Advances in Intelligent and Soft Computing

Editor-in-Chief

Prof. Janusz Kacprzyk
Systems Research Institute
Polish Academy of Sciences
ul. Newelska 6
01-447 Warsaw
Poland
E-mail: kacprzyk@ibspan.waw.pl

T0138138

Jia Luo (Ed.)

Soft Computing in Information Communication Technology

Volume 1

 Springer

Editor
Jia Luo
National Kinmen Institute of Technology
Kinmen
Taiwan

ISSN 1867-5662   e-ISSN 1867-5670
ISBN 978-3-642-29147-0   e-ISBN 978-3-642-29148-7
DOI 10.1007/978-3-642-29148-7
Springer Heidelberg New York Dordrecht London

Library of Congress Control Number: 2012935669

Printed on acid-free paper

Springer is part of Springer Science+Business Media (www.springer.com)

Preface

The main theme of SCICT2012 is set on soft computing in information communication technology, which covers most aspect of computing, information and technology. 2012 International Conference on Soft Computing in Information Communication Technology is to be held in Hong Kong, April 17–18.

Soft computing is a term applied to a field within computer science which is characterized by the use of inexact solutions to computationally-hard tasks such as the solution of NP-complete problems, for which an exact solution cannot be derived in polynomial time.

Soft Computing became a formal Computer Science area of study in early 1990's. Earlier computational approaches could model and precisely analyze only relatively simple systems. More complex systems arising in biology, medicine, the humanities, management sciences, and similar fields often remained intractable to conventional mathematical and analytical methods. That said, it should be pointed out that simplicity and complexity of systems are relative, and many conventional mathematical models have been both challenging and very productive. Soft computing deals with imprecision, uncertainty, partial truth, and approximation to achieve practicability, robustness and low solution cost.

Generally speaking, soft computing techniques resemble biological processes more closely than traditional techniques, which are largely based on formal logical systems, such as sentential logic and predicate logic, or rely heavily on computer-aided numerical analysis (as in finite element analysis). Soft computing techniques are intended to complement each other.

Unlike hard computing schemes, which strive for exactness and full truth, soft computing techniques exploit the given tolerance of imprecision, partial truth, and uncertainty for a particular problem. Another common contrast comes from the observation that inductive reasoning plays a larger role in soft computing than in hard computing.

The conference receives 480 manuscripts from the participants. After review process, 149 papers are finally accepted for publication in the proceedings.

The Editors would like to thank the members of the Organizing Committee for their efforts and successful preparation of this event.

We would like to thank all colleagues who have devoted much time and effort to organize this meeting. The efforts of the authors of the manuscripts prepared for the proceedings are also gratefully acknowledged.

Jia Luo
Publication Chair

Contents

X   Contents

Gaussian-Heimite in the Atom Population Solution[*]

Liu Xin[**]

College of Information and Computer Engineering, Northeast Forestry University,
Harbin 150040, China

Abstract. To solve the population of the multi-electron atoms, it is calculated corely with the method of Gaussian-Heimite integral. It is evidenced that the Hamiltonian Matrix reduced in the system of multi-electron atoms. Dynamics of atomic population can be clearly predominated by means of the population with distributed radial and angle. It is very important for micrometric experiment. This calculation method works little time and higher accuracy.

Keywords: Atom Population, Gaussian-Heimite, Hamiltonian Matrix.

0   Introduction

The computer has always been a major tool in physics calculation since it came into the world, it was applied to calculate all sorts of atom parameters in the atomic physics field, atom spectrum field and so on. With the development and improvement of hardware and software, the atom population solution had been improved and updated accordingly. Current software program showed a huge advantage and great value in many factors, especially for quick calculation, short programming language, accurate analog data and others, like proving the correctness of theory and guiding the experiment.

The atom population is an absolutely necessarily data in the physics, chemistry, biology, microcosmic engineering field, we are starving for a more prompt and effective method and program in order to understand the atom population data rapidly. This article project is to use Gaussian-Heimit integral program to calculate atom radial wave function. So predigest the calculated method and enhanced precision.

1   Theory Model

In polyatomic system, electron can solve the stationary Schrodinger equation [1] under the population of various energy levels, then the wave functions can be separated relative to time variables and spatial variable, but this is a special cases. In most of the atomic system, especially when we cannot neglected the interaction between the atoms

---

[*] State natural sciences fund 71003020: Based on Data Mining Citation Network Diagram Hot Technology Field Prediction Research.

[**] Author :Liu Xin (1976 -), female, docent, Heilongjiang Mu Ling people, Master degree, the main research direction: intelligent information processing and query database optimization, data mining technology and Web data management, etc.

with the environment, the wave function of atomic system which relative to the time variable "t" and spatial variable "r" is inseparable, namely Ψ (r, t). Now to understand the movement rule of the system need solution for the basis of timed Schrodinger equation.

$$i\frac{\partial}{\partial t}\Psi(r,t) = \hat{H}\Psi(r,t) \tag{1}$$

$\hat{H}$ in Equations (1) is Hamiltonian functor, AKA Hamiltonian, it acts directly on the system of wave function Ψ (r, t), when a wave function Ψ (r, t) of spatial variable "r "and time variable "t" cannot separate variable directly from wave function Ψ (r, t), the above equation will become complicate accordingly. In considering the interactions between adjacent original electronic, also it is a just coulomb interaction, Hamiltonian operator can be represented as below:

$$\hat{H} = -\sum_{i=1}^{Z}\frac{Z\nabla_i^2}{2\mu} - Ze^2\sum_{i=1}^{Z}\frac{1}{r_i} + \frac{e^2}{2}\sum_{\substack{i,j \\ i\neq j}}^{Z}\frac{1}{|r_i - r_j|} \tag{2}$$

Z is atomic number, e is electronic charge quantity, μ is a converted quality through electronic and coulomb interaction, ∇_i^2 is a quadratic differential vs. wave function Ψ (r, t) spatial variable:

$$\nabla_i^2 = \frac{\partial^2}{\partial x_i^2} + \frac{\partial^2}{\partial y_i^2} + \frac{\partial^2}{\partial z_i^2} \tag{3}$$

I, j is the electrons ordinal number of atom.

2   Methodology

If we solve the model (1) directly based on wave function Ψ (r, t), the consumed time is long, computing accuracy can not reach the physical demands. Now let assume the wave function will be:

$$\Psi(r,t) = \sum_{i=1}^{N}a_i(t)\Phi_i(r) \tag{4}$$

In this model, a_i (t) is a spread coefficient, usually indicates electronic odds on specific level. According that the Hamiltonian operator is hermite-gaussian operator ,we can do the conjugate equation (1) of wave function expressed (4) , do a integral for both ends to make integral equations (1) into a matrix equation, and the Hamiltonian operator matrix cell will be:

$$H_{ij} = \int\limits_{-\infty}^{\infty} d\tau \int\limits_{-\infty}^{\infty} d\tau' \Psi_i(r,t) \Psi_j(r',t') \qquad (5)$$

$d\tau$ and $d\tau'$ is total effect of integral vs spatial variable r and time variable t.

$$H = \begin{bmatrix} H_{11} & H_{12} & \cdots & H_{1n} \\ H_{21} & H_{22} & \cdots & H_{2n} \\ \vdots & \vdots & \vdots & \vdots \\ H_{m1} & H_{m2} & \cdots & H_{mn} \end{bmatrix} \qquad (6)$$

For any I, j, according to the characteristic of coulomb's and the emi characteristics of Hamiltonian descriptor, the density matrix cell in model (6) will be: Hji Hij = *, this can reduce the operational time.

First step: in terms of initial conditions, set up a temptation wave functions , calculate a wave function with iterative method which available for the accuracy requirement.

Second Step: solve the matrix elements in model (5), wave function in model (4) can be divided into a new model :radial wave function R (R) multiply sphere harmonic function Y (theta, Φ) (theta and Φ is spatial variable), then calculate radial wave function R (R) and sphere harmonic function Y (theta, Φ) [4,5,6]respectively.

$\int_{-\infty}^{\infty} e^{-x^2} f(x)dx$ calculation is based on Gaussian - Elmer infinite numerical integral formulas and its function program of error estimation in the MATLAB main program:

```
function [GH,Y, RHn]=GaussH2(fun,X,A,fun2n)
n=length(X);n2=2*n; Y=feval(fun,X);
GH=sum(A.*Y); sun=1; su2n=1; su2n1=1;
for k=1:n
sun=sun*k;
end
for k=1:n2
su2n=su2n*k;
end
mfun2n =max(fun2n); potential field interaction
RHn = ( sun*sqrt(pi))* mfun2n /( 2^n*su2n);
```

3 Analyse

Atomic population is correlated with radial wave function. A proposed algorithm which used Gaussian-Heimite integral can solve radial wave function R (r), and main program repeatedly use the above function and replace its operational result in the equation (6), solve a matrix characteristic values, atomic population relates to matrix character, we can get the sagittal character once determine the matrix character, that is wave function. Every iterated algorithm of wave function show a number on the interface, give an estimation of wave function which is close to accuracy requirements after comparing with a fixed number. On this basis, put forward a kind of rapid iteration algorithm in order to reduce the computational complexity, by introducing a probability data connection method, we shall avoid the integral calculation of wave function estimation of standards and multiplicative algorithm. Numerical simulation results show that the iterative wave function arithmetic on using Gauss-heimite integral algorithm can more approach optimal performance of the algorithm, and the rapid iteration can close to precision more faster demanded by physical, also reduce dramatically r computational complexity at the same time.

4   Summary

To solve electrons radial wave function of atoms with Gaussian – Heimite, this method reduced iterative frequency and shortens the operation times to ensure to get the population state of atom quickly.

References

1. Burkov, A.A., Hawthorn, D.G.: Spin and charge transport on the surface of a topological insulator. Phys. Rev. Lett. (105), 066802 (2010)
2. Wei, Z.Y., Dong, Q.W., Mi, W.: A LU key distribution project with the certification. Computer Application (29), 161–164 (2009)
3. Xue, X., Shi, H.Z.: A hierarchical classification method research of china text based on vector space model. Computer Application (26), 1126–1133 (2006)
4. Nan, D., Min, X.H., Yue, P.: A simulated times choice as calculating reliability by Monte Carlo method. Mining Machinery (3), 13–14 (2002)
5. Xiang, M., Fu, Z.J., Feng, Y.H.: Heuristic Attributed reduction algorithm for distinguishing matrix. Computer Application (30), 1999–2037 (2010)
6. Akoplan, D.: Fast FFT based GPS Satellitc acpuisition methods. IEE Proceedings of Rada Sonar and Navigation 154(4), 277–286 (2005)

The Study of Financial Management Information for Business Process Reengineering under the Low-Carbon Economy

Xuesong Jiang[1], Yan Xu[2], and Nanyun Xiao[3]

[1] Accounting Institute, Harbin University of Commerce, Harbin, Heilongjiang, China
No. 1 Song Xue-Hai Street, District Harbin China, 150028, 13313628811
[2] Economics and Management Institute, Northeast Agricultural University,
Harbin, Heilongjiang, China
No. 59 Mucai Street, Xiangfang District Harbin China, 150030, 15045649765
[3] Counselor, Harin Finance University,Harbin, Heilongjiang, China
No. 65 Diantan Street, Xiangfang District,Harbin China, 150030,13796821567

Abstract. The low-carbon economy is not only a low power, low emissions, low pollution-based economic model, but also a new transformation model of economic development, energy consumption, its essence is to improve energy efficiency and create a clean energy structure to pursue the green GDP. The core of low-carbon economy is technological and system innovation. With economic globalization and national comprehensive national heightening, China's enterprises are facing unprecedented competition from abroad. Allocation of resources through the optimization of enterprise, enterprise information can greatly improve the management level and competitiveness, is the only way for China enterprises to the world. As an important part of enterprise information, financial management information can bring huge economic benefits for the enterprise. Therefore, actively research information related to financial management theory and practice, either financial management of enterprises or improving the competitiveness of enterprises, have a positive and practical significance.

1 Introduction

The generation of financial management information development was beginning from 1950, mainly in the United States took the lead in General Electric Company as a symbol on a computer be used to calculate wages; nowadays to establish a low energy consumption, low pollution, low-emission-based economy has become world economic development trend. Integrated MRPIJ, ERP software, financial management system, a new generation ERP system has been completely changed from the accounting information in the reflection of the financial management of information processing to multi-level, global the financial management support services .

1.1 The Meaning of Financial Management Information

Financial management information is that accountants through the use of modern technology, carry out business process reengineering, to establish corresponding financial organization model, to muster financial information on the potential of human resources, to developed corporate financial information resources and improve the efficiency of financial activities, to better organization of corporate finance activities, processing of financial relations, business stakeholders in order to achieve financial goals to maximize benefits.

1.2   Content of Financial Management Information

Implementation of financial management information relies on the integration of several information systems; in general, the financial management information should include five parts: the accounting transaction processing information systems, financial management information systems and financial decision support system, financial managers and the organization of interconnected information systems information system. Accounting transaction processing information systems is to provide accurate, timely information to improve financial efficiency and success rate; financial management information system, financial decision support systems and financial manager of information systems from different perspectives, different levels of financial management solution plan, control, decision making and other issues; organizations to solve interconnect information systems within the enterprise and between organizations, enterprises and associated enterprises transfer information between issues. The successful establishment of these systems and the integration between financial management, information management is the successful embodiment of the inextricable relationship between them.

2   Traditional Financial Accounting Process and Shortcomings

2.1   Traditional Financial Accounting Processes

Traditional financial accounting process is based on hundreds of years ago Pacioli's accounting theory developed, the core idea is that classification system, data is provided by the classification of aggregated data, which collected data from business processes, the data is the main carrier original documents, accounting strict accordance with the " Fill in the certificate - registration books - prepared statements "the order of the original certificate after generating all kinds of books for data processing, and finally to the books, accounting vouchers, based on reports submitted to the preparation of internal

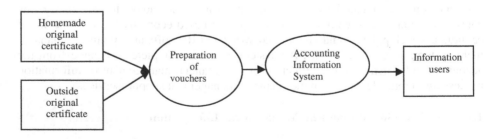

Fig. l. Traditional financial accounting process

and external Investors, creditors, government departments and managers and so on. Specific structure shown in Figure 1:

2.2   Shortcomings of Traditional Financial Accounting Processes

Information sharing can not be achieved. In the period of the traditional manual accounting system, financial accounting processes are based on the division of labor under a sequence of business processes. Period to the computerized accounting system, only simple hand soil imitation and copy the accounting process, although the use of modern information technology environment in the IT established a number of independent subsystems, such as material accounting systems, accounts processing system , but only implement the financial accounting process automation, real change only means of operation, and the structure does not change the nature of traditional information systems. Between the various subsystems or modules each separated subsystem is based on the formation of one unit of information for the island. The accounting subsystem provides data and information, can only meet the needs of the financial sector, but can not meet the other functions associated with the need for sharing of financial information can not be present.

Difficult to meet the needs of information age management. The whole process of enterprise business activities associated with capital flow, logistics and information flow. The traditional accounting system structure, ideas and technology constraints, business process accounting does not capture all the data, but by an economic business activities to determine which data of the financial statements, which comply with accounting definitions which capture the flow of funds information, the process for the operational activities associated with logistics and information flow, such as information customers need, such as productivity, performance, reliability, like the other information, but not be considered. Results of operations related to the same economic data is stored separately in the hands of accountants and business people.

Unable to meet the needs of real-time control. Any business cash flow is associated with the logistics flow, but the traditional financial accounting processes reflect the cash flow information is often lagging behind the material flow of information, financial management processes are often in control after the control, which allows companies not from the perspective of efficiency production and business activities of real-time control. This is because the accounting data is usually in the business after the acquisition, but not in the business of real-time acquisition occurs; accounting data processing is delayed posting the data collection, compilation, reconciliation, etc.; financial report can not be directly used, must be a number of Road links in order for users to use processing; traditional financial management is due to technical limitations and economic constraints make financial control only after the control. In today's rapidly changing economic environment, to improve the usefulness of financial information and control efforts need to achieve real-time information, but the lag of the traditional financial process so that the financial manager can not obtain the necessary information from which can not meet the financial real-time control needs.

3   Financial Management Business Process Reengineering

For the smooth implementation of the network of financial, to achieve financial information management, traditional financial accounting processes must be reconstructed. We should be based on the theory of business process reengineering, carefully study the specific content of the financial processes and all aspects of the defect from the traditional financial process starting to rebuild the financial accounting processes to achieve the centralized financial management and achieve financial and business integration, the overall budget management and funding of the dynamic control.

3.1   Reengineering Targets

Data sharing. As the internet continues to develop, in the basic theory of traditional accounting, work procedures, organizational methods, baking has undergone tremendous changes, in accordance with the requirements of information processing, only the full use of modern information technology in order to adapt to the changing regulatory requirements. Financial accounting process reengineering is no longer the traditional accounting process simulation system, but in modern information technology as the basis, with the current social, economic and technological environment suited to a new system, is the true sense of data sharing. In the original data, based on the data coding of the standard simple processing, the formation of the source data to meet internal and external requirements of all users of information, financial data is truly from the same source shared.

Establish a centralized management system. The one hand, a centralized network server and data server, the other will focus on administrative rights to the company headquarters. This centralized services and data sharing, but also to ensure the safety of operation of the system. Meanwhile, the parent body can do real-time from your computer inquiries, the audit, the company's internal control system strictly, and strengthen financial management.

Enhanced system capacity. Full use of the automatic processing computer system, the procedures set by human, to achieve real-time business automatically generate accounting documents, financial data and automatically reporting, and strengthen data collection, consolidation and analysis capabilities, the liberation of the workforce, improve working efficiency.

Establish an effective feedback mechanism. Service decision-making is the fundamental purpose of financial management, how to timely and accurate financial reporting feedback to policy makers is the ultimate goal of business process reengineering.

3.2   Process Reengineering

Established based on business event-driven integration of financial and business information processing. "Event-driven" is the information users need information in accordance with the intent to use is divided into several different events. Since the introduction of data warehouse, the number of origins, information focus, avoiding duplication of data is incomplete and the occurrence of the situation, to maximize the enterprise-wide data sharing, simplifying the process to achieve real-time access to information, real-time processing information, real-time reporting information, so that all data from a common use, managers at all levels in real time, dynamic access to information, support decision-making. Figure2 shows the process of specific processes:

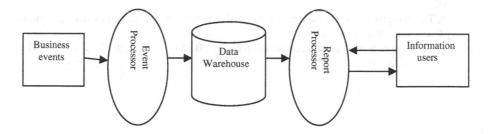

Fig. 2. Event-driven structure

Embedded real-time information processing in the financial management process. Embedded real-time information to financial management processes to the process, the enterprise implementation of operational activities, the business-related event information into financial management information systems, financial decision-making information system, through the implementation of business rules and information processing rules to generate integrated Information to achieve integrated financial management. This will change the existing financial staff management, the financial sector extended to the various business units, direct attention to the actual business process to achieve real-time control of things in business and process risks.

Financial officers change from Information processer to business Manager. The help of information technology, finance staff from the daily processing of financial information to get out, companies can better focus on business processes and achieve their management responsibilities. Traditional financial accounting processes and business process of phase separation, and only deal with business in the course of events in a subset of the financial accounting and financial management is important provider of data needed, so this situation will lead to financial personnel and business management personnel out of touch, unable to play the management of financial management functions.

4 Conclusion

From accounting to the accounting information and then to the financial management information, the decision is not only a noun or a concept changed, it represents a revolutionary concept, a new generation of accounting thought is the inevitable trend development of financial management. In the production of financial management information, although the development process there is a variety of problems, it is still growing. We firmly believe that the financial management of information technology will be developed under the science and technology to overcome various problems and healthy development.

References

1. Zhang, R.: e times of financial management. China Renmin University Press, Beijing (2002)
2. Yao, T.: Enterprises to implement a comprehensive budget management practices. Accounting Communications (December 2004)
3. Li, G.: BRP re-use of financial accounting processes. Business Research 67 (March 2004)

Measures of Building Tax Information

Dongling Li[1], Ying He[2], Yunpeng Xu[3], and Yadong Fan[4]

[1] Economics and Management Institute,
Northeast Agricultural University,Harbin, Heilongjiang, China
No.59 Mucai Street Xiangfang District Harbin China, 150030, 15846526946
364777398@qq.com
[2] Economics and Management Institute,
Northeast Agricultural University,Harbin, Heilongjiang, China
No.59 Mucai Street Xiangfang District Harbin China, 150030, 13945002528
398289336@qq.com
[3] Economics and Management Institute,
Northeast Agricultural University, Harbin, Heilongjiang, China
No.59 Mucai Street Xiangfang District Harbin China, 150030, 13946082997
305222623@qq.com
[4] Economics and Management Institute,
Northeast Agricultural University, Harbin, Heilongjiang, China
No.59 Mucai Street Xiangfang District Harbin China, 150030, 13359716008
Fanyadong1966@hotmail.com

Abstract. With the continuous development of global information technology, cmputer applications become increasingly common and in-depth. The application of information technology are widely used in ativities of all areas, it also facilitates the process of world development.Information technology of the tax authorities in many countries have made great achievements. In our case, speeding up the construction of tax information is an important task for the current tax. This paper seeks to tax information on the construction of the current situation and problems, and giving recommendations to the tax information of the future development.

Keywords: tax information management, Information, Tax Administration.

1 Introduction

Tax information technology started from the early 80s of last century, construction of tax information has gone through more than 20 years in our country. Especially since the 90s of twentieth century, after the government having launched the process and the national tax authorities, tax information technology comes into a more rapid stage of development. It has become an important infrastructure in tax collection, and played a growing role in the developing of tax administration and promoting tax. However, there are some problems in the construction of tax information, the research on these issues has great significance in speeding up the construction of tax information.

J. Luo (Ed.): Soft Computing in Information Communication Technology, AISC 158, pp. 11–15.
springerlink.com © Springer-Verlag Berlin Heidelberg 2012

2   Current Situation of Tax Information

The earliest tax information started from the 80s of the twentieth century. In 1983, several tax authorities used computers in tax accounting, tax data, statistics and analysis. Since then, the construction of tax information is in a continuous development in the twists and turns. Some domestic experts holed that the tax information of the development process is divided into the following three stages: the first stage is the initial stage of tax information (1983-1989); the second stage is the initial application of the tax information phase (1990-1994); the third stage is the application of tax information development stage (since1995).

The main features of the first stage is to use a simple computer program simulating manual business process. The idea is one computer application platform, using accounting, statistics and information processing applications as starting point. So that tax officials can get free from the onerous statistics and statistical reports, and have enough awareness on the tax staff of the computer application. All the country has developed report processing and statistics, collection and management, key sources of revenue management and revenue management, and to some extent it has had a great promotion and application. In the initial stage, the tax system has 3,400 personnel computer applications, 800 professional and technical personnel people, 5300 computer machines following 286. Development platform generally built on DOS operating system and DBASE database basis.

The main features of the second stage is that computer application in the tax system has a certain promote. All the country are beginning to develop their own tax collection and management software. Collection and management have been computerized. We progressively carried out a variety of individual software development and applications starting from the establishment of the system and the implementation of standard. In order to meet the collection and the implementation of tax reform, the State Administration of Taxation also focus on export tax rebates and management software. But in Administration of Taxation, there is no unified software development management. As a result ,all regions have developed their business needs according to their own software, which has laid a foreshadowing in lacking of unified planning on current tax information for construction.

The main features of the third stage is that there has been big margins in the organization and leadership, equipment configuration, software application, network building and team building and other areas. At this stage, the national tax system has conducted a variety of large-scale infrastructure and implementated the "Golden Tax" engineering. It has also developed and promoted a national unified tax collection and management software which is called TAXS and CTAIS tax system. The information technology has begun to take shape.

At this stage, the construction of tax information has made a rapid development, and effectively promoted the work of raising the level of taxation. First, the application system is building steadily. Second, the level of information application is increasing. Third, information technology infrastruction is developing rapidly. While fully affirming our achievements, we should also see the problems in the construction of tax information.

3   Problems in the Process of Building Tax Information

3.1   Construction of the Tax Information Is Not Standardized

The most basic requirements is that the information processing should be standard and uninfected. It is the only way to achieve information sharing. Tax information is currently working in China, though the State Administration of Taxation management integration program are undertaken within the system. In addition to the provincial director of information technology software, the business sector in the higher tax authorities will be introduced from time to time to some specialized management software. There are also grass-roots tax authorities which are needed to develop their own according to their own software. The result is the application software inconsistency between the code and data interfaces are not unified, so as to running can not be compatible with other platforms, which resulting in a unified tax collection and management software lost its unity and seriousness.

3.2   Construction of the Tax Information Is Not Standardized

As applications continue to expand and conduct, the application system of data is even more prominent. On the one hand, the performance of the data is not comprehensive, some relevant information is not included in the systems management, which would weaken the application software; the other hand, the data is not accurate enough. The information is not precise enough to provide the taxpayer, staff operator error or system problems caused by technical data errors and data collection standards are not unified, thereby affecting the establishment of the application on this basis the accuracy of business system is difficult to function properly. Information technology through the years, accumulated a large amount of tax data, the decentralized applications resulted in a large number of tax management information has not been fully utilized.

3.3   Lack of Funds Restricted the Construction of Tax Information

Lack of funds restricted the tax administration process of information construction, the state tax information for construction funds and other investment funds for the same cast over and spread, leading to the provinces, municipalities and autonomous regions, formed a large situation , no one would like, and no one wanted to rob the distance, but few can really achieve good results.

3.4   The Quality of Tax Officials Need to Be Improved

Tax information is not only to establish a tax collection of information, but also to its widely used in tax collection and a part of the work. So the majority of the tax authorities and the business skills of the staff has proposed new requirements. Line at the grassroots level, e additional personnel are needed very year, but new recruits computer professionals in the real is few. Tax professionals understand the knowledge is almost none.

4   Some Suggestions on How to Solve the Construction

4.1   We Should Build a Unified Application and Improve Information Sharing Systems

In the construction process of tax information, we must stand highly in a long-term development and construct planning according to industry standards as well as the idea of system theory of information-oriented. Tax departments must follow uniform standards, and actively promote the steady pace of building tax information. They should also enhance the integration of business and norms in accordance with requirements of the construction of China's information. At the same time they should design and develop the software and implement the standards and norms related technology and information technology in the implementation of the process seriously. Internal revenue tax department should be to promote business process reengineering and integrate comprehensively tax services. It is also require us to built business communication and exchange of good interaction between cross-sectored coordination and use information network system to achieve the greatest degree of data sharing and utilization.

4.2   The Data Management Should Be Strengthened

We should establish and improve data collection and management system, so as to strengthening the management of data quality. The quality of the data reflects the value of the data. Only by ensuring data authenticity and integrity, can we improve the utilization of data resources and promote collection and quality improvement. To improve the quality of the data ,we can follow these aspects: First, we can strengthen data collection and other aspects of the management and monitoring. Second, we can establish information collection assessment system,. As we know, a clear responsibility to review the accuracy of the information could improve the quality of data and information. Third, we can   broaden the scope of basic data collection and relevant departments as soon as possible with real-time data exchange and sharing.

4.3   We Should Improve the Security System

The advance tax information is a systematic project, involving all aspects. A large number of proven information has promoted that the work of the primary issue is not just technical issues and the coordination is particularly important in the county, so the organization must have a strong security system. From the first leadership attention, we can see that leadership can not attach great importance to the construction of information. It is difficult to break through in a variety of difficulties to obtain good results. It is necessary to have an independent leadership responsible for managing information body responsible for the overall planning. In addition it is high time to have a coordination mechanisms, the construction of information related with various departments is fully to do a good job with the department of information coordination between business units. The information technology sector position must be accurate, all levels of technical departments should be located in the bridge business and maintenance work.

4.4   Strengthen the Tax Administration Personnel

Information in the tax system has been everywhere, and all department staff are inextricably linked, you need to tax officials and full participation. The only way is that all the people involve into the construction of the tax information. To achieve this goal we should do as following aspects:

First, we can improve the tax system of personnel training and strengthen the targeted training. To select easily understandable training materials and supporting Courseware, workers should be trained for professional skills. Each training should be rigorous testing of the participants, and the situation assessment should be recorded.

Second, we can establish a scientific point of view of information on performance index, to assessment the ability to work on the post comprehensive of personnel. The key is to identify performance appraisal entry point and the use of scientific methods. Tax officials should make a correct evaluation of the work and try to improve their skills.

Third, we can introduce incentives to improve the tax staff motivation. Award a fair and objective is to stimulate and mobilize the human potential of the most effective means. For those who can master the skills required for the job and related performance evaluation scores, Tax officials will be able to increase enthusiasm for learning, the formation of good study and work atmosphere, so that everyone can assume their responsibilities and ensure that the tax information system can play its due effect.

5   Conclusion

Information is not only an important driving force to promote economic growth, but also the development trend. The work of tax authorities is the requirements of information age. We should take the initiative to meet the challenges of the information age and to improve the efficiency of tax administration. Tax information will bring the regulation of tax work, process, organization of major change and tax policy. It will also create a sound basis and bring a great reduction in tax costs of economic development.

References

1. Tan, R., Wang, J.: Integration: A new stage of China's tax information development. China Tax (2) (2005)
2. Tong, L.: Information on Tax Collection Management, vol. (2), Zhongnan University (2004)
3. Tan, Y.: Information Age Strategy of China Tax Information, vol. (4) (2006)
4. Fang, L., Chen, F.: Problems and solutions of tax information technology. Information and Network (5) (2006)
5. Jiang, L.: Tax problems in the construction of information and countermeasures. Huxiang Forum (1) (2007)
6. Yang, G.: Information on Tax Thoughts on the Construction. Finance Study (2) (2009)

A New Type of Using Morphology Methods to Detect Blood Cancer Cells

Yujie Li[1,*], Lifeng Zhang[1], Huimin Lu[1], Yuhki Kitazono[2],
Shiyuan Yang[1], Shota Nakashima[3], and Seiichi Serikawa[1]

[1] Department of Electrical Engineering and Electronics,
Kyushu Institute of Technology, 1-1 Sensui-cho, 804-8550, Japan
[2] Department of Electronics and Control Engineering,
Kitakyushu National College of Technology, 5-20-1 shii, 802-0985, Japan
[3] Department of Electircal Engineering, Ube National College of Technology,
2-14-1, Tokiwadai, 755-8555, Japan
{gyokukeitu.ri,keibin.riku,nakashima}@boss.ecs.kyutech.ac.jp,
{zhang,yang,serikawa}@elcs.kyutech.ac.jp
http://www.boss.ecs.kyutech.ac.jp/

Abstract. In order to resolve the problem of recognizing blood cancer cells accurately and effectively, an identifying and classifying algorithm was proposed using grey level and color space. After image processing, blood cells images were gained by using denoising, smoothness, image erosion and so on. After that, we use granularity analysis method and morphology to recognize the blood cells. And then, calculate four characterizes of each cell, which is, area, roundness, rectangle factor and elongation, to analysis the cells. Moreover, we also applied the chromatic features to recognize the blood cancer cells. The algorithm was testified in many clinical collected cases of blood cells images. The results proved that the algorithm was valid and efficient in recognizing blood cancer cells and had relatively high accurate rates on identification and classification.

Keywords: image processing, mathematical morphology, image denoising, granularity detection, cell clustering, cell recognition.

1 Introduction

This Cell recognition is the hot field of medical image processing and pattern recognition. It has a very wide range of applications in the biomedical field [1]. The shape, number, distribution of these cells, which are reflected a variety of important exceptions to the information of disease. Cells in bone marrow smear as location, identification and differential counting is a complex and time-consuming work. Therefore, the using of computer aided automated for analysis the cell image is of

---

* The author with the double degree student in Department of Electrical Engineering and Electronics of Kyushu Institute of Technology in Japan and College of Information Engineering of Yangzhou University in China.

great significance to reduce the intensity of labor of doctors and to improve the diagnostic accuracy [2-4].

Many beneficial explorations have been carried out in this field. Many researchers adopt Genetic Algorithm to optimize and identify image, some scholars use statistical model and other methods to segment and process images from different aspects. Some researchers use the principle component analysis and natural network for detect the cells image. But these methods have some defects to different extents, such as, complexity of arithmetic, difficulty to ensure parameters and so on.

In this paper, we proposed a new simple method which can be adapted to detect image regularity. The outline of the article is as follows. Section 2 covers the preprocessing of blood cells image processing. Given this, the cancer cells detection method by granularity detection, morphological structure, and chromatic features is derived in Section 3. In Section 4, the experimental results are shown, at the same time, we discuss the results. We conclude this paper in Section 5.

2 Related Works

The system is constituted by Olympus microscope, Panasonic color camera, color image acquisition card and computer. The smears of bone marrow cells, which are catch by the optical microscope, then converted into the computer image through the CCD camera. After that, the images are transmitted to the acquisition card and converted to digital images for computer processing, storage, display, print and so on.

2.1 Eliminate the Background of the Original Image

The influence of illuminance in original image may be affecting the final processing results, therefore, to modify the whole image, smoothness and denoising is necessary. Using the subtraction method in the original image and the background one can eliminate the uneven background effectively. This is defined as

$$g_i(x, y) = f_i(x, y) - h(x, y) + k \tag{1}$$

where $g_i(x, y)$ is the image after processing, $f_i(x, y)$ is original image. $h(x, y)$ is the background. k is the correction conference.

2.2 Principles of Denoising by Using Morphology

The basic operations of mathematical morphology have dilation, erosion, opening and closing [6]. Based on the type of the prepared image and requirement, the mathematical morphology can also be divided into the binary morphology, gray morphology and color morphology. The basic operations in gray morphology are introduced as followings.

Assuming that $f(x, y)$ expresses input image, $b(x, y)$ stands for structure element. D_f and D_b are the definition domain of image $f(x, y)$ and structure element $b(x, y)$ respectively. The expression of gray dilation is

$$(f \oplus b)(s,t) = \max\left\{ f(s-x,t-y) - b(x,y) \mid (s-x),(t-y) \in D_f; (x,y) \in D_b \right\} \tag{2}$$

The expression of gray erosion is

$$(f\Theta b)(s,t) = \min\left\{ f(s+x,t+y) - b(x,y) \mid (s+x),(t+y) \in D_f; (x,y) \in D_b \right\} \tag{3}$$

The opening and closing operations of gray image have the identical form with the corresponding operations of binary image.

Opening operation carries out simple corrosion on image f with structure element b, and then carries out dilation operation on previous result with the b. But closing operation is right opposite with opening operation. It carries out simple dilation on image f with structure element b, and then carries out erosion operation on previous result with structure element b.

The expression of opening operation is

$$f \circ b = (f\Theta b) \oplus b \tag{4}$$

The expression of closing operation is

$$f \bullet b = (f \oplus b)\Theta b \tag{5}$$

Opening operation is often used to remove less bright detail relative to the size of structure elements, and keep relatively the whole gray level and bright large-area invariable at the same time. Closed operation is often used to remove dark detail part in the image, but to keep relatively the bright part not being affected.

Image denoising is a fundamental link in image pre-processing. It removes interference in image to extract useful information in complicated image. Although a series of traditional filtering methods can get good effect on de-noising, some useful image information will lose simultaneously. For instance, the image border will be blurred accompanied by de-nosing. De-noising based on morphology uses the structure element to handle the image, to extract useful information and reserve the border information well.

Gray morphology is, in the aspect of eliminating noise, similar to binary morphology, but it handles gray image [5]. Therefore it has the unique characteristic.

2.3   Granularity Detection

After preprocessing, we can extract corresponding characteristic information in image. Here we mainly detect image granularity by using "probe" structure element to carry through morphology operation with image [6-8]. On the basic of applying one-dimensional structure element, two-dimensional structure element is expanded to detect objects.

Given a test Image, we need to select right structure element, then the original image is eroded and dilated, which is equivalent to completing opening operation on the original image. Some less noise spots are reduced. Therefore, as long as choosing appropriate structure element and doing opening operation on the image, we could detect the granules in image.

We use granule distribution function to describe image granularity. The distribution function is given as following definition.

$$f(\lambda) = \frac{A(X \circ \lambda B) - A(X \circ (\lambda + 1)B)}{A(\lambda B)} \tag{6}$$

where X represents the original image, B represents structure element with radius 1, λ represents radius, $A(X \circ \lambda B)$ expresses the area of reserved granules in the image which complete opening operation by using λB. It is easy to see that if λ gets bigger; the reserved granules in image will get less.

2.4   Entropy of One-Dimensional Histogram

In this work, we use entropy as a way to segment the image. Entropy is an important concept in information theory. Theoretically, it is an element of the average amount of information. The formula is

$$H = - \int_{-\infty}^{\infty} p(x) \cdot \lg p(x) dx \tag{7}$$

where, $p(x)$ is the probability of occurrence x. In an original image, set T as the threshold. If the pixel grey level less than T, it is a object. Otherwise, it is the background. p_i is the probability of grey level at pixel i, the probability of objects and background respectively as,

$$p_o = \sum_{i=0}^{T-1} p_i, \quad i = 1,2,\cdots,T-1; \tag{8}$$

$$p_b = \sum_{i=T}^{255} p_i, \quad i = T,T+1,\cdots,255; \tag{9}$$

Then, we get the entropy of object and background as

$$H_o(T) = -\sum_i (p_i / p_o) \lg(p_i / p_o) \quad i = 0,1,2,\cdots,T-1; \tag{10}$$

$$H_b(T) = -\sum_i (p_i / p_b) \lg(p_i / p_b) \quad i = T,T+1,\cdots,255; \tag{11}$$

The entropy of image is

$$H(T) = H_o(T) + H_b(T) \tag{12}$$

3   Features Analysis

3.1   Morphological Feature Extraction

Area of Cytoplasm. Area of cytoplasm is defined as all of the area of the sum pixels of the cytoplasm. Usually, the area of lymphocytes is small; the area of plasma cells is larger. The function is given as following,

$$S = \sum_{i=1}^{n} a_{ix}(y_{i-1} + a_{iy}/2) \tag{13}$$

$$y_i = \sum_{k=1}^{i} a_{ky} + y_0 \tag{14}$$

where a_{ix} is the x direction of the direction chain a_x, a_{iy} is the y direction of the direction chain a_y. (x_0, y_0) is the coordinate of initial point.

Circumference of Cell Compartment. The circumference of cell compartment is

$$L = n_e + n_o\sqrt{2} \tag{15}$$

where n_e express the even numbers of the cell compartment, n_o is odd numbers of the cell compartment.

Shape Factor. Shape factor is defined as

$$R = 4\pi A / L^2 \tag{16}$$

where R is the shape factor, L express the perimeter of the cytoplasm. A is the area of the cytoplasm. Roundness is used to describe the deviation between cell region and the circle. At the same condition, the cell area boundary smooth and round, the circumference of the cell is shorter, and the circularity $R = 1$. Otherwise, R is much smaller.

Rectangle Factor. The rectangle factor is defined as

$$I = A/(W \times H) \tag{17}$$

where A is the area of the cytoplasm, W is width of cell compartment, H is height of cell compartment. Rectangle factor used to describe the degree of deviation between the cell compartment and rectangular. When the cell compartment is rectangle, $max\{I= 1\}$.

Elongation. The elongation is

$$E = \min(W,H)/\max(W,H) \tag{18}$$

The more slender cell region, E is smaller. As the cell region is round, $E = 1$.

Ratio of Nucleus and Cytoplasm. The ratio of nucleus and cytoplasm is

$$B = S_N / S_C \tag{19}$$

If the value of B is large, the cell may be cancer cell.

3.2   Recognize of Chromatic Features

After the morphological identification, majority of lesions are correctly detected, but some normal cells are incorrectly classified as disease cells. In order to improve the recognition rate, we use chromatic features for further identification. Compared with normal cells, nuclear color of lesion cell is darker than that of normal cell. Therefore, we calculate mean and variance of each color component. After that, we determine the threshold, to reclassification of the wrong cells, to improve the recognition rate.

4   Results and Analysis

Using the above morphology and chromatic recognition algorithm, we processed 100 samples. Research experiment is Pentium 3.2 G Duo CPU, 2G memory. Computed the total of 100 images detection takes 2.5 minutes, the average piece of image detection takes about 1 ~2 seconds. The results are shown in Fig.1.

We can see from the Fig.1 that, Fig.1(a)is the original cell image. Fig.1(b) is the blood cells contained with Gauss noise and pepper and salt noise. Fig.1(c)-(f) we compared our method with Sobel detection and Gauss Laplace detection. From the right image, we can see that, our proposed method can effectively extract the silhouette of the nucleus and cytoplasm. Because of this, we can calculate the ratio of nucleus and cytoplasm more correctly. Through 100 clinic examples, we get the result that, our method can highly effectively detect the blood cancer cell.

5   Conclusion

In this paper, we consider the drawbacks of the traditional morphology method, firstly adopts morphology filters to segmentation and granularity detection to process image by researching characteristics of blood cells images, which not only eliminates original image mostly, but also detects the main edges of image from different angles. These methods can well detective the cancer cells. On the one hand, the introduction of useful method can solve the unself-adjust function of image sensors. It is easily to find the color clustering center. And we also solve the difficult problem in medical image processing, which is the errors caused by differences in light. On the other hand, we use the advanced algorithms of morphology. Then, we can obtain the region of a single cell. After that, we extract their characteristic parameters. Combining the two advantages, it can be relatively identification of various cell components in complex background and certify them effectively.

After that, we analysis the redacted cell and blush cell in Fig. 1.(a), we set the threshold parameters by granularity detection in Fig. 3, that is, $A \in (800,2000)$, $L=120$, $I=0.76$, $E=0.8$, $R=0.5$, $B=0.015$. Then, we get the normal cells and the cancer cells accurately. Results are shown in Fig. 2.

We can see from the result that, the proposed methods is better than other methods, like Sobel detection and Gauss Laplace detection. Thus a preferably recognition effect is gained.

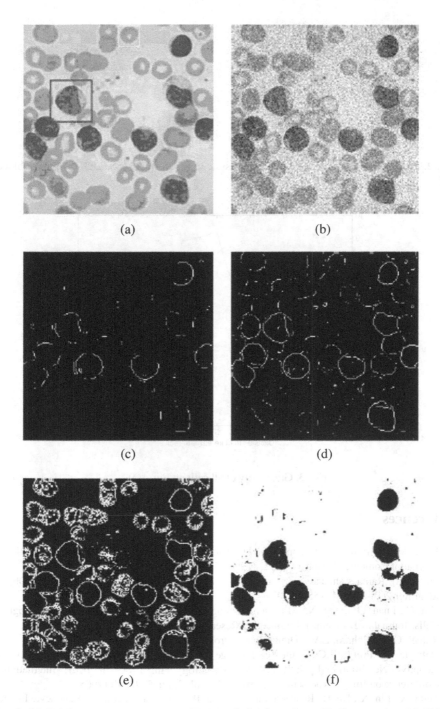

Fig. 1. (a) Original image with noise. (b) Granularity detection with Gauss noise. (c) Using Sobel filter to detection. (d) Using Gauss-Laplace filter to detection. (e) Using traditional morphology filter to detection. (f) Our proposed method to detection.

(a)   (b)

Fig. 2. Results of detection cancer cells in our system. (a) Normal blood cell. (b) Cancer blood cell.

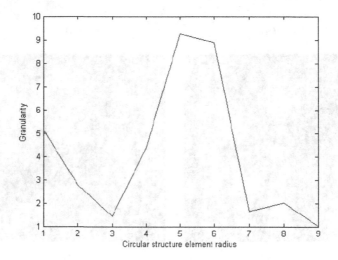

Fig. 3. Granularity of all cells in Fig. 1(a).

References

1. Kass, M., Witkin, A., Terzopoulos, D.: Sankes: Active Contour Models. International Journal of Computer Vision 1(4), 321 (1998)
2. Sinha, N., Ramakrishnan, A.G.: Automation of Differential Blood Count. Digital Object Identifier 2(15-17), 547–551 (2003)
3. Yin, C., Luan, Q., Feng, N.: Microscopic Image Analysis and Recognition on Pathological Cells. Journal of Biomedical Engineering Research 28(1), 35–38 (2009)
4. Rafael, C.G., Richard, E.W.: Digital Image Processing, 2nd edn. Prentice Hall (2002)
5. Debayle, J., Pinoli, J.C.: Multi-scale Image Filtering and Segmentation by Means of Adaptive Neighborhood Mathematical Morphology. In: Proc. of IEEE International Conference on Image Processing, Genova, Italy, vol. 3, pp. 537–540 (2005)
6. Tang, X., Lin, X., He, L.: Research on Automatic Recognition System for Leucocyte Image. Journal of Biomedical Engineering 24(6), 1250–1255 (2007)

7. Funt, B., Barnard, K., Martin, L.: Is machine colour constancy good enough? In: Burkhardt, H.-J., Neumann, B. (eds.) ECCV 1998. LNCS, vol. 1406, pp. 445–459. Springer, Heidelberg (1998)
8. Huimin, L., Lifeng, Z., Seiichi, S.: A Method for Infrared Image Segment Based on Sharp Frequency Localized Contourlet Transform and Morphology. In: IEEE International Conference on Intelligent Control and Information Processing, Dalian, China, pp. 79–82 (2010)

An Automatic Image Segmentation Algorithm Based on Weighting Fuzzy C-Means Clustering

Yujie Li[1,2], Huimin Lu[1], Lifeng Zhang[1], Junwu Zhu[2], Shiyuan Yang[1], Xuelong Hu[2], Xiaobin Zhang[2], Yun Li[2], and Bin Li[2], and Seiichi Serikawa[1]

[1] Department of Electrical Engineering and Electronics,
Kyushu Institute of Technology, E7-404,
Kitakyushu 804-8550, Japan
[2] College of Information Engineering, Yangzhou University,
Yangzhou 225009, China
{gyokuketu.ri,keibin.riku}@boss.ecs.kyutech.ac.jp,
{zhang,yang,serikawa}@elcs.kyutech.ac.jp,
{zhujunwu,huxuelong,zhangxiaobin,liyun,libin}@yzu.edu.cn

Abstract. Image segmentation is an important research topic in the field of computer vision. Now the fuzzy C-Means (FCM) algorithm is one of the most frequently used clustering algorithms. Although a FCM algorithm is a clustering without supervising, the FCM arithmetic should be given the transcendent information of prototype parameter; otherwise the arithmetic will be wrong. This limits its application in image segmentation. In this paper, we develop a new theoretical approach to automatically selecting the weighting exponent in the FCM to segment the image, which is called Automatic Clustering Weighting Fuzzy C-Means Segmentation (ACWFCM). This method can reduce the disturbance of noise; get the segmentation numbers more accurately. The experimental results illustrate the effectiveness of the proposed method.

Keywords: Image segmentation, Fuzzy C-Means clustering, Weighting Fuzzy C-Means algorithm, Clustering analysis, Automatic Clustering Weighting Fuzzy C-Means Segmentation.

1 Introduction

Image segmentation is one of the important image analysis technologies. One of the most basic definitions is: divide image into each with distinct features of regional and extract interested goals. Many different segmentation approaches [1-5] have been proposed such as: thresholding, region extraction and so on. But all of them cannot be generalized under signal scheme. In this paper, we bring forward an approach based on clustering.

Clustering is a process for grouping a set of objects into classes or clusters, so that, the objects with a cluster or class have high similarity, but they are very dissimilar to other objects [6]. There are two main clustering approaches: crisp clustering and fuzzy clustering. In the crisp clustering method the boundary between clusters is

J. Luo (Ed.): Soft Computing in Information Communication Technology, AISC 158, pp. 27–32.
springerlink.com   © Springer-Verlag Berlin Heidelberg 2012

clearly defined. However, in practice, the boundaries between clusters cannot be clearly defined. Some objects may belong to more than one cluster. In these cases, the fuzzy clustering methods provide better and more useful methods to cluster these objects [7].

The Fuzzy C-Means algorithm is a popular fuzzy clustering method. Traditional FCM clustering is dependent of the measure of distance between objects. In most situations, standard FCM uses the common Euclidean distance which supposes that each feature has equal importance. There are variable selection and weighting methods in approve the standard FCM.

Note that we propose a segmentation method in this paper, named Automatic Clustering Weighting Fuzzy C-Means (ACWFCM), which uses gradation-gradient-two dimensional histogram for automatically to obtain the number of clustering. The paper is organized as follows. A new objective function of automatic clustering we are considering is presented in section 2. The steps of ACWFCM are presented in section3. In section4 we examine the behavior of the algorithm on a variety of images, and compare the results with Otsu method and stand FCM algorithm. The conclusions of this paper are summarized in section 5.

2   Proposed ACWFCM Algorithm

2.1   Gradation-Gradient 2D Histogram

If the gray level of the target and background is highly correlated, the gradient value in each target region and background region is smaller than that at edge region. That is to say, the pixel value of target and background is close to the gradation axis in the gradation-gradient-two dimensional coordinate plane. While the edge region stay away from the gray axis. Furthermore, the edge regions are definitely between the background and targets or between different targets [8].

We can abandon the larger gradient value pixels, which can filter the image. See Figure 1, the gradation-gradient coordinate plane.

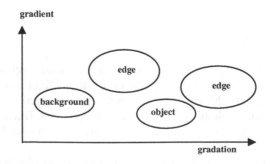

Fig. 1. Gradation-gradient two-dimensional histogram

2.2   Automatic Cluster Number Method

In the clustering algorithm, the number of clusters is often given as the initial conditions, which can realize the unknown data categories and give a clustering algorithm for estimating the numbers of objects. If inappropriate choice of cluster number, it will make the mismatch between the results of the division data sets and real data sets, leading to a failure of cluster.

In this paper, we propose the gradation-gradient-two dimensional histogram coordinate plane for deciding the objects in an image. In this histogram, according to the number of peaks to determine the number of clusters to complete clustering unsupervised.

Let $f(x,y)$ as an image, x, y represent the location of the image. The gradient value of each pixel is defined as

$$\nabla f = \left[G_x^2 + G_y^2 \right]^{1/2} = \left[\left(\tfrac{\partial f}{\partial x} \right)^2 + \left(\tfrac{\partial f}{\partial y} \right)^2 \right]^{1/2} \tag{1}$$

Set the pixel number as N_{AB}, which with a gray level A and gradient value B. Then the gradation-gradient histogram is

$$H_{AB} = N_{AB} / N \tag{2}$$

where N is the all pixel number of the image. We set the gradation-gradient threshold to remove the high gradient pixels in the gradation-gradient histogram. And then, make a projection to gradation, forming one-dimensional histogram $h(x)$.

After that, make a convolution of $h(x)$ and $g(x)$.

$$\Phi(x) = h(x) * g(x) = \int_{-\infty}^{\infty} h(u) \frac{1}{\sqrt{2\pi}\sigma} \exp[-\frac{(x-u)^2}{2\sigma^2}] du \tag{3}$$

where, $g(x) = \dfrac{1}{\sqrt{2\pi}\sigma} \exp[-\dfrac{(x-u)^2}{2\sigma^2}]$. The number of cluster C can be automatically calculated by the aggregate $\{x_i \mid \Phi'(x_i) = 0, \Phi''(x_i) < 0\}$. The centroids v_k are $\{x_i\}$.

$$C \mapsto \{x_i \mid \Phi'(x_i) = 0, \Phi''(x_i) < 0\} \tag{4}$$

2.3   ACWFCM Algorithm

In this section, we propose a new objective function for obtaining fuzzy segmentations of images with automatically determine the cluster number C.

The ACWFCM algorithm for scalar data seeks the membership functions u_k and the centroids v_k, such that the following objective function is minimized:

$$J_{ACWFCM} = \sum_{(x,y), i \in N} \sum_{k=1}^{C} w_i u_k (x, y)^q \parallel I(x, y) - v_k \parallel^2 \tag{5}$$

where $u_k(x, y)$ is the membership value at pixel location (x,y) for class k such that $\sum_{k=1}^{C} u_k(x, y) = 1$, $I(x,y)$ is the observed image intensity at location (x,y), and v_k is the centroid of class k. The total number of class C is automatically calculated by function (4). The parameter q is a weighting exponent on each fuzzy membership and determines the amount of fuzziness of the resulting classification. For simplicity, we assume for the rest of this paper that $q=2$ and the norm operator $\| \cdot \|$ represents the standard Euclidean distance. The main role of weighting parameter w_i is adjusting the cluster centers. w_i can be calculated by source image $f(x,y)$ and smooth image $\overline{f}(x,y)$. n expresses the times of pixel (x,y) appeared in the image. We form a two dimensional gradation histogram by $f(x,y)$ and $\overline{f}(x,y)$. In the two dimensional gradation histogram, $H(s,t)$ expresses the joint probability density, which with the gray level s at source image $f(x,y)$ and with gray level t at smooth image $\overline{f}(x,y)$.

$$w_i = H(s,t) \tag{6}$$

where, $\sum_{i=1}^{N} w_i = 1$, N is the all pixel number of the image [9]. When $w_i=1/N$, that is to say, the same effect to the classification of each class, ACWFCM is degenerated to the standard FCM.

The ACWFCM objective function (4) is minimized when high membership values are assigned to pixels whose intensities are close to the centriod for its particular class, and low membership values are assigned where the pixel data is far from the centroid.

The advantage of ACWFCM is that if a pixel is corrupted by strong noise, then the segmentation will be only changed with some fractional amount .But in hard segmentations, the entire classification may change. Meanwhile, in the segmentation of images, fuzzy membership functions can be used as an indicator of partial volume averaging, which occurs where multiple classes are present in a single pixel. Furthermore, the improved algorithm can remove noise, get the number of clusters more accurately, and obtain a better segment result.

3   Algorithm Steps

The objective function J_{ACWFCM} can be minimized in a fashion similar to the standard FCM algorithm. The steps for our ACWFCM algorithm can be described as follows:

step 1. Provide initial values for centroids, v_k, $k=1,...,C$, and C is automatically defined by function (4).

step 2. Compute new memberships as follows:

$$u_k(x, y) = \frac{\| I(x, y) - v_k \|^{-2/(q-1)}}{\sum_{l=1}^{C} \| I(x, y) - v_l \|^{-2/(q-1)}} \tag{7}$$

for all (x,y), $k=1,...,C$ and $q=2$.

step 3. Computer new centroids as follows:

$$v_k = \frac{\sum_{x,y} w_i u_k(x, y)^2 I(x, y)}{\sum_{x,y} w_i u_k(x, y)^2} \tag{8}$$

$k=1,...,C$ and $0 < i \le N$.

step 4. If the algorithm has converged, then quit. Otherwise, go to Step 2. In practice, we used a threshold value of 0.001.

4   Experimental Results and Discussion

We implemented ACWFCM on a Core2 2.0GHz processor using MATLAB. Execution time for ACWFCM ranged from 2 seconds to 5 seconds for a 256×256 image. For comparison, the first experiment applies the Otsu and FCM methods to segment the image. In the second experiment, we take the ACWFCM for segmenting. Fig.2 (b) to (d) shows the results of applying the Otsu method, FCM and ACWFCM algorithms to segment. The noise of Otsu and FCM are appeared clearly, but there is less noise in ACWFCM. The ACWFCM segment result is better than other methods.

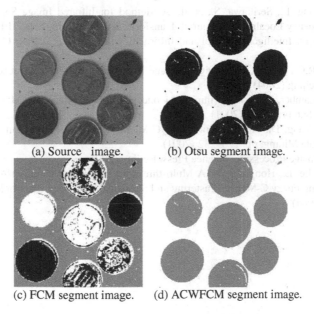

(a) Source   image.   (b) Otsu segment image.

(c) FCM segment image.   (d) ACWFCM segment image.

Fig. 2. Results of different segment methods

5   Conclusions

In this paper, we present ACWFCM, a new FCM algorithm that gives the number of cluster automatically for image segmentation. At the same time, using two-dimensional gradation-gradient histogram method, remove the high value of histogram, which can more effectively eliminate the influence of strong noise, enhance the segment performance and can automatically obtain the number of cluster. A large number of experiments show that the new fuzzy clustering image segmentation algorithm makes significantly reduced the number of iterations. But this new algorithm is not suitable to gray uniform variation images. This problem needs further study. However, it is significant meaning in automation segmentation of image with obvious changes in background and objects.

Acknowledgments. The research described in this paper are supported by the Project for the Chinese Jiangsu Province Nature Science Research Plan Projects for Colleges and Universities (Grant No. 08KJD120002) and Open fund Projects of Jiangsu Province Key Lab for Image Processing and Image Communication (No. ZK206008).

References

1. Chen, C.W., Luo, J., Parker, K.J.: IEEE Trans. on Image Processing 7(12), 1673–1683 (1998)
2. Parvati, K., Prakasa, R.S., Mariya, D.M.: Discrete Dynamics in Nature and Society, pp. 1–8 (2008)
3. Otman, B., Hongwei, Z., Fakhri, K.: Fuzzy Based Image Segmentation. Springer, Berlin (2003)
4. Lu, H., Zhang, L., Serikawa, S., et al.: A Method for Infrared Image Segment Based on Sharp Frequency Localized Contourlet Transform and Morphology. In: IEEE International Conference on Intelligent Control and Information Processing, Dalian, China, pp. 79–82 (2010)
5. Gonzalez, R.C., Woods, R.E.: Digital Image Processing. Publishing House of Electronics Industry, Beijing (2004)
6. Han, J., Kamber, M.: Data mining: Concepts and Techniques. Morgan Kaufmann Publishers, San Francisco (2001)
7. Hung, M., Yang, D.: An Efficient Fuzzy C-Means Clustering Algorithm. In: IEEE Intel. Conf. on Data Mining, pp. 225–232 (2001)
8. Sun, J.X.: Image Processing. Science Press, Beijing (2005)
9. Xinbo, G., Jie, L., Hongbing, J.: A Multi-threshold Image Segmentation Algorithm Based on Weighting Fuzzy C-Means Clustering and Statistical Test. Acta Electronica Sinica 32(4), 661–664 (2004)

Secure DV-Hop Localization against Wormhole Attacks in Wireless Sensor Networks[*]

Ting Zhang[1], Jingsha He[2], and Yang Zhang[2]

[1] College of Computer Science and Technology,
[2] School of Software Engineering
Beijing University of Technology
Beijing 100124, China
{zhangting06,gordenzhang}@emails.bjut.edu.cn, jhe@bjut.edu.cn

Abstract. Secure localization has become a serious issue along with increasing security threats in wireless sensor networks (WSNs). In this paper, we address the problem of making sensors in WSNs to locate their positions in an environment in which there exists the threat of wormhole attacks. We propose a secure localization scheme based on DV-Hop algorithm, named secure DV-Hop localization scheme (SDVL), which can fight against wormhole attacks in wireless sensors networks without increasing hardware complexity or too much computational cost. In SDVL, sensors ensure the security of their localization by using an effective detection mechanism. Simulation results show that the proposed scheme is capable of fighting against wormhole attacks while improving accuracy compared to the basic DV-Hop algorithm.

Keywords: secure localization, distance vector-hop, wormhole attack, wireless sensor networks.

1 Introduction

With the increasing functionality and decreasing cost of sensor nodes, wireless sensor networks (WSNs) have started to be deployed. Since the data collected by the sensors are usually accompanied by the position of the sensors, localization has become a significant issue in WSNs. So has the security of localization. Current localization algorithms can be divided into range-based algorithms and range-free algorithms [1, 2]. In resent years, some secure localization algorithms are been proposed [3, 4]. DV-Hop [5] is one of the typical range-free algorithms which is a distributed positioning algorithm based on distance vector routing. However, the algorithm is vulnerable to some threats and attacks. Wormhole attack is one of the most common forms of attacks. Therefore, some localization algorithms based on DV-Hop have been proposed [6] in which the trust mechanism is introduced into DV-Hop to improve the security of localization.

In this paper we propose a secure localization scheme based on DV-Hop, called secure DV-Hop localization scheme (SDVL), which achieves secure localization

---
[*] The work in this paper has been supported by funding from Beijing Education Commission.

against wormhole attacks. We also provide simulation to show that SDVL can locate sensors with a higher level of accuracy in an un-trusted environment compared to basic DV-Hop localization. The rest of this paper is organized as follows. In the next section, we describe the wormhole attack model in DV-Hop localization. In Section 3, we present the SDVL scheme and describe some implementation details. In Section 4, we present some simulation results for SDVL to verify its performance. Finally, in Section 5, we conclude this paper in which we also describe our future work.

2   Wormhole Attack Model in DV-Hop Localization

2.1   The Problem Statement

We consider secure localization in the context of the following design goals: robustness against wormhole attacks; improving accuracy; acceptable computational cost and communication overhead.

In the discussion, we assume that the network consists of beacon nodes, sensors and attackers. The beacon nodes are capable of positioning themselves while the sensor nodes need to locate their own positions. We also assume that all beacon nodes are credible, all the nodes in the sensor network are capable of performing clock synchronization, and each node has a unique identifier ID.

2.2   Wormhole Attack Model for DV-Hop

DV-Hop is the most basic scheme in an ad hoc positioning system (APS) with three non-overlapping stages. In the first stage, each node in the network gets a hop-count to the beacon nodes. In the second stage, beacon nodes calculate the average size for one hop to other beacon nodes. In the third stage, sensors estimate their positions by using trilateration after calculating the distance to the beacon nodes. Wormhole attack is one of the typical threats in DV-Hop, which results in the packets propagating through further paths. Hence, we analyze the secure issue of wormhole attacks in DV-Hop localization for WSNs.

Wormhole attack is usually implemented by two nodes in conspiracy to establish a wormhole link between the two attacking nodes. Attackers can reduce the hop counts easily through wormhole link to acquire the right of message propagation. Wormhole attacks are difficult to detect since they can be launched without compromising any sensors. As shown in Fig. 1, the normal path between the two beacon nodes B1 and B2 is B1-S1-S2-S3-S4-S5-B2. But the two attackers W1 and W2 can make B1-S6-W1-W2-S7-B2 the shorter path through the wormhole link than the normal one. In DV-Hop localization, the error in estimation of the hop count between two nodes can result in wrong hop size, consequently a lower accuracy of localization.

To achieve our goals, in SDVL, we introduce the wormhole detection mechanism into DV-Hop to equip the network with the capability of detecting wormhole attacks and segregating the attackers from the sensors before the stage of position estimation to ensure the accuracy of localization even in the presence of wormhole attackers.

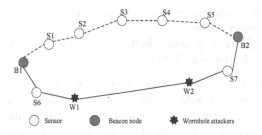

Fig. 1. Wormhole attack model

3   The Secure DV-Hop Localization Scheme

SDVL solves the problems of wormhole attack detection in localization for WSNs, which helps improve the accuracy of localization and make the localization system more secure. Main steps of the algorithm are described as follows.

3.1   Routing Initialization

Beacon nodes broadcast messages that include the positions of themselves and the initial hop-count value of 0. A node that receives such a message would acquire the position of a beacon node, write its own ID into the message and send it to its neighbors with an increment of the hop-count value. Each beacon node can thus get the positions of other beacon nodes and the shortest distance in hops between itself and the other beacon nodes with this method. Then, each beacon nodes calculates the average size of one hop using Equation (1) in which c_i denotes the average hope size of beacon node i to other beacon nodes and h_i denotes the hope-count of beacon node i to other beacon nodes. In the equation, we assume that the two-dimensional coordinate of beacon node i is (X_i, Y_i), all the beacon nodes around i are denoted by j, and their coordinates are denoted as (X_j, Y_j).

$$c_i = \frac{\sum \sqrt{(X_i - X_j)^2 + (Y_i - Y_j)^2}}{\sum h_i}, i \neq j \quad \cdot \tag{1}$$

3.2   Detection of Attacks

Wormhole attacks are mainly implemented through the approaches of packet encapsulation and proprietary channel. In the first approach, the attackers envelop the routing packets to make hop-counter fail to work normally. In the second approach, the attackers transfer routing packets through a proprietary channel with high bandwidth. We can thus classify wormhole attackers into two categories: hidden identity and un-hidden identity. In the first category, a normal sensor in the routing path adds its own ID and increments the hop-count. However, the attacker doesn't add itself into the packet, nor does it change the hop-count. In the second category, the attacker receives and sends packet normally, but makes the hop-count smaller by going through a proprietary channel.

The goal of our scheme is to detect both categories of wormhole attackers. Main steps of detection are described as follows:

1. Each node in the routing path checks whether the previous node has added itself into the packet. The previous node will be labeled as an attacker unless it adds its ID into the packet and increments the hop-count. If the node determines that its previous node is a malicious sensor, it will broadcast an alarm to remove the attacker. Un-hidden identity attackers will be detected in this way.

2. If every node in the routing path detects and thus ensures that its previous node receives and sends routing packet normally, an integrated routing path will be constructed between beacon nodes. Beacon nodes will get the minimum hop-count between each other and estimate the average hop size using Equation (1). A beacon nodes will then compare the average hop size c with transmission radius R. if $c>R$, wormhole attackers exist in the path.

3. After the presence of wormhole attackers are detected, the network should determine the attacker nodes and remove them. The network compares the actual distance d between each pair of nodes with transmission radius R using Equation (2) in which d denotes the actual transmission distance of signal, v denotes the propagation velocities of the signal and t denotes the propagation time between the nodes. If $d>R$, the nodes at both ends of this distance are a pair of wormhole attackers.

$$d = v \cdot t .$$
(2)

3.3   Position Estimation

Sensors estimate their positions through maximum likelihood estimation after getting the positions around beacon nodes and the distance between each other. Suppose the number of beacon nodes around an unknown node is n and their coordinates are $(x_1, y_1), \ldots (x_n, y_n)$, and the distance between unknown node $U(x, y)$ and the beacon nodes are $d_1, d_2 \ldots d_n$, respectively. Using Equation (3), the unknown node's position can be inferred. In addition, n distance equations about U and n beacon nodes are listed with each of the first $n-1$ equations minus the last equation. The results are show in Equation (4). $U(x, y)$ can then be calculated using Equations (5) and (6).

$$(x - x_i)^2 + (y - y_i)^2 = d_i^2 , i = 1, 2, \ldots, n .$$
(3)

$$\begin{cases} x_1^2 - x_n^2 - 2(x_1 - x_n)x + y_1^2 - y_n^2 - 2(y_1 - y_n)y = d_1^2 - d_n^2 \\ \qquad\qquad \ldots \\ x_{n-1}^2 - x_n^2 - 2(x_{n-1} - x_n)x + y_{n-1}^2 - y_n^2 - 2(y_{n-1} - y_n)y = d_{n-1}^2 - d_n^2 \end{cases} .$$
(4)

$$U = A^{-1}b .$$
(5)

$$A=2\begin{bmatrix} x_1-x_n & y_1-y_n \\ \ldots \\ x_{n-1}-x_n & y_{n-1}-y_n \end{bmatrix}, \quad b=\begin{bmatrix} x_1^2-x_n^2+y_1^2-y_n^2-d_1^2+d_n^2 \\ \ldots \\ x_{n-1}^2-x_n^2+y_{n-1}^2-y_n^2-d_{n-1}^2+d_n^2 \end{bmatrix}. \tag{6}$$

4 Simulation

We have done some simulation to show the performance of SDVL as described below.

1. The network configuration of the simulation is set up as follows: 50 sensor nodes, 10 beacon nodes and 6 attacking nodes are deployed randomly in a 40×60 m^2 area. The transmission radius of each beacon and sensor node is 10 m. We compare the performance of SDVL to DV-Hop localization algorithm in the presence of wormhole attacks. We show the improvement on the average location error when SDVL is used in the WSN. The localization error is calculated using Equation (7) and the average localization error of all the nodes is calculated using Equation (8) in which (x_i, y_i) denotes the measurement coordinate of sensor i, (x_i', y_i') denotes the actual coordinate of sensor i and N denotes the number of sensors. We repeat the two localization algorithms 10 times in the network. The results of the simulations are shown in Fig. 2.

$$e_i = \frac{\sqrt{(x_i - x_i')^2 + (y_i - y_i')^2}}{R}. \tag{7}$$

$$\bar{e} = \sum_{i=1}^{N} e_i / N. \tag{8}$$

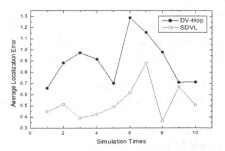

Fig. 2. Comparison of relative localization errors

2. For performance evaluation, we randomly distribute 50 sensors within an area, which generates a topology as shown in Fig. 3, and randomly select 6 wormhole attackers. Then, we compute the average location errors for different numbers of beacon nodes. Fig. 4 shows the average localization errors of sensors in the presence of wormhole attackers with varying numbers of beacon nodes.

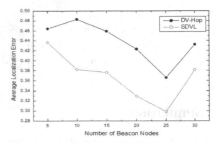

Fig. 3. Test topology

Fig. 4. Localization errors for varying number of beacon nodes

5 Conclusion

In this paper, we first analyzed wormhole attacks in DV-Hop localization. We then proposed a secure localization scheme called SDVL to detect and fight against wormhole attacks in WSN, which helps to improve localization accuracy in the presence of wormhole attackers. We described how SDVL detects the attackers in the network to get more accurate positions of sensors. Simulation results showed that SDVL can achieve higher localization accuracy in the presence of wormhole attacks than the basic DV-Hop. In our future work, we will extend our secure localization scheme to improve the security of localization in the case of other kinds of malicious nodes without incurring too much computational overhead and communication cost.

References

1. Savvides, A., Han, C.C., Strivastava, M.B.: Dynamic Fine-Grained Localization in Ad-Hoc Networks of Sensors. In: Proc. 7th Annual International Conference on Mobile Computing and Networking, Rome, Italy, pp. 166–179 (2001)
2. Bulusu, N., Heidemann, J., Estrin, D.: GPS-Less Low Cost Outdoor Localization for Very Small Devices. IEEE Personal Communications 7(5), 28–34 (2000)
3. Lazos, L., Poovendran, R.: SeRLoc: Secure Range-Independent Localization for Wireless Sensor Networks. In: Proc. 2004 ACM Workshop on Wireless Security, Philadelphia, pp. 21–30 (2004)
4. Lazos, L., Poovendran, R.: HiRLoc: High-Resolution Robust Localization for Wireless Sensor Networks. IEEE Journal on Selected Areas in Communications 24(2), 233–246 (2006)
5. Niculescu, D., Nath, B.: DV Based Positioning in Ad Hoc Networks. Journal of Telecommunication Systems 22(1-4), 267–280 (2003)
6. Wu, J.F., Chen, H.L., Lou, W., Wang, Z., Wang, Z.: Label-Based DV-Hop Localization against Wormhole Attacks in Wireless Sensor Networks. In: Proc. 2010 IEEE International Conference on Networking, Architecture, and Storage, Macau, China, pp. 79–88 (2010)

An Analysis on the Application and Development Trend of CAD Technology of Our Country

Huanzhi Gao[1], Beiji Zou[2], and Xiaoying Wang[3]

[1] China Nonferrous Metals Industry Association, Beijing, China
[2] Central South University, School of Information Science and Engineering,
Changsha, China
[3] Institute of Disaster-prevention Science and Technology,
School of Disaster Information and Engineering, Sanhe Heibei, China
ghzygc_1976@163.com

Abstract. This paper elaborates the development course and application status of CAD technology in our country, gives an analysis on problems of CAD application and discusses the development trend of CAD technology from about five aspects.

Keywords: CAD technology, CAD system, technological revolution, development trend.

1   Introduction

Computer Aided Design (CAD), which came into being about in recent 40 years, is a new subject combining computer science with engineering design methods. It is a technology which takes advantage of computer system to assist create, modify, analyse and optimize design. With the popularization of the Internet/Intranet network, parallel and high-performance computering and transaction processing, the synergy virtual design of different places and real-time simulation are also widely used.

As an important part of information technology, the CAD technology integrates the computer capability of high-speed mass data storage, processing and mining with the human ability of comprehensive analysis and creative thinking. It plays an important role in aspects such as accelerating development of project and products, shortening designing and manufacturing cycle, improving the quality, reducing the cost and promoting the market competitiveness and innovation ability of enterprises.

Being limited by computer software and hardware technologies in the early period, the CAD technology progressed slowly. It is until the middle of 1970s when the development of the CAD technology was accelerated because of the rapid growth of computer software and hardware technologies. Thus, the development and application level of CAD becomes a major criterion of industrial modernization of a nation during a certain period and, to some extent, reflects the comprehensive strength of a nation. So far, the development of CAD technology has mainly gone through four phases.

The first technological revolution of CAD--- surface modeling

In order to cope with large quantities of free surface problems in airplane and automobile manufacturing, developers of French Dassault Aviation create a method of free surface modeling characterized by surface model and develops the three-dimensional surface modeling system CATIA on the basis of the two-dimensional mapping system CADAM, marking the emancipation of CAD technology from the view mode of sheer imitation of engineering drawing. It first realizes the possibility of complete describing the main information of products part by computer. Meanwhile, it lays the realistic foundation for the development of CAD.

The second technological revolution of CAD--- solid modeling

Based on the exploration of the integrated technology of CAD/CAE, the SDRC company issued I-DEAS, the first large-sized CAD/CAE software of the world in 1979, which is completely based on the solid modeling. It represents the future development direction of CAD technology.

The third technological revolution of CAD--- parametric technology

In the middle of 1980s, the CV company proposed an algorithm that is more innovative and better than free modeling--- parametric solid modeling. This algorithm is characterized by feature-based, full-size constrained, data correlation, dimension-driven design and modify.

The fourth technological revolution of CAD--- quantity technology

Developers of the SDRC company proposed a solid modeling technology being guided by the chief source of parametric technology, which is quantity technology. It is more advanced than the parametric technology. The quantity technology not only retains original advantages of the parametric technology but also overcomes many disadvantages.

2   The Application Status of CAD Technology in Our Country

The CAD technology was introduced about in 1960s in our country. It started relatively late. However, the development of recent years is fairly rapid. It is gradually used in industries such as machinery, electronics, architecture, automobile, costume, etc. And it is also in the probationary period of many other areas. In recent years, the gradually declining price of computer also promotes the development of CAD technology, mainly for the following aspects.

2.1   Sizable Enterprises Have Established Relatively Complete CAD System

Most sizeable enterprises have successfully applied the CAD technology to their own two-dimensional engineering drawing. Some enterprises try to store data files in floppy disk, being archived by archives administration, and then conduct the collection of relevant data. It saves greatly manpower, physical resources and the time of product development. There are even some units with strong technical strength and proficient skills of three-dimensional CAD application who conduct solid modeling of graphic model by three-dimensional design software. During the preliminary stage of the scheme, complete a variety of images and animations and conduct preview and report of scheme with three-dimensional modeling. Carry out finite element analysis and optimal design by taking advantage of finished model in the design process. These enterprises have accomplished the "drawing board throwing" and have shorten the production circle.

The establishment of complete CAD system in sizable enterprises indeed improves the quality of products, shortens the production circle and achieves favorable social and economic benefits.

2.2   Small and Medium-Sized Enterprises Gradually Use the CAD Technology

The CAD technology has been more widely popularized and applied in 1990s, especially in sizable enterprises where CAD system brought obvious economic benefits. Thus, it enables small and medium-sized enterprises to realize the importance of appling CAD technology in production. They establish one after another their own CAD system by limited hardware and network environment under the premise of classifing existing CAD software, replacing the traditional manual design with CAD graphic design and greatly improving the efficiency of enterprises.

2.3   Colleges and Universities Develop Numerous Practical CAD Softwares by Themselves

During the middle and later period of 1970s, some domestic colleges, scientific research institutions and sizable enterprises conducted a great deal of researches in the filed of CAD technology and developed some practical CAD software by themselves according to their own requirements. These softwares are relatively cheap and easier for maintenance and training so that they are more convenient for application and popularization.

2.4   Introduction Foreign Unitized CAD Equipment

Since the reform and opening up, the CAD technology of our country has gain a rapid development. However, it is still far from satisfing design requirements of enterprises. In order to be geared to international standards, many enterprises start to spend huge sums of money in introducing unitized CAD equipment from abroad to enhance the competitiveness of enterprises.

At present, Chinese enterprises are widely using CAD technology, which improves design quality of products and shortens design circle. Yet it is undeniable that there are still a lot of problems in the application of CAD, which are mainly as follows:

Low utilization of CAD equipment. In practical use of enterprises, there are few designers who can completely grasp such an advanced designing tool for serving the research task. What's more, inadequate tasks leave equipments unused.

Looking up to "hardware" and down on "software". Instead of buying legal copy of CAD software, enterprises prefer to spend huge sums of money in purchasing computer, printer and plotting instrument. "Pirate softwares" are bound to be not guaranteed in terms of functions and technical services.

Insufficient understanding of the secondary development of software. Commercial CAD softwares on the market currently are basically development platforms of a certain profession with common features. Yet they fail to satisfy various needs of different enterprises. This is doomed to requiresecondary development according to their own actual needs. However, this kind of secondary development is not highly concerned by enterprises.

3　The Development Trend of the CAD Technology Application

It is the path of enterprise informatization that further deepen the application of CAD on the basis of current situation. At present, a favorable idea is to use database technology, which treats normative design information as data object, so as to conduct real-time change. The CAD technology is an important part of advanced manufactuing technology. With the realization of manufacturing process and information collection becoming the goal, many enterprises have embarked on the establishment of CIMS system of the enterprise level in order to achieve system integration and information sharing.

3.1　High Openness of CAD System

CAD system is now widely established on platforms of open operating system XP/NT and UNIX. There are also products of CAD on platforms like Java and Linux. In addition, CAD system provides final users with secondary development environment, of which the kernel source code is accessible, so that users are able to customize their own CAD system. The openness of CAD system is the foundation of its practicability and the possibility of being transformed into actual productivity. The openness of CAD system is mainly reflected in such aspects as working platform, customer interface, environment of application and development, and information interchange with other systems.

3.2　Hereditability of CAD System

During the overall process of production, information of the product can be transited among different links which is beneficial to developers in terms of comprehensively considering cost, quality, progress and user demand of products.

In the updating progress of products, it is required that all historical data should be acquired in order to take full advantage of product information that have been practiced in production. During the process of new product's development, a brand new product can be acquired only by modifing and reshaping a very small part of components. It not only shortens the development cycle and saves the cost of research and manufacturing, but also improves the standardization of products and ensures an one-time success of products.

3.3　High Intelligence of CAD System

Intelligent CAD is the inevitable trend of its development. Intelligent CAD is not just a simple integration of intelligence technology and CAD technology. It requires to carry out deeper researches in thinking model of human design, being expressed and simulated by information technology. In this way, an efficient CAD system will be established and will definitely provide the filed of artificial intelligence with new theories and approaches. The development trend of CAD is going to have a profound impact on the development of information science.

3.4 Standard of CAD System

CAD softwares are usually integrated on a heteroid working platform. In order to support the environment of heteroid cross-platform, CAD system should be open which is mainly solved by the technology of standardization.

Support software of CAD has gradually actualized the ISO standard and industrial standard. In addition, application-oriented standard elements and standardized methods also became essential contents of CAD system and are developing towards the application of rationalized engineering design.

Currently there are two types of CAD standard. One is common standard proceeding from national or international standard formulation unit; the other is market standard, namely industrial standard, which is private.

3.5 Three-Dimensional Design Is a Reform of Design Concept

Information expressed by planar drafting are not unique nor complete in many cases. Only by constant modification and perfection can information be expressed clearly. What's more, it fails to formulate the design idea and has difficulty in representing three-dimensional entities of thinking such as materials, shapes and sizes of products and relevant products. Three-dimensional CAD system enables people to start design directly from the three-dimensional model of thinking, makes it possible to express all geometric parameters and design concepts, and conducts analysis and researches of the whole design process in three-dimensional mode with unified data. Thus, the design concept can be improved and idealized. There is a full correlation of three/two dimensions in the system of three-dimensional CAD. In the product design of three-dimensional CAD, some basic attributes of products designed by rendering can be adjusted so as to achieve the effect of rendering the outlook of products. Establish an ample, comprehensive and unified design database, on the basis of which various analysis can be carried out on global design or components. From the perspective of long-term development, the three-dimensional CAD technology will inevitably replace the planar drafting of CAD. With the constant increasing of the performance price ratio of computer and the development of popularization of network communication, intelligentization of information processing and practical multimedia technology, popularization and application of CAD technology are becoming wider and deeper. CAD technology is developing in the direction of open, integrated, intelligent and standardized. It is of far-reaching significance to grasp the development trend of CAD technology, not only for developing readily marketable products of our CAD software industry but also for the correct lectotype and program of their own CAD application system. It is unavoidable that the ultimate way of products design is three-dimensional design. Three-dimensional CAD technology is the inevitable development trend of CAD application. The earlier the three-dimensional design is put into practice, the more economic benefits and technical benefits will be achieved.

References

1. Zheng, G.: Three stages of development of CAD. China Information World (October 2003)
2. Sun, S., Sun, Y.: Introduction of the original equipment manufacturing. Machinery Industry Press (2000)
3. Hao, J.: Computer aided engineering. Aerospace Industry Press (2000)
4. Tong, B.: Basis of the mechanical CAD technology. Tsinghua University Press (January 2011)
5. Tan, G.: Basis of the mechanical CAD technology. Harbin Institute of Technology (July 2006)

Na⁺ Distribution and Secretion Characters from *Tamarix Hispida* under Salt Stress

Tingting Pan[1], Weihong Li[2], and Yapeng Chen[2]

[1] Xinjiang Agricultural University Urumqi, China
pantingting125@126.com
[2] Xinjiang Institute of Ecology and Geography,
Chinese Academy of Sciences, Urumqi, China

Abstract. Taking Tamarix hispida as test materials, a group of different concentrations of NaCl were then added to the pots and the salinity was maintained at5, 10, 15 and 20g/L. Na+ content of different samples were analyzed by ICP-AES. This paper studied Na+ distribution in various organs and secretion characters from Tamarix hispida under the stress of different salt concentration, discuss the selective absorption and Na+ transport mechanisms. The result show that Na+ content of Tamarix hispida were increasing prominently with increased salinity, the order of Na+ accumulation on Tamarix hispida is leaf>root>stem, leaf is the main part of Na+ accumulation and secretion. The secretion of Na+ is also enhanced with the increase of salt concentration. The excreting rate of salt gland was higher in the morning than that in the afternoon and higher in daytime than that at night. Because salt secretion has a high positive correlation with atmospheric humidity, at the same time it has a higher negative correlation with atmospheric temperature. Such high selectivity of salt gland secreting salt is the self-adjustment mechanism of Tamarix hispida to adapt to the habitat, maintaining the proper salt and nutrient contents.

Keywords: salt stress, Tamarix hispida, Na⁺ secretion.

1   Introduction

The influence of salt stress on the plant growth is very complex[1], which has been related to the ion toxicity[2], the osmotic stress, the lack of mineral elements and some others[3,4]. The plants under salt stress in the short term will accumulate too much salt, which can form the osmotic stress in plants and restrain the plant growth. And the long-term salt stress will cause the formation of ion toxicity in plants. Because of the secretion of salt glands on the ions, the salt secretion plants have played an important role in reducing salt's damage on the active tissues and adjusting the balance of mineral elements. The *Tamarix hispida* is one of the Tamarix L plants that belong to the Tamaticaceae and have the best adversity stress-tolerant abilities. It is the main shrub in desert riparian of Tarim River. The ecological adaptability of them is widely, such as the drought-resistance, the salinity-resistance and other physical characteristics. Meanwhile, they can also curb the desertification, prevent the wind and fix the sand.

J. Luo (Ed.): Soft Computing in Information Communication Technology, AISC 158, pp. 45–52.
springerlink.com   © Springer-Verlag Berlin Heidelberg 2012

Moreover, they can grow in extreme drought areas in which the rainfall is scare, the evaporation is high and the wind is big. These plants are widely distributed in saline-alkali soils around the Tarim Basin, which has formed the salt shrubs that regard *Tamarix hispida* as the main species and have had important ecological functions.

For a long time, *Tamarix hispida* has aroused the attention of scholars depending on its significant ecological and economic benefits as well as the unique ecological and biological characteristics. Nowadays, there have been many reports about the community characteristics of *Tamarix hispida*, the salt mechanism and the relationship between distribution and salinity[6,7]. And there are also some in-depth studies in the taxonomy of *Tamarix hispida*[8,9]. However, the studies about the salt secretion functions of salt stress on *Tamarix hispida* are involved few. This paper has employed *Tamarix hispida* as materials, mainly focusing on the Na+ accumulation and the secretion characteristics under salt stress. It has analyzed the ion transportation and the optional absorption mechanism of plants in salt stress environment, which aims to provide theoretical basis for the repair of desertification as well as the references for restoring and reconstructing the impaired ecosystem along the lower reaches of Tarim River.

2 Materials and Methods

2.1 Plant Materials

The test is carried out in the ecology and restoration monitoring test station located in lower reaches of Tarim River. The sandy loam is regarded as the potted medium. On April 2010, the *Tamarix hispida* seedlings which grow well and have relatively consistent growth are transplanted to a PVC pipe with a diameter of 30cm and 100cm high. The pot experiment has employed the complete randomized experimental design. And the salt (NaCL) stress is handled in early July. The NaCL is configured into a certain concentration of solution according to the dry weight of soil and the soil is manured into it once. There are four levels of salt treatment: 5.0g/L、 10.0 g/L、 15.0 g/ and 20.0 g/L and six replicates are set per treatment. The samples won't be collected every five days until the end of stress and there are four times for collecting the samples. The *Tamarix hispida* that will be tested in this paper is taken from the 4th sampling.

1) The measuring methods of Na+ content and salt excretion of plants

Na$^+$ secretion is respectively collected according to Marilyn's method in the morning, in the afternoon and at night. Before the sampling, *Tamarix hispida* that will be tested should be washed by deionized water. Then they need to be cut leaves and branches in the intervals which are from 20:00 on August 9 to 10:00 on August 10 and the intervals of 8:0---14:00, 14:00---20:00 and 8:00---20:00 on August 10. Na$^+$ secreted by salt gland should be washed and Na$^+$ secretion are collected in the morning, in the afternoon and at night. The collecting of secretion should be repeated three times for each time period. After collecting the salt secretion solution, the plants should be naturally dried. Then the samples are taken to the lab and washed by tap water in order to remove the surface dirt. At the same time, they need to be washed three times with deionized water. After drying they should be put into the drying oven with a temperature of $60 \sim 80\,°C$ for $8 \sim 10h$. The dried samples are grinded and the crushed samples are put into the ethylene bottle for use through 200 mesh sieve. 0.5000g($\pm$0.0001) Powder samples are weighed and put into the 50ml porcelain crucible. They are firstly carbonized in the furnace with low temperature for $1 \sim 2h$ and then incinerated in muffle furnace for $10 \sim 12h$ ($600\,°C$). After incinerating the powder samples, 1:1 nitric acid solution is added into them with 5ml for sample's dilution and extraction. Then the treated samples are put into 500ml graduated cylinder and the deionized water is used to volume. Moreover, the inductively coupled plasma emission analyzer (ICP-AES) is employed to test Na^+ content in samples. The ion secretion rate is represented by Mg number of Na^+ secreted by per gram fresh weight in one hour.

Based on the experimental data, the single factor analysis (ANOVA) has been employed to significantly analyze the differences between different salt stress levels and different growth parts. And the least significant difference (LSD) multiple comparison method has been employed to comparatively describe the differences between different salt stress levels and different growth parts under 95% reliability.

3   Results and Analysis

3.1   The Comparison of Stem, Leaf and Na⁺ Content

The Na^+ content in roots, stems and leaves of *Tamarix hispida* under salt stress with different levels is showed in Table 1. Form this Table, we can find that the basic trend of Na^+ content in different parts of *Tamarix hispida* is leaf $>$ root $>$ stem. With the increase of NaCL concentration, the Na^+ content in *Tamarix hispida* is also rapidly increased. However, the increase magnitudes of Na^+ content in different parts are of difference. The Na^+ content in leaves is highest, when the NaCL concentration is 20g/L, the content will be from 13.22 to 14.20. The Na^+ content in roots is from 10.30 to 10.72. And the Na^+ content in stems is from 9.13 to 9.31. The Na^+ content in leaves is significantly higher than the one in roots, which indicates that the sodium-resistant ability of Tamarix leaves under salt stress is greater than the roots and the Tamarix leaves are the main part for accumulating the sodium ions. At the same time, the Na^+ content in Tamarix leaves is also significantly higher than the one in stems, which proves that the stem is only a Na^+ transportation channel and there is no process of accumulation. So the content is the smallest.

The Na^+ content of *Tamarix hispida* leaves have showed significant differences in different levels of salt stress ($F4,24=5.93$, $p > 0.001$). By NaCL irrigation, the Na^+ content of *Tamarix hispida* leaves has been increased. When the salt stress is 5, 10g/L, the Na^+ content of *Tamarix hispida* leaves hasn't showed significant difference. When the salt stress is 15, 20g/L, the Na^+ content of *Tamarix hispida* leaves is significantly higher than the one that the salt stress is 5, 10g/L. However, the Na^+ content of leaves hasn't showed significant difference when the salt stress is 15, 20g/L. The Na^+ content of roots has showed extremely significant differences under different levels of salt stress ($F4,24=20.26$, $p < 0.001$). The different levels of salt stress have all significantly increased the Na^+ content of roots. When the salt stress is 10, 15g/L, the Na^+ content of roots hasn't showed extremely significant difference. And when the salt stress is 20g/L, the Na$^+$ content of *Tamarix hispida* roots is significant higher than the one that the salt stress is 5, 10g/L. The Na$^+$ content of stems has also showed extremely significant differences under different levels of salt stress (F4,24=39.75, p<0.001). After the NaCL irrigation, the Na$^+$ content of *Tamarix hispida* leaves has been also significantly increased. With the strengthening of salt stress, the Na$^+$ content of *Tamarix hispida* stems has showed a gradually increasing trend.

Table 1. Na$^+$ contents in roots, stems and leaves of tamarix hispida

NaCl (g/L)	sampling organization (mg·g^{-1})		
	root	stem	leaf
5	9.49±0.13a	7.59±0.36a	12.88±0.48a
10	10.05±0.18b	8.08±0.28b	12.99±0.05a
15	10.23±0.38bc	8.41±0.23c	13.56±0.57b
20	10.51±0.21c	9.22±0.09d	13.76±0.54b
F Value	20.26***	39.75***	5.93**

Note: the data in the Table is mean ± S.D.. The same letters show that the Na$^+$ content has showed some significant differences under different salt stress (the least significant difference multiple comparison, p>0.05). *** shows that the results of variance analysis have showed extremely significant differences (p<0.001). ** has showed the results of variance analysis have showed significant differences (0.001<p<0.01).

3.2   The Rhythmic Changes of Salt Excretion

The Tamarix hispida's secretion of Na$^+$ has showed that the one in the morning is higher than the afternoon and the one in daytime is higher than the night. That is because the secretion change of *Tamarix hispida* has high positive correlation with the trends of atmospheric humidity changes. At the same time, it has high negative correlation with the atmospheric temperature changes. From the Figure 1, we can find that with the increase of NaCL concentration, the secretion rate of leaf salt gland on Na$^+$ has been gradually increased. The secretion rate of *Tamarix hispida* on Na$^+$ in daytime is higher 49%, 55% and 51% than night, which may be relative to the phenomenon that the temperature during the day is higher than night and the humidity during the day is lower than night. The differences between day and night have reached a significant level (p < 0.001). However, the differences between the treatments are not significant.

In Table 2, the secretion rates of Na$^+$ in the morning are respectively 1.05 times, 1.10 times, 1.07 times and 1.01 times faster than the ones in the afternoon. Form the perspective of secretion rate, although the secretion of leaf salt gland on Na$^+$ is reduced, the variation amplitude is slightly. The secretion of Na$^+$ between the morning and the afternoon is no significant difference, but there are some significant differences between the treatments (p<0.001).

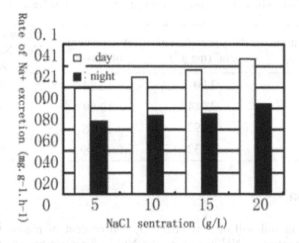

Fig. 1. Comparison of Na+ secrition from Tamarix hispida leaves between day and the night

Table 2. Na+ secretion rates from Tamarix hispida in a.m. (8:00—14:00) and p.m. (14:00—20:00)

sampling time	the secretion rate of Na+ under different salt stress (mg·g^{-1}·h^{-1})			
Sampling time	5g/L	10g/L	15g/L	20g/L
8:00—14:00	0.046	0.074	0.078	0.083
14:00—20:00	0.044	0.067	0.073	0.082

3.3 The Relationship between the Secretion of Tamarix Hispida on Na+ and the Na+ Content of Leaves

From the Table 3, the secretion ability of *Tamarix hispida* salt gland on Na+ is enhanced with the increase of salt concentration and the differences are extremely significant ($p < 0.001$). Through the further comparative analysis of the daily secretion of Na+ and the Na+ content of leaves, the daily secretions of *Tamarix hispida* on Na+ have respectively reached 15.6%, 16.8%, 16.9% and 18.4% of the Na+ content of leaves. Na+ absorbed by roots are mostly discharged in vitro through the salt glands, which is relative to the phenomenon that *Tamarix hispida* is a kind of salt excretion plant. As the salt glands are mainly distributed in aquamous leaves, the salt which is brought into body by transpiration stream is discharged through salt glands so as to reduce the salt content of leaf tissue and produce a certain adaptation to the salt habitat[10,11]. The salt secretion of *Tamarix hispida* is lower than the trigyna, which shows *Tamarix hispida* is a plant with strong salt secretion ability[12]. The leaves of *Tamarix hispida* have a higher selectivity for the secretion of ions, which may be related to maintaining appropriate salt and nutrient concentration in plants for meeting the saline habitats.

Table 3. Comparison of excreted Na$^+$ contents in leaves from *Tamarix hispida*

NaCL concentration (g·L^{-1})	daily secretion of Na$^+$ (mg·g^{-1})	Na$^+$ content of leaves	daily secretion of Na$^+$ / Na$^+$ content of leaves
5	2.004	12.88	0.156
10	2.184	12.99	0.168
15	2.292	13.56	0.169
20	2.532	13.76	0.184

4   Discussion

The salt content in soil will directly affect the ion content of plants, which mainly shows the high content of Na$^+$ in soil, so the Na$^+$ in *Tamarix hispida* under salinity environment are enriched. In physiological terms, although the ion components in *Tamarix hispida* and salt glands secretion have agreed with the ones in root environment, the Na$^+$ contents of different parts in plants are different. The Na$^+$ content is highest in *Tamarix hispida* and the content in leaves is higher than the one in roots and stems. As the halophytes mainly maintain the ability of osmotic absorption of water of plants through the osmotic adjustment action that the inorganic and organic osmosis regulate the substance, the high salt content in *Tamarix hispida* is beneficial to adjust the osmotic potential of cells so as to make them absorb water in arid and saline environment and keep the moisture content of plants. In particular, the leaves have the priority to absorb and accumulate a large number of ions in order that the total net accumulation of inorganic ions is higher than the roots. The upper part of ground has accumulated a large number of ions, which is conducive to increasing the osmotic potential difference between the upper part of ground and the roots. It can promote the transportation of water from root to upper part of ground, improve the water status in upper part of ground and promote the plant growth, which is similar to the Suaeda salsa leaves. *Tamarix hispida* has the multi-cellular salt glands, which can secrete a variety of ions. In order to meet the habitat salt and the water changes, *Tamarix hispida* has produced a clear ecological adaptation in theory and form. For example, there is a thick cuticle covering the ektexine, which can maintain the water, prevent the moisture loss in dry and heat environment, adapt to the osmotic stress and ion toxicity of plants in high salt environment so as to make itself live in saline soils or in saline. There are some salt glands distributed in leaves and the salt glands have discharged the excess salt that invade the body out of it by means of transpiration potential so as to regulate the ion content in cells and maintain the relatively stable salt concentration in plants. In addition, as for the plants, except that the salt glands of *Tamarix hispida* is related to the Na$^+$ content of plants for Na$^+$ secretion, the selective actions have been taken place in the two processes that the salt is transported from root to leaf and the leaf discharges the secretion through the salt glands. *Tamarix hispida* has high selectivity for Na$^+$ and this high selectivity for salt secretion is related to its long-term adaptation of high salt content social environment[13]. At the same time, it has also avoided the plant mineral nutrition imbalance caused by non-selective secretion. The strong enrichment for soil salt ions and the physiological adaptability of the living environment of *Tamarix hispida* have a certain value in the soil improvement of saline environment and the regetation recovery.

The secretion of salt secreting plants on ions has been affected by external environmental factors, such as the salt concentration of root environment, the humidity, the light, the temperature and so on. The salt secretion rate of the red sand in arid deserts has been increased with the improvement of salt concentration in root growth environment. As for the Limonium sinense Kuntze grown in different salt concentrations, we can find that the low concentration NaCL treatments mainly improve the salt secretion rate of leaves by increasing the number of leaf salt glands, which has nothing to do with the salt secretion rate of single salt gland. The increase of salt secretion rate under high concentration NaCL treatments is the common result of the improvement of single salt gland's salt secretion rate and the number's increase of salt glands. A certain concentration NaCL treatment can promote the development of leaf salt glands. The daily changes of salt secretion have a positive correlation with the atmospheric humidity and have a negative correlation with the daily transpiration intensity. Ramadan's findings have showed that the proper humidity is conducive to secreting ions for salt glands. The soil and the water in atmosphere can inhibit the secretion process of red sand (Reaumuria hirtella). The light promotes the salt transportation by increasing the transpiration and then is contributing to the deer hair salt secretion. However, the photosynthesis inhibitor DCMU has reduced the salt secretion, which indicates that ATP has been involved in salt secretion. The salt has reached the leaves with the transpiration stream in daytime and it can be secreted for a whole day. The salt secretion can occur for a whole day, which has showed that it isn't directly affected by the light and the internal rhythms of plants. As the concentration of secretion Na$^+$ is usually higher than the soil environment, so the salt secretion, the salt drainage way, is regarded as the active energy physiological process which is related to metabolism. This secretion will be adjusted by the temperature.

The results of this study have showed that the secretion of Na$^+$ is enhanced with the increase of salt concentration. The excreting rate of salt gland was higher in the morning than that in the afternoon and higher in daytime than that at night. Because salt secretion has a high positive correlation with atmospheric humidity, at the same time it has a higher negative correlation with atmospheric temperature. As for the trend of Na$^+$ secretion, the daily transpiration intensity and the light intensity, Chen Yang has analyzed that there is no significant correlation among the Na$^+$ secretion, the daily transpiration intensity and the light intensity. It is not clear that the increase of Na$^+$ secretion of *Tamarix hispida* is caused by the improvement of salt secretion of single salt gland or due to the increase of salt gland's number. And the molecular mechanism of the improvement of salt secretion of single salt gland needs to be further studied.

Acknowledgment. Foundation Item: National Natural Science Foundation of China (40871059 ; 91025025).

References

1. Wang, G., Cao, F.: Effects of Soil Water and Salt Contents on Photosynthetic Characteristics. Journal of Applied Ecology 15(12), 2396–2400 (2004)
2. Yamg, C., Li, C., Zhang, M., et al.: PH and Ion Balance in Wheat-wheatgrass under Salt-alkali Stress. Journal of Applied Ecology 19(5), 1000–1005 (2008)
3. Munns, R.: Comparative physiology of salt and water stress. Plant, Cell & Environment 25, 239–250 (2002)
4. Cater, C.T., Grieve, C.M.: Mineral nutrition, growth, and germination of Antirrhium majus L (Snapdragon) when produced under increasingly saline conditions. HortScience 43, 710–718 (2008)
5. Chen, Y., Li, W., Xu, H., et al.: The Influence of Groundwater on Vegetation in the Lower Reaches of Tarim River, China. Acta Geographica Sinica 58(4), 542–549 (2003)
6. Li, Q.: Study on Salt Tolerance of Tamarix.spp in Xinjiang, pp. 23–32. Xinjiang University (2002)
7. Yang, J., Zhang, D., Yin, L., et al.: Distribution and Cluster Analysis on the Similarity of the Tamarix Communities in Xinjiang. Arid Zone Research 19(3), 6–11 (2002)
8. Zhang, D., Yin, L., Pan, B.: A Review on the Systematics Stady of Tamarix. Arid Zone Research 19(2), 41–45 (2002)
9. Feng, Y., Yin, L.: Study on Organs Morphology and its Taxonomic Significance of Tamarix L. Arid Zone Research 17(3), 40–45 (2000)
10. Barhoumi, Z., Djebali, W., Smaoui, A., Chaibi, W., Abdelly, C.: Contribution of NaCl excretion to salt resistance of Aelurppus littoralis (Willd) Parl. Journal of Plant Physiology 164(7), 842–850 (2007)
11. Xue, Y., Wang, Y.: Study on Characters of Ions Secretion from Reaumuria trigyna. Journal of Desert Research 28(3), 437–442 (2008)
12. Zhang, H.: A Study on the Characters of Content of Inorganic Ions in Salt-stressed Suaeda Salsa. Acta Botanica Boreali-Occidentalia Sinica 22(1), 129–135 (2002)
13. Ding, F., Wang, B.: Effect NaCL on Salt Gland Development and Salt-secretion Rate of the Leaves of Limonium Sinense. Acta Botanica Boreali-Occidentalia Sinica 26(8), 1593–1599 (2006)

The Design of Function Signal Generator Based on Delphi and ARM

Zhao-yun Sun, Xiao-bing Du, Feng-fei Wang, and Peng Yan

School of Information Engineering, Chang'an University, Xi'an, 710064 China
zhaoyunsun@126.com, keleyisheng2006@126.com,
lwwangfengfei@126.com, yan.gaocai@163.com

Abstract. Traditional signal sources have a large body. It can not meet the needs of Special occasions and is difficult to operate by computers. A function generator need to be designed which has such features as Multi-wave pattern, large-scale, high-precision, stable, portable and general. The design of new generator combined ARM and EDA technologies together using ADS1.2 development environment and Delphi 7.0 OOP development environment. We choose chip LPC2103 as ARM processor, chip MAX038 as signal generator unit, chip MAX232 as Level shifter and chip AD811 as signal amplifier. The signal generator we designed is operating normally and satisfying in the testing experiments based on LPC2103 development board .

Keywords: Signal Generator Module, ARM, LPC2103, Delphi, MAX038.

0   Introduction

Function signal generator is one of the necessary instruments in electronic measurement, electronic equipment development and electronic engineering courses. It was not only widely used in communication, instrument and signal acquisition or processing, automatic control system test, but also in other blame electric measurement fields. This topic is put forwarded for the wide application of signal generator in electronic design field, combining with function signal generated module, ARM technology and object-oriented programming tools of Delphi. This design completed a set of multi-function, portable and simple operating of the function signal generator design. This topic is mainly implement hardware design of generator and the development of upper computer control software.

1   System Design

According to the design requirements, signal generator can produce various waveforms such as triangle wave, square-wave and sine wave under the control of peripheral circuit. It can also realize fine-tuning of frequency and duty ratio and amplification of signal independently. This kind of function generator is potable with advantages of big adjust range, high precision and stability. It can be used in all kinds of situations where need signal sources. Overall system design is shown in the Fig.1.

J. Luo (Ed.): Soft Computing in Information Communication Technology, AISC 158, pp. 53–60.
springerlink.com   © Springer-Verlag Berlin Heidelberg 2012

Fig. 1. System solution diagram

This system include LPC2103 ARM processor module circuit, PC control interface module, multilevel conversion circuit module, signals generated circuit module and signal amplifier circuit module. LPC2103 was selected as the center of system control, MAX038 as the core of function generator. We design and compiled a circuit driver, thus a completed system was form that can produce various waveforms.

2   System Realization

LPC2103 ARM processor as the system control center is used to handle data and instructions from PC, and then send signals to the signal produce chips; MAX038 receives data from ARM processor, and then send the corresponding waveforms to signal amplifier circuit as the core of function generator. The fine-tuning of frequency and duty ratio can be controled by corresponding tube independently. MAX232 chip as multilevel conversion part can provide various power levels, such as +5V,+12V and -12V, to meet the needs of 232 serial port. The software programmed under Delphi 7.0 programming environment can generate control signals for Lower level computer. AD811 as amplifier amplify the signals produced by MAX038 processing, so as to meet the technical indicators requirements of amplifier signal peak -peak values.

2.1   Processor Circuit Design

The design of Processor circuit includes the connection of external pins on LPC2103 chip and design of crystal oscillator circuit, decoupling circuit, reset circuit, JTAG circuit and power conversion circuit. LPC2103 is the core of system control. It is a 32-bit microprocessor, which is suit for industrial control and medical systems, with the advantages of small size, low power consumption, cheap price and so on[1]. So it is very suitable for the applications that take miniaturization as the main requirements. Processor and peripheral circuit design is shown in Fig.2.

Decoupling circuit place number of capacitors near power and ground on each chip, It is usually choose 0.1UF (10^4) ceramic capacitor. Each power pin and ground pins on processor chip should be connected to achieve a good decoupling effect and to improve the overall reliability of the circuit board.

The basic function of reset circuit: provide reset signal when the system's power is available. Until power is stable, the reset signal was canceled. CAT1025 reset chip was used to design the reset circuit in this paper. As Fig.3, when click the button, the CAT1025 of MR/pin is low power, reset output pin is activated. This output is connected to the reset pin of processor. Reset the LPC2103 need low power, so this circuit can be completed manual reset button.

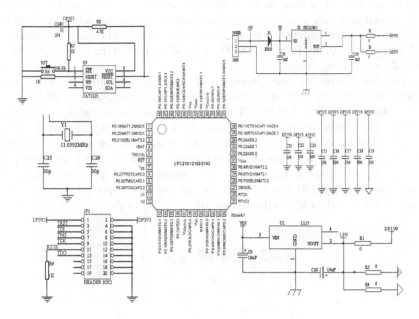

Fig. 2. Processor hardware and principle diagram

The role of crystal oscillator circuit is to provide basic system clock. Crystal oscillator adapts crystal oscillator circuit [2]. The two-port of crystal oscillator circuit is connected to the LPC2103 chip's OSC1 port and OSC0 port. 6MHz is standard of a crystal oscillator, and this frequency shared by system. Crystal oscillator circuit generates the clock signal to synchronize the various parts.

JTAG is an international standard testing protocol that is used for chip testing and simulation of the system. Special testing circuit TAP tests the internal nodes through a dedicated JTAG testing tool.

LM1117-3.3V is used to obtain the necessary 3.3V voltage. Power in fig.3 is the external power input. D1 is a Schottky diode. When the power is reverse polarity, it can protect system circuit. C0 and C_{10} are isolation capacitance, R_2 and R_3 is 0 Euro resistance. They are used to isolate analog and digital power supply. LM1117-1.8V is used to obtain the necessary 1.8V voltage. C_9 and C_{20} is protected capacitors, R_1 is 0 Euro resistance, which is used to isolate the power signal. R_4 and R_5 is 0 Euro resistance, they are used to separate the digital ground, analog ground and the earth signals.

2.2 Software System Design

Software system combining integrated function signal generator chip, ARM processor and software programming tools to complete the production of the function signals. The main functions of the software that include upper and lower machine software is to control the work of the waveform signal module.

(1) PC software is written using Delphi7.0, using to control signal generation module. It used to run the user interface program, provide the user interface control the send of instructions according to various interface protocols. The main function of the control program is control the MAX038, CD4051, TLC5618, LM324 module to generate the waveform signals.

(2) The lower software was programmed in C programming language under integrated ADS development environment. We write the program in LPC2103 chip through serial port to control MAX038 and other modules work together to generated waveform. The lower software is responsible for identifying the corresponding response control procedures, reading the device status data in real time (typically analog) and converting them into digital signal then feedback to the host computer. MSComm control transmit and receive data through serial port, providing serial communication for application to achieve upper and lower computer's communication, It coordinates the work of the hardware and software systems[3].

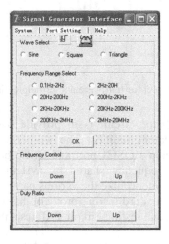

Fig. 3. PC control interface

2.3   Signal Generator Circuit Design

Signal generator circuit chooses MAX038 chip. MAX038 is signal generator with high frequency, high accuracy. It can produce accurate triangle wave, sawtooth, sine wave, square wave, pulse wave. The output frequency range can be controlled in the 0.1Hz to 20MHz[4]. Duty ratio and frequency control can be adjusted separately. In MAX038, when the A1 port is high level, the output waveform is sine wave; when the A1, A0 port at the same time is low level, the output waveform is square wave; when the A1 port is low level, A0 port is high level, the output waveform is triangular. A0 port is connected with P0.5 port of processor LPC2103, A1 port is connected with the P0.6 port of processor LPC2103. COSC port is used to adjust the band, port IIN is used to adjust the frequency. Signal generator circuit is shown in Fig. 4.

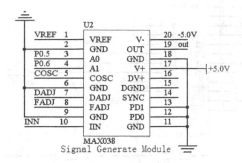

Fig. 4. Signal generate circuit principle diagram

2.4 Band Selection Circuit Design

We use CD4051 chip to select different capacitance C_f, then we can get the frequency range (ie, band). There are eight bands this system for switching. Circuit design is shown in Fig.5.

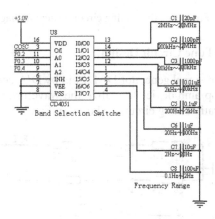

Fig. 5. Band selection circuit principle diagram

2.5 Frequency Fine-Tuning Circuit Design

The frequency of MAX038 output waveform is determined by current INN of the input pin IIN, the capacitance C_f of COSC pin and the voltage FADJ of VFADJ. The fundamental frequency of the output waveform is determined by the current IIN and the C_f. By TLC5618 and LM324 quad op amp device, we can control signal generator FADJ to realize the frequency fine-tuning. Frequency tuning circuit is shown in Fig.6.

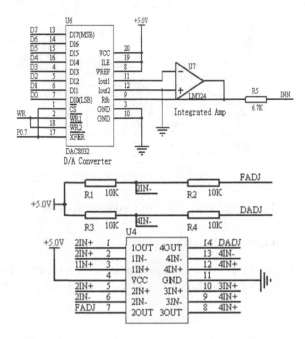

Fig. 6. Frequency fine-tuning circuit principle diagram

2.6   Duty Ratio Fine-Tune Circuit Design

We control the voltage of signal generator DADJ pin by D/A converter TLC5618 chip and chip LM324 quad op amp device to realize the waveform duty ratio adjustment. Duty ratio tuning Duty ratio tuning circuit is shown in Fig.7.

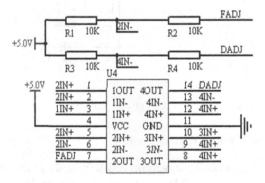

Fig. 7. Duty ratio circuit principle diagram

2.7   Signal Amplifier Circuit Design

AD811 chip is a broadband high-speed current feedback type operational amplifier. The output amplitude of MAX038's waves are 2V (Peak-Peak). To meet technical indicators requirements, in this topic, AD811's voltage amplifier gain is 2, and it plays the role of power amplifier. Signal power amplifier circuit is shown in fig.8.

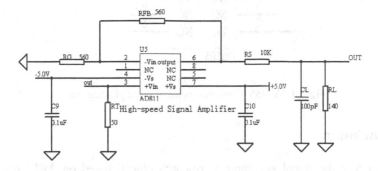

Fig. 8. Signal amplifier circuit principle diagram

2.8 Multilevel Conversion Module Design

Multilevel conversion module chose MAX232 chip. MAX232 chip is an RS232 standard chip[5]. It is used to realize conversion between microprocessor 5V TTL signals and computer RS232 signal. MAX232 chip's serial port is 9-needle. we just need connect serial 2, 3, 5 tube (receive, send, ground) and MAX232, then the PC can communicate with low level computer. Multilevel conversion circuit design is shown in Fig.9.

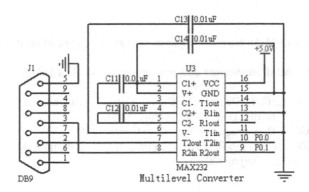

Fig. 9. Multilevel conversion circuit principle diagram

2.9 Power Choice Circuit Design

Power system adopts USB 5V provides voltage, using B0505 chip to obtain -5V voltage that MAX038 chip and AD811 chips need, B0505's output is a voltage difference. If B0505's output + VO connect the earth, then the B0505's 0V output is -5V. Its principle circuit design is shown in fig.10.

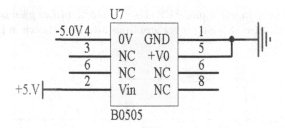

Fig. 10. 5V~-5V Converting circuit principle diagram

3   Conclusion

This paper take the signal generator as research object, based on ARM processor, ADS1.2 development environment, object-oriented programming environment Delphi7.0 and serial interface communication programming control MSComm, EDA technology, completing a set of function signal generator design. The function signal generator can produce various waveforms, including triangle wave and square-wave, sine wave, etc; The signal frequency between 0.1 Hz ~ 20MHz (eight band). Signal amplitude is 0 ~ 5V, signal duty ratio and frequency can be continuously adjusts. It has the signal amplification effect. This system is a practical, adjustable range, high precision and portable signal generator, can be applied in various situations where need signal source.

References

1. ARM Inc. ARMv7-M Architecture Application Level Reference Manual. first beta release (2008)
2. Du, C.-L.: The ARM architecture and programming. Tsinghua University Press, Beijing (2003)
3. Wang, Y.-T., Yang, G.-J.: Delphi + MSComm control development serial communication process. Journal of Industrial Control Press (July 2008)
4. Max, M.: MAX038 High-Frequency Waveform Generator Rev. (2004)
5. Zhao, X.-H., Zhou, C.-L., Liu, T.: MAX232 principle and application. Beijing University of Aeronautics and Press, Beijing (2006)

The Optimal Joint Power Allocation Scheme and Antennas Design in AF MIMO Relay Networks

Yuanfeng Peng and Youzheng Wang

Wireless Multimedia Communication Laboratory, Tsinghua University, Beijing, China
pengyf@wmc.ee.tsinghua.edu.cn, yzhwang@mail.tsinghua.edu.cn

Abstract. In this paper, a closed expression of the optimal joint PA is obtained by applying a zero-forcing (ZF) receiver at the destination node to recover the original signal. But the MSE performance of the system would have a great deterioration when applying a ZF receiver at the destination node. The smallest singular value distribution in the theory of matrix analysis is used to explain this MSE deterioration. On this basis, this paper proposes an antennas design at the relay node to eliminate this MSE deterioration. And numerical results show that by applying a ZF receiver at the destination node and the proposed antennas design at the relay node, we not only obtain a closed expression of the optimal joint PA, but also make the MSE performance of the system almost as best as it can be.

Keywords: MIMO relay network, optimal joint power allocation, zero-forcing receiver, antennas design, smallest singular value distribution.

1 Introduction

The AF MIMO relay network, with one relay node and amplify-and-forward (AF) relaying strategy, is focused in this paper. The AF MIMO relay network would provide an excellent information service with high data rate and ubiquitous coverage [1]-[4]. However, power allocation (PA) at the source node and PA at the relay node are of crucial importance to the performance of the AF MIMO relay network.

Some power allocation schemes have been proposed for the MIMO relay network in existing literatures [5]-[7]. In [5], an AF MIMO relay network was designed by maximizing the instantaneous channel capacity, leading to a number of parallel SISO subchannels and a water-filling PA scheme for the subchannels. In this method, the PA was considered for the relay node solely, rather than the source and relay jointly, so the channel capacity of the entire MIMO relay network has not been maximized. The authors of [6][7] have attempted to optimize the channel capacity by considering a joint PA for both source and relay node. But e xisting literatures, which deal with joint PA optimization in MIMO relay networks, always apply a linear minimum MSE (LMMSE) receiver at the destination node to recover the original signal. This method makes the joint PA optimization problem without any closed expression. Then the distributed iteration approach has to be applied to get the optimal solution. But the authors pointed out that the resulting channel capacity may be only a local maximum rather than the global maximum depending on the chosen starting values for the

J. Luo (Ed.): Soft Computing in Information Communication Technology, AISC 158, pp. 61–70.
springerlink.com © Springer-Verlag Berlin Heidelberg 2012

proposed iteration approach. And it is known that the distributed iteration has following defects: firstly the iteration approach may not converge or may need lots of steps to converge; secondly computational complexity of the iteration approach is very high and it may cost most of hardware resources.

Instead of the LMMSE receiver, this paper applies a zero-forcing (ZF) receiver at the destination node to recover the original signal, and then a closed expression of the optimal joint PA is obtained. But compared to the LMMSE receiver application, the ZF receiver application has a great deterioration on the MSE of the MIMO relay network when least antennas are deployed at the relay, which means that the relay node has the same number of antennas as that at the source node. In order to eliminate this deterioration, this paper proposes that double number of antennas could be deployed at the relay node. Simulation results show that on the premise of optimal joint PA at the source and relay node, the MSE of ZF receiver application with double antennas at the relay node is much smaller than the MSE of ZF receiver application with only least antennas at the relay node, and also smaller than the MSE of LMMSE receiver application with only least antennas at the relay node. Even compared to the MSE of LMMSE receiver application with double antennas at the relay node, the MSE performance of ZF receiver application with double antennas is almost as good as that. Therefore, with a ZF receiver applied at the destination node and double antennas deployed at the relay node, this paper not only obtains a closed expression to the joint PA optimization problem, but also makes the MSE performance as good as that in the LMMSE receiver application.

2   System Model

The MIMO relay network model used in this paper is comprised of a source node, a relay node and a destination node as shown in Fig. 1. Throughout this paper, it is assumed that the network operates in a Rayleigh flat fading environment. The source, relay and destination nodes have M_S, M_R and M_D antennas respectively. In order to obtain spatial multiplexing in the network, it requires $M_R \geq M_S$ and $M_D \geq M_S$ such

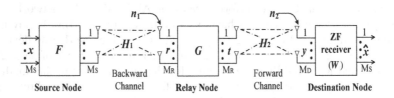

Fig. 1. The MIMO relay network

that the compound channel can support M_S independent substreams. But in reality the source node also serves as a destination node, or vice versa. So it should always have $M_S = M_D \triangleq M$ to maintain the consistency of RF equipments. $x \in \mathbb{C}^{M \times 1}$ denotes the transmitted vector, assuming $E\{xx^H\} = I$. F denotes the precoder matrix at the source node and has $trace\{FF^H\} \le P_S$ where P_S is the total transmit power at the source node. $n_1 \in \mathbb{C}^{M_R \times 1}$ refers to the noise vector received at the relay node with zero mean and covariance matrix $E\{n_1 n_1^H\} = \sigma_R^2 I$. Then $G \in \mathbb{C}^{M_R \times M_R}$ is the signal processing matrix at the relay node. $t \in \mathbb{C}^{M_R \times 1}$ denotes the relayed vector, assuming $E\{t^H t\} \le P_R$ where P_R is the total transmit power at the relay node. And $n_2 \in \mathbb{C}^{M \times 1}$ refers to the noise received at the destination node with zero mean and covariance matrix $E\{n_2 n_2^H\} = \sigma_D^2 I$. $H_1 \in \mathbb{C}^{M_R \times M}$ denotes the backward channel matrix, and $H_2 \in \mathbb{C}^{M \times M_R}$ denotes the forward channel matrix. It is always assumed that both H_1 and H_2 are in full rank. So the signal vector received at the destination node is

$$y = H_2 G(H_1 Fx + n_1) + n_2 = H_2 G H_1 Fx + H_2 G n_1 + n_2 .$$ (1)

Perform the singular value decomposition (SVD) of H_i as,

$$H_i = U_i \Lambda_i V_i^H, \ i = 1, 2 ,$$ (2)

where U_i and V_i are unitary matrices and Λ_i is a diagonal matrix composed of the singular values of H_i. Here we assume that $\Lambda_1 = \text{diag}\{\sqrt{a_1}, \sqrt{a_2}, ..., \sqrt{a_M}\}$ with $a_1 \ge a_2 \ge ... \ge a_M > 0$ and $\Lambda_2 = \text{diag}\{\sqrt{b_1}, \sqrt{b_2}, ..., \sqrt{b_M}\}$ with $b_1 \ge b_2 \ge ... \ge b_M > 0$. It has been proved in [7] that the minimal MSE between the source node and the destination node is obtained when the overall channel is decomposed into a number of parallel uncorrelated subchannels. So the structure of F and G is as follows,

$$F = V_1 \cdot \Lambda_F \qquad G = V_2 \cdot \Lambda_G \cdot U_1^H ,$$ (3)

where Λ_F and Λ_G are two diagonal matrices to be designed. The diagonal elements of Λ_F and Λ_G represent the power allocation coefficients for parallel data streams at the source and relay node, respectively. Without loss of generality, we assume that $\Lambda_F \triangleq \text{diag}\{\sqrt{c_1}, \sqrt{c_2}, ..., \sqrt{c_M}\}$ with $(c_1 + c_2 + \cdots + c_M) \le P_S$ and $\Lambda_G \triangleq \text{diag}\{\sqrt{d_1}, \sqrt{d_2}, ..., \sqrt{d_M}, \underbrace{0, 0, ..., 0}_{M_R - M}\}$ with $\sum_{i=1}^{M}(d_i a_i c_i + \sigma_R^2 d_i) \le P_R$. So from (1)(3), we get

$$y = U_2 \Lambda_2 \Lambda_G \Lambda_1 \Lambda_F x + U_2 \Lambda_2 \Lambda_G n_1 + n_2 .$$ (4)

At last, a receiver matrix W should be applied to the signal y to recover the original transmitted signal x. Although the LMMSE receiver is the optimal linear receiver [8], this paper applies a ZF receiver at the destination node in order to obtain a closed expression of the optimal joint PA. So the estimated signal vector is

$$\hat{x} = W \cdot y ,$$ (5)

where $W = \underbrace{(\Lambda_2\Lambda_G\Lambda_1\Lambda_F)^{-1}}_{\Phi}\bullet U_2^{\mathrm{H}}$. So when the overall channel is totally decomposed,

the network model is transformed from Fig. 1 to Fig. 2. Note that $(n_1)_i$, $(n_2)_i$ and $(\Phi)_{i,i}$ are i-th element of vector n_1, n_2 and (i,i)-th entry of matrix Φ respectively.

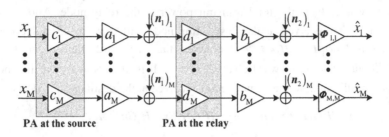

Fig. 2. The diagonal MIMO relay network model

3   The Optimal Joint Power Allocation Based on MSE Criterion

3.1   Formulation of the Joint PA Optimization Problem

From (4) and (5), the estimated signal vector is as follows,

$$\hat{x} = x + (\Lambda_1\Lambda_F)^{-1}n_1 + (\Lambda_2\Lambda_G\Lambda_1\Lambda_F)^{-1}U_2^{\mathrm{H}}\bullet n_2 , \tag{6}$$

So the MSE of the MIMO relay network is measured as,

$$
\begin{aligned}
\mathrm{MSE} &= E\left\{\|\hat{x}-x\|^2\right\} = trace\left\{E\left\{(\hat{x}-x)\bullet(\hat{x}-x)^{\mathrm{H}}\right\}\right\} \\
&= trace\left\{E\left\{(\Lambda_1\Lambda_F)^{-2}\|n_1\|^2 + (\Lambda_2\Lambda_G\Lambda_1\Lambda_F)^{-2}\|n_2\|^2\right\}\right\} \\
&= trace\left\{(\Lambda_1\Lambda_F)^{-2}\sigma_{\mathrm{R}}^2 + (\Lambda_2\Lambda_G\Lambda_1\Lambda_F)^{-2}\sigma_{\mathrm{D}}^2\right\} \\
&= \sum_{i=1}^{M}\left(\frac{\sigma_{\mathrm{R}}^2}{a_i c_i} + \frac{\sigma_{\mathrm{D}}^2}{a_i b_i c_i d_i}\right)
\end{aligned}
\tag{7}
$$

So our goal is to design the power allocation matrices Λ_F and Λ_G (or c_i and d_i; $i=1,...,M$) jointly to minimize the MSE. Mathematically, the problem of joint PA optimization under the MSE criterion can be formulated as follows,

$$
\begin{aligned}
(\hat{c}_1,...,\hat{c}_{\mathrm{M}};\hat{d}_1,...,\hat{d}_{\mathrm{M}}) &= \underbrace{arg\ min}_{(c_n,d_n;n=1,...,\mathrm{M})}\ \mathrm{MSE} \\
&= \underbrace{arg\ min}_{(c_n,d_n;n=1,...,\mathrm{M})}\left\{\sum_{i=1}^{M}\left(\frac{\sigma_{\mathrm{R}}^2}{a_i c_i} + \frac{\sigma_{\mathrm{D}}^2}{a_i b_i c_i d_i}\right)\right\} .
\end{aligned}
\tag{8}
$$

s.t. $(c_1 + c_2 + \cdots + c_{\mathrm{M}}) \le P_S;\ \sum_{i=1}^{M}(d_i a_i c_i + \sigma_{\mathrm{R}}^2 d_i) \le P_R;$

$c_{\mathrm{n}} \ge 0,\ d_{\mathrm{n}} \ge 0,\ \mathrm{n}=1,...,\mathrm{M};$

3.2   The Closed Expression to the Optimal Joint PA Problem

Because there are two sets of allocation factors c_n and d_n to design to minimize the MSE, we firstly assume that the allocation factors c_n are optimal already, and then find out the relationship between d_n and c_n. So we simplify the original problem (8) to

$$
\begin{aligned}
(\hat{d}_1,...,\hat{d}_M) &= \underset{(d_n;n=1,...,M)}{arg\ min} \left\{ \sum_{i=1}^{M} (\frac{\sigma_R^2}{a_i c_i} + \frac{\sigma_D^2}{a_i b_i c_i d_i}) \right\} \\
&= \underset{(d_n;n=1,...,M)}{arg\ min} \left\{ \sum_{i=1}^{M} (\frac{1}{a_i b_i c_i d_i}) \right\}
\end{aligned}
\tag{9}
$$
$$
s.t.\ \sum_{i=1}^{M} (d_i a_i c_i + \sigma_R^2 d_i) \le P_R;\ d_n \ge 0,\ n=1,...,M;
$$

From Karush-Kuhn-Tucker (KKT) conditions [9], the closed expression of d_n is

$$
d_n = \sqrt{\frac{1}{a_n b_n c_n (\sigma_R^2 + a_n c_n)}} \cdot \frac{P_R}{\sum_{i=1}^{M} (\frac{\sqrt{\sigma_R^2 + a_i c_i}}{\sqrt{a_i b_i c_i}})};\ n=1,...,M;
\tag{10}
$$

Then we should find out the closed expression of the allocation factors c_n. Combining (10) and (8), we could transform the original optimization problem to

$$
(\hat{c}_1,...,\hat{c}_M) = \underset{(c_n)}{arg\ min} \left\{ \sum_{i=1}^{M} \left[\frac{\sigma_R^2}{a_i c_i} + \frac{\sigma_D^2 \sqrt{\sigma_R^2 + a_i c_i}}{P_R \sqrt{a_i b_i c_i}} \cdot (\sum_{i=1}^{M} \frac{\sqrt{\sigma_R^2 + a_i c_i}}{\sqrt{a_i b_i c_i}}) \right] \right\}.
\tag{11}
$$
$$
s.t.\ (c_1 + c_2 + \cdots + c_M) \le P_S;\ c_n \ge 0,\ n=1,...,M;
$$

By using the KKT method again, the closed expression of c_n is

$$
c_n = \frac{P_S \sqrt{\sigma_R^2 (\frac{1}{a_n} + \frac{\sigma_D^2}{2P_R a_n \sqrt{b_n}}) \cdot \left[(\sum_{i=1}^{M} \frac{1}{\sqrt{b_i}}) + \frac{1}{\sqrt{b_n}} \right]}}{\sum_{i=1}^{M} \sqrt{\sigma_R^2 (\frac{1}{a_i} + \frac{\sigma_D^2}{2P_R a_i \sqrt{b_i}}) \cdot \left[(\sum_{i=1}^{M} \frac{1}{\sqrt{b_i}}) + \frac{1}{\sqrt{b_i}} \right]}};\ n=1,...,M;
\tag{12}
$$

So the combination of expressions (12) and (10) is the closed expression of the optimal joint PA based on the criterion of MSE minimization. Because $a_1 \ge a_2 \ge ... \ge a_M$ and $b_1 \ge b_2 \ge ... \ge b_M$, we can get $c_1 \le c_2 \le ... \le c_M$ and $d_1 \le d_2 \le ... \le d_M$ from (12) and (10). It means that in the optimal joint PA solution of ZF receiver application, the subchannel whose channel gain is smaller should be assigned more power both at the source and relay node.

Therefore, unlike the traditional application using a LMMSE receiver at the destination node, this section obtains a closed expression of the optimal joint PA when applying a ZF receiver at the destination node. However, the ZF receiver application has its own defect, which would be shown out and solved in next section.

4   Antennas Design at the Relay Node

From Fig. 1 and Fig. 2, it is known that the number of antennas at the relay node should be more than the number of antennas at the source node, which means $M_R \geq M$. And from the hardware efficiency perspective, the least number of antennas is the best choice, which means $M_R = M$. However, from the simulation results in next section, we find that the number of antennas at the relay node has a great influence on the MSE performance of the system. So antennas design at the relay node is also a research focus in our paper. Fig. 5 shows that the MSE of ZF receiver application has an enormous deterioration compared to the LMMSE receiver application when only least number of antennas ($M_R = M$) is deployed at the relay node. The theory of matrix analysis could explain this deterioration.

From (7), we have

$$\text{MSE} \geq \sum_{i=1}^{M} (\frac{\sigma_R^2}{a_i c_i}) \geq \frac{\sigma_R^2}{a_M c_M} \geq \frac{1}{P_S/(\sigma_R^2)} \cdot \frac{1}{a_M} \; . \tag{13}$$

So the smallest singular value $(\sqrt{a_M})$ of the backward channel matrix H_1 determines the lower bound of the MSE. When least number of antennas is deployed at the relay node, H_1 is a $M \times M$ complex Gaussian matrix in full rank whose components are independent standard normal variables. Paper [10] derived the pdf of a_M as follows,

$$f(a_M) = \frac{M}{2} e^{-Ma_M/2} \; . \tag{14}$$

Because H_1 is in full rank, we assume $a_M \geq 10^{-3}$. Then,

$$E\left\{\frac{1}{a_M}\right\} = \int_{0.001}^{+\infty} \frac{1}{a_M} \cdot \frac{M}{2} \cdot e^{-Ma_M/2} \, da_M \; . \tag{15}$$

Taking Fig. 5 for example, when $M = 4$,

$$E\left\{\frac{1}{a_M}\right\} = \int_{0.001}^{+\infty} \frac{1}{a_M} \cdot \frac{4}{2} \cdot e^{-4a_M/2} \, da_M \approx 11.3 \; , \tag{16}$$

So

$$\text{MSE} \geq \frac{11.3 \times M}{P_S/(M\sigma_R^2)} \; , \tag{17}$$

which is the lower bound curve as shown in triangle dashed curve in Fig. 5. But in the LMMSE receiver application, paper [7] derived that the worst case of MSE is $\text{MSE} = M_S = 4$. So from the above analysis, it is known that the ZF receiver application has a great deterioration compared to the LMMSE receiver application when $\text{SNR} = P_S/(M\sigma_R^2) < \frac{11.3 \times 4}{4} \approx 10.5(\text{dB})$.

In the ZF receiver application, from (13) it is known that the smallest singular value of H_1 has a decisive effect on the MSE. So in order to improve the MSE performance, we should increase the average smallest singular value of H_1. From the perspective of the SVD of H_1, there are two ways to increase the smallest singular value: one is to increase the antenna number at the source node, the other is to increase the antenna number at the relay node. However, it is usually much more difficult and complicated to increase the antenna number at the source node. Because the source node, which is always a mobile terminal such as a telephone set, is too tiny to deploy many antennas without correlation. But the relay node is always a base station, which is fixed and carefully designed. It means that we can deploy much more antennas at the relay node without worrying the correlation between antennas. So we choose the second way to increase the smallest singular value of H_1.

For simplicity, the solution in this paper to eliminate the deterioration between ZF and LMMSE receiver application is to duplicate the antenna number at the relay node. So H_1 is a $2M \times M$ rectangular Gaussian matrix, and Paper [10] also gave out the pdf of a_M in this case as follows,

$$
\tilde{f}(a_M) = \pi^{1/2} \cdot 2^{-(M+1)/2} \cdot \Gamma(\frac{2M+1}{2}) \cdot a_M^{(M-1)/2} \cdot
$$
$$
e^{-a_M/2} \Big/ \Big\{ \Gamma(\frac{M}{2}) \cdot \Gamma(\frac{M+1}{2}) \cdot \Gamma(\frac{M+2}{2}) \Big\} \tag{18}
$$

When $M = 4$,

$$
E\Big\{ \frac{1}{a_M} \Big\} \approx 0.2 . \tag{19}
$$

So

$$
\text{MSE} > \frac{0.2 \times M}{P_S / (M\sigma_R^2)} . \tag{20}
$$

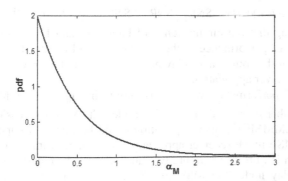

Fig. 3. Pdf of the square smallest singular value in a 4×4 Gaussian matrix

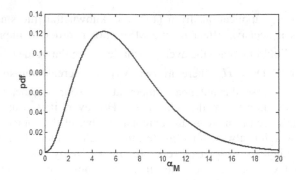

Fig. 4. Pdf of the square smallest singular value in a 8×4 Gaussian matrix

Fig. 3 and Fig. 4 are pdf curves of the square smallest singular value (a_M) corresponding to (14) and (18). We can find that after duplicating the antenna number at the relay node, the average smallest singular value has greatly increased. And comparing (17) and (20), we can get about $10 \times \log_{10} \dfrac{11.3}{0.2} \approx 17.5 \text{dB}$ improvement in SNR $= P_S / (M \sigma_R^2)$ due to deploying double number of antennas at the relay node from the perspective of lower bound of MSE. Although the lower bound of MSE in ZF receiver application is not tight to real MSE in ZF receiver application, simulation results also show that about 20dB improvement in SNR could be obtained by comparison of the real MSE in ZF receiver application between Fig. 5 and Fig. 6.

5   Numerical Results

In this section, the MSE performance of an AF MIMO relay network is evaluated over Rayleigh flat fading channels. The signal-to-noise ratio of the backward channel is denoted as $\text{SNR}_1 = P_S / (M \sigma_R^2)$, and the signal-to-noise ratio of the forward channel is denoted as $\text{SNR}_2 = P_R / (M \sigma_D^2)$. $\text{SNR}_1 = \text{SNR}_2 \triangleq \text{SNR}$ is assumed in the simulation for simplicity, meaning that the circumstance of backward and forward channels is the same. And the MSE performance in the simulation is all obtained on the premise of optimal joint PA at the source and relay node, whether in the ZF receiver application or the LMMSE receiver application.

First, the MSE performance vs. SNR is shown in Fig. 5 when $M_S = M_D = 4$ and least antenna number is applied at the relay node. The MSE expression and optimal joint PA in the LMMSE receiver application could be found in paper [7]. And the lower bound of MSE in ZF receiver application is obtained from the theory of matrix analysis in section IV. From this figure we can find that when least antenna number is applied at the relay node, the MSE of ZF receiver application has an enormous deterioration compared to that of LMMSE receiver application, especially when SNR is low.

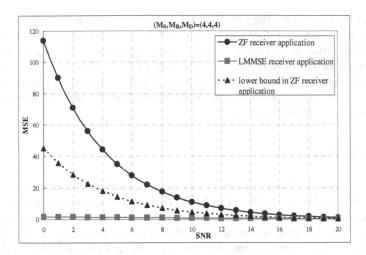

Fig. 5. MSE performance when the relay has least antennas

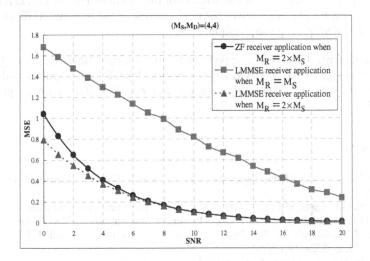

Fig. 6. MSE performance comparison when the relay has least antennas and double antennas

But as mentioned above, the ZF receiver application has a great deterioration in MSE performance. So the paper proposes to duplicate the antenna number at the relay node to eliminate this deterioration. Fig. 6 shows the great improvement in MSE performance when double antenna number is deployed at the relay node. Compared to the MSE of ZF receiver application in Fig. 5, we know that the MSE of ZF receiver application has greatly decreased when duplicating the antenna number at the relay node. It means about 16dB improvement in SNR. What's more, from Fig. 6 we can find that the MSE of ZF receiver application with double antennas at the relay node is much smaller than the MSE of LMMSE receiver application with only least antenna number at the relay node. The circle solid curve and the triangle dashed curve in Fig. 6 show that even compared to the MSE of LMMSE receiver application with double antennas at the relay node, the MSE of ZF receiver application with double antennas at the relay node almost performs as well as that. From this figure we know that when SNR>6dB these two MSE curves almost overlap together. And it is known that the LMMSE receiver is the optimal linear receiver [8], so this conclusion shows that when double antennas are deployed at the relay node, the MSE performance in ZF application is almost as good as the MSE performance when the optimal linear receiver is applied at the destination.

6   Conclusions

In order to obtain a closed expression to the optimal joint PA problem, this paper applies a ZF receiver instead of the LMMSE receiver at the destination node to recover the original transmitted signal. But based on the MSE minimization criterion, the ZF receiver application has a great deterioration compared to the LMMSE receiver application when least antennas are deployed at the relay node. So this paper proposes a scheme with a ZF receiver applied at the destination node and double antennas deployed at the relay node, which not only obtains the closed expression of the optimal joint PA, but also makes the performance of the minimal MSE as good as that in the LMMSE receiver application.

Acknowledgment. This work is supported by the National Basic Research Program of China under grant No. 2007CB310601, National S&T Major Project under grant No. 2009ZX03006-007-02 and the Natural Science Foundation under grant No. 60928001.

References

1. Teltar, I.E.: Capacity of multi-antenna Gaussian channels. European Trans. Tel. 10, 585–595 (1999)
2. Foschini, G.J., Gans, M.J.: On limits of Wireless Communications in a Fading Environment when Using Multiple Antennas. Wireless Personal Communications, 311–335 (1998)
3. Cover, T.M., Gamal, A.A.E.: Capacity theorems for the relay channel. IEEE Trans. Inf. Theory 25(5), 572–584 (1979)
4. Host-Madsen, A., Zhang, J.: Capacity bounds and power allocation for wireless relay channel. IEEE Trans. Inf. Theory 51(6), 2020–2040 (2005)
5. Tang, X., Hua, Y.: Optimal design of non-regenerative MIMO wireless relays. IEEE Trans. Wireless Commun. 6(4), 1398–1407 (2007)
6. Hammerstrom, I., Wittneben, A.: Power allocation for amplify-and-forward MIMO-OFDM relay links. IEEE Trans. Wireless Commun. 6(8), 2798–2802 (2007)
7. Li, C., et al.: A joint source and relay power allocation scheme for a class of MIMO relay systems. IEEE Trans. on Signal Processing 57(12), 4852–4860 (2009)
8. Palomar, D.P., Cioffi, J.M., Lagunas, M.A.: Joint Tx-Rx Beamforming Design for Multicarrier MIMO Channels: A Unified Framework for Convex Optimization. IEEE Trans. on Signal Processing 51(9) (2003)
9. Boyd, S., Vandenberghe, L.: Convex Optimization. Cambridge University Press (2004)
10. Alan, E.: Eigenvalues and condition numbers of random matrices. SIAM J. Matrix Anal. Appl. 9(4), 543–560 (1988)

A Novel Adaptive Digital Pre-distortion Method in the UHF RFID System

Dan-Feng Li, Chun Zhang, Ziqiang Wang, Jing-Chao Wang, Xu-Guang Sun

Institute of Microelectronics, Tsinghua University, Beijing 100084, China
zhangchun@tsinghua.edu.cn

Abstract. The linearity of the radio frequency identification (RFID) system has a great effect on its performance and power dissipation. A novel adaptive digital pre-distortion (ADPD) method based on Look Up Table (LUT) is proposed in this paper, which can effectively improve the linearity of the RFID system. With the help of the ADPD method, the harmonic component is eliminated about 40 dB, and the three-order inter-modulation distortion (IMD3) is eliminated about 47 dB. By using the interpolation module, the size of the LUT is reduced to 3% of the full size. The rate of the ADPD method's convergence is speeded up by 55%, owing to the acceleration module.

Keywords: RFID, inter-modulation distortion, ADPD, LUT.

1 Introduction

RFID technology develops rapidly in the last few years. This technology can be used in security, positioning, communication and other areas of high utility value. RFID system is typically constituted of the forward link and the backscatter link entailing RFID tags and RFID readers. In the forward link, the reader performs as the interrogation, transmitting a Radio Frequency (RF) wave to tag. In the backscatter link communication, passive tag generates the reverse RF wave just by reflecting back a portion of interrogating RF wave in a process known as backscatter [1].

The nonlinearity of the RFID system will cause many noises, such as the distortion of harmonic components and intermediation, the Adjacent Channel Interference (ACI). All of these would raise the Bit Error Rate (BER). Also, the isolation from the transmitter to the receiver is generally less than 20 dB, which makes the information disturbed by the leakage carrier, so the RFID system must be linear enough [2].

The Output Backoff method is usually used to improve the linearity of the wireless communication systems. But this method makes the cost and the power increasing, and then its performance is unsatisfactory. Now, some new methods are also proposed to solve this problem, such as envelope elimination and restoration, feed forward, feed back, polar transmitter and digital pre-distortion. With the rapid development of modern digital signal processing technology, digital pre-distortion is growing to be a more general way for its effectiveness and high efficiency.

The LUT method [3] and the polynomial method [4] are the two most commonly used digital pre-distortion methods. The effect of the LUT method is better, but the rate of convergence is low and the method costs more hardware resources. The polynomial method costs much less hardware resources, but its effect is worse than the LUT method.

In this paper, a novel adaptive digital pre-distortion method is proposed, which combines the advantages of the LUT method with those of the polynomial method. The ADPD method can obviously eliminate the harmonic component and IMD3 at lower consumption of hardware resources.

2 Design of the ADPD Method

2.1 Theory of the Digital Pre-distortion

The key of the digital pre-distortion method is to design a pre-distorter in the forward link to compensate the output characteristic of the system. The output characteristic of the ideal pre-distorter is completely opposed to that of the original system, which ensures the new system is linear, as Fig.1 shows.

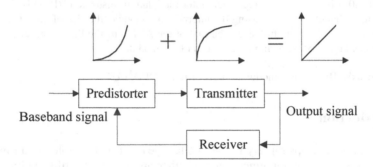

Fig. 1. Diagram of the adaptive digital pre-distortion

Traditionally, the LUT method is used in the common wireless communication systems in order to carry out the digital pre-distortion, for its better performance. Assume the bit width of the DAC is N, and the full size of the LUT will be 2^N. Such huge hardware consumption is difficult to realize in the RFID system. So, this ADPD method is proposed.

2.2 Architecture of the ADPD Method

The nonlinearity of the RFID system is mainly caused by the power amplifier, so some traditional methods are just designed to improve the linearity of the PA. But, this ADPD method is designed to improve the linearity of the whole RFID system.

An adaptive pre-distorter is designed in the transmitter. Some typical values in the output characteristic of the system are chosen linearly to quantify the linearity of the RFID system. The pre-distortion weights related to the typical values are saved in the LUT. The whole system is a feedback architecture, which ensures its stability.

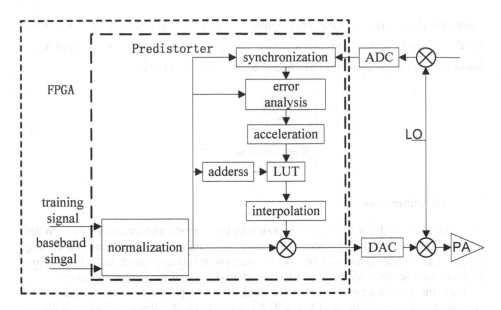

Fig. 2. Architecture of the ADPD in the RFID system

The Architecture of the proposed ADPD in the RFID system is shown in Fig.2. Both the baseband signal and training signal are normalized by the normalization module to ensure they are within the valid range of the pre-distorter. And the synchronization module is designed to cancel the phase error caused by the system delay. In the error-analysis module, a linearity index Lx is defined to describe the linearity of the RFID system, shown in Eq. (1), in which p_i is the amplitude of the input signal and p_o is that of the output signal.

$$Lx = \frac{(p_o - p_i)}{p_i} \qquad (1)$$

In traditional digital pre-distortion method, a fixed step is used to update the weights in LUT, which makes a slow rate of convergence. Related to Lx, an adaptive step Δp, shown in Eq. (2), is designed in the acceleration module to speed up the rate of the method's convergence.

$$\Delta p = f(Lx) \cdot p_i \qquad (2)$$

Another unique of this ADPD method is the design of interpolation module, which can describe the linearity of the system with a LUT in quite small size. The interpolation module calculates the pre-distortion weights of the non-typical values by

$$w_x = \alpha * w_{x_k} + \beta * w_{x_{k+1}}, \qquad (3)$$

where x is the amplitude of the input signal, x_k and x_{k+1} is the nearest values to x in the LUT, w_{x_k} and $w_{x_{k+1}}$ is the pre-distortion weights corresponds to x_k and x_{k+1}. Naturally, α and β can be set as linear interpolation, denoted by

$$\begin{cases} \alpha = \dfrac{x_{k+1} - x}{x_{k+1} - x} \\ \beta = \dfrac{x - x_k}{x_{k+1} - x_k} \end{cases} \tag{4}$$

2.3   Algorithm Flow

The ADPD method has two operate modes, training mode and linearity mode. When it is powered on, the RFID system will enter the training mode and send the training signal. The output signal will be sent to the synchronization module and the error-analysis module, in order to calculate the system delay and the amplitude error.

Then the acceleration module will calculate the step by the linearity index. The pre-distortion weights in the LUT will be updated by the linearity index, with the address from the address module. When the linearity index reaches the value we set, the pre-distortion weights will be fixed temporarily. At the same time, the training mode stops and the linearity mode starts. In the linearity mode, the system is nearly linear, and the base band signal modulated by OOK could be send to the transmitter.

3   Verification of the ADPD Method

3.1   Simulation Environment

The nonlinear model of PA is important to estimate the performance of the pre-distortion methods. Based on the measurement results, the RAPP model [5] is adopted, and its transfer function can be expressed as

$$f(A) = \frac{A}{(1 + A^{2*p})^{1/2*p}}, \tag{5}$$

where p is the index to describe the output characteristic of the PA, A denotes the magnitude of the input signal, and $f(A)$ denotes the magnitude of the output signal. Setting p to be 2, it can describe a more actually model according to our tests. The phase distortion of the RAPP model is small [6], so only AM/AM conversion is needed to research.

The bit widths of the DAC and the ADC are set to 10, the size of LUT is set to 32, and the target of the training is to make Lx less than 1%. To avoid the Gibbs effect, the training signal is a 40 KHz triangular wave signal. A 40 KHz sinusoidal wave signal is used to test the harmonic components. Then a two-tone signal consisting of 30 KHz and 50 KHz is used to test the IMD3. Data rate of Pulse-Interval Encoding (PIE) is set to be about 40 KHz, which is modulated with OOK.

3.2 Simulation Results

Simulation results show that the ADPD method is effective for compensating the non-linearity of the RFID system. The harmonic components are eliminated about 40 dB (Fig.3), and the improvement of IMD3 is about 47 dB, shown in Fig.4. Optimization components also show excellent performance. The rate of convergence is improved by 55%, and the size of LUT is reduced to 3% of its full size.

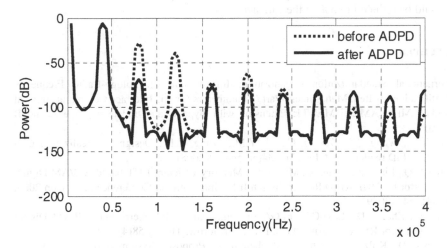

Fig. 3. Improvement of the harmonic components

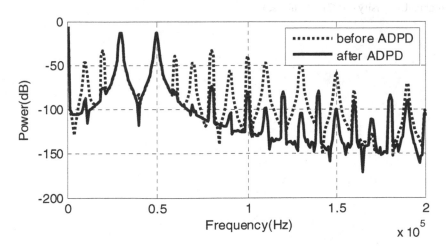

Fig. 4. Improvement of the IMD3

4   Conclusion

In this paper, a novel ADPD method is proposed to improve the linearity of the RFID system. It integrates the advantages of the LUT method and the polynomial method, which is simple and efficient. Simulations indicates that the method is effective, the three-order inter-modulation distortion is eliminated about 47 dB, and the harmonic component is eliminated about 40 dB. With the help of the optimization modules, the rate of convergence in the RFID system was speeded up by 55%, and the size of the LUT could be reduced to 3% of the full size.

References

1. International Standardization Organization: Information Technology-Radio Frequency Identification for Item Management-Part 6: Parameters for Air Interface Communications at 860-960 MHz. AMENDMENT1: Extension with Type C and update of Type A and B (2006)
2. Gao, T.B., Wang, J.C., Zhang, C., Li, Y.M., Wang, Z.H.: Design and realization of a portable RFID reader. Appl. Int. Circ. 34(5), 56–58 (2008)
3. Yang, Y.Q., Liu, Z.M., Tan, X., Min, H.: A Memory-Efficient LUT-based AM/AM Digital Predistortor for UHF RFID Reader. In: 8th IEEE International Conference on Asicon 2009, pp. 541–544 (2009)
4. Guan, L., Zhu, A.D.: Low-Cost FPGA Implementation of Volterra Series-Based Digital Pre-distorter for RF Power Amplifiers. IEEE T. Microw. Theory 58(4), 866–872 (2010)
5. Falconer, D., Kolze, T., Leiba, Y., Liebetreu, J.: Proposed System Impairment Models. IEEE 802.16.1pc-00/15 (2000)
6. Zhang, T.T.: Research of Pre-distortion of High Power Amplifier. Thesis for Master degree, Tsinghua University (2007) (in Chinese)

Design of UHF RFID Tag with On-Chip Antenna

Xijin Zhao, Chun Zhang, Yongming Li, Ziqiang Wang, and Yuhui He

Institute of Microelectronics, Tsinghua University, Beijing, 100084, China
zhao-xj08@mails.tsinghua.edu.cn

Abstract. This paper presents a UHF RFID tag with on-chip spiral antenna based on the principle of inductive coupling. The Whole chip die area is 1mm² and the integrated antenna is fabricated with the top metal layer in 0.18μm standard CMOS process. When the reader generating an output power of 20dbm, the fully integrated tag can achieve 4mm communication distance. An off-chip meandered shape antenna, magnetically coupled to the chip die, has been designed to overcome the reading range limitation. The measured reading range is raised to about 3 m in free space with 1 W RF power output of the reader. As there is no physical connection between the chip die and the off-chip far-field antenna, this structure can be used as an economical and convenient method for RFID tag packaging.

Keywords: RFID Tag, On-chip Antenna, Inductive Coupling.

1 Introduction

Radio frequency identification (RFID) technology provides the capability of wirelessly identifying and tracking objects in warehouse, supply chain, control system and automation process [1]. RFID applications operate at low frequency (30~300KHZ) and high frequency (3~30MHZ) bands using inductive coupling for power and data transfer. UHF and higher frequency band systems are usually coupled using electro-magnetic field for long readable range. In recent years, near-field UHF RFID systems also received a lot of attention in item-level RFID applications due to its huge market [2], [3]. Electrically small antenna used in near-field communication can be integrated on silicon substrates in UHF band [8]. RFID tags with on-chip antennas (OCA) not only save antenna fabrication and packaging costs but also reduce the tag size to adapt to the harsh size requirements for item tagging, monetary anti-counterfeiting, et al. The communication range of the RFID system with OCA is small. A UHF RFID system which employs a magnetic coupling between the die of a tag and its external antenna has been presented to overcome the reading range limitation in [4]. However, it only develops a structure for near field applications which the off-chip antenna is single-turn circular spire and the communication distance is only 4cm.

In this study, we present a UHF RFID tag integrated on-chip antenna in 0.18μm CMOS technology. A near-field loop reader antenna magically coupled to the integrated tag antenna is utilized to conduct power and data transfer. The magnetic coupled RFID system has a communication distance of less than 1cm. In order to overcome the reading range limitation in RFID systems with OCA, an off-chip

J. Luo (Ed.): Soft Computing in Information Communication Technology, AISC 158, pp. 77–83.
springerlink.com © Springer-Verlag Berlin Heidelberg 2012

radiator, inductively coupled to the OCA has been designed. This structure can also be used as an economical and convenient method for RFID tag packaging.

The paper is organized as follows: In section 2, the near field UHF RFID system with integrated antenna is explained; in section 3, the analysis and implementation of the inductively coupled antenna is presented; Conclusions are given in section 4.

2   Near-Field RFID System with OCA

2.1   System Architecture

The proposed inductively coupled near-field UHF RFID system is illustrated in Fig.1. The system based on a transformer-type coupling between the primary coil in the reader and the secondary coil integrated in the tag. A high efficiency rectifier harvests energy from the magnetic alternating filed of the reader's antenna to power up the circuits in the tag. A current controlled ring oscillator generate clock for data communication. The reset circuit is designed to provide a reset pulse when the supply voltage reaches 0.8V (the minimum voltage required by the circuits).

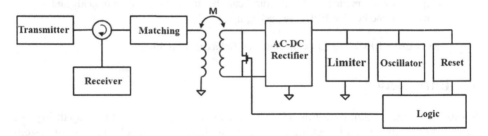

Fig. 1. Architecture of the inductively coupled RFID system

The OCA tag transmits data to the reader by captive load modulation. Due to weak coupling between the reader antenna and the integrated tag antenna, high modulation depth is selected to enhance the signal received by the reader. The logic circuit is just design to verify the system by outputting a frequency of 110 KHz pulse to modulate the on-chip antenna.

2.2   Voltage Multiplier Rectifier

In passive UHF RFID systems, an AC-DC charge pump rectifier is applied to convert AC power received by the antenna to a DC voltage. In standard CMOS technology, it's hard to integrate a high Q and high inductance inductor. The coupling between OCA and reader's antenna are weaker than off-chip antennas. Designing a voltage multiplier rectifier with high power conversion efficiency and low turn-on voltage is an effective approach to improve communication range.

A cross-connected CMOS rectifier circuit is chosen here. Fig. 2 shows the units of the rectifier. It includes a five-stage full-wave voltage multiplier. This kind of circuit topology is known as a high power efficiency rectifier for UHF RFIDs [5].

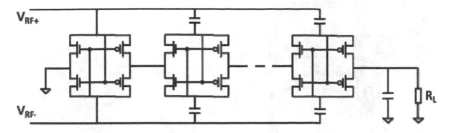

Fig. 2. Five-stage of voltage multiplier rectifier

In this study, low-threshold voltage MOSFETs are chosen to obtain low turn-on voltage of the multiplier rectifier. The simulated power conversion efficiency as a function of the input RF power and the output DC voltage as a function of the RF amplitude are shown in Fig.3, 4 respectively.

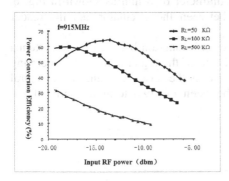

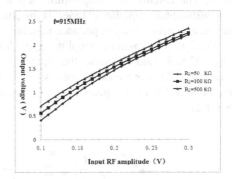

Fig. 3. Power conversion efficiency as a function of input RF power

Fig. 4. Output voltage as a function of input RF amplitude

2.3   Reader Antenna and OCA Co-design

Though numerous uses of on-chip antennas fabricated in silicon IC technology have been proposed, most of the operating frequencies are around 10 GHz or higher [6]. In the UHF band, for passive OCA RFID system, designing electrically-small loop antenna and using inductive/near-field coupling is a feasible method [7]. With silicon-based integrated spiral inductors as chip antennas, the working principle is very similar to LF and HF RFID systems.

We design an on-chip spiral antenna with the top layer metal in 0.18μm CMOS process. The low electrical resistivity and large thickness of the top layer contribute to designing high Q factor inductor. In order to limit the chip die area and reduce fabricate costs; the chip circuit is surrounded by the OCA. The whole chip area is 1mm × 1mm where the microphotograph is shown in Fig. 5.

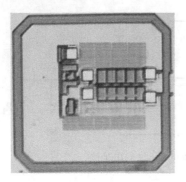

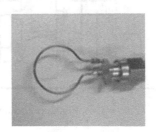

Fig. 5. Microphotograph of the RFID tag with OCA

Fig. 6. Reader antenna

A near field electrically small reader antenna is manufactured by insulated copper wire. The antenna is a single-turn loop with diameter of 8 mm as shown in Fig. 6. An automatic impedance tuner is inserted between the circulator and the reader antenna to give a well impedance matching.

Fig.7 is the magnetic coupling equivalent circuit diagram between reader antenna and OCA. R_L and C_L model the input impendence of the tag circuits at the operating frequency. Ls represents the intrinsic inductor. Rs is the series resistance of the spiral metal and Cp is the parasitic capacitance between the two terminal ports of the antenna.

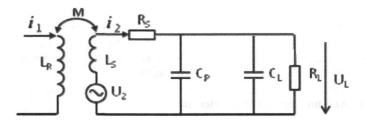

Fig. 7. Magnetic coupling equivalent circuit diagram between reader antenna and OCA

A time variant current i_1 in reader loop antenna generate a time variant magnetic flux. A voltage U_2 is induced in OCA according to Faraday's law.

$$U_2 = i_1 \cdot j\omega M \tag{1}$$

The voltage delivered to the chip circuit can be calculated as

$$U_L = \frac{U_2}{1 + \left(j\omega L_s + R_s \right) \cdot \left(\frac{1}{R_L} + j\omega \left(C_P + C_L \right) \right)} \tag{2}$$

The magnitude of U_L is

$$|U_L| = \frac{|U_2|}{\sqrt{\left(\frac{\omega L_S}{R_L} + \omega R_S C_2\right)^2 + \left(1 - \omega^2 L_S C_2 + \frac{R_S}{R_L}\right)^2}} \qquad (3)$$

Where $C_2 = C_P + C_L$.

It has been explained and proved that the U_L achieves maximum when the OCA resonates with the tag circuit [7], [8]. So the big challenge of the OCA design is optimizing the electrical and geometric parameter to guarantee the maximum power delivered to tag circuits. Inductance modeling method and electromagnetic field simulation tools is useful to analyze the power link between the reader antenna and OCA.

2.4   Measurement Results

The single loop reader antenna is fine tuned to good matching by the automatic impendence tuner, with its return loss measurement to be less than -25dB at 915MHZ. Measurement result show that a distance of 4mm read range is achieved under 100mW RF power delivered from the reader transmitter to the reader antenna. The tag response received by spectrum analyzer is show in Fig.8.

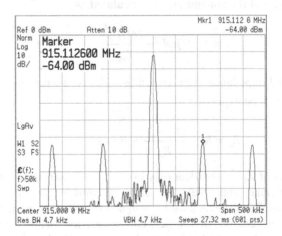

Fig. 8. The spectrum of tag response

3   Inductively Coupled Antenna for RFID Tag with OCA

Many off-chip inductively coupled far field antennas in different structures have been proposed [9], [10]. These antennas are composed of a feeding loop and a radiating body, which are coupled inductively [10].

As introduced in section 2, the OCA work principle is based on inductive coupling. An off-chip radiating body, inductively coupled to the OCA, can be designed to achieve far field communication. Fig.9 shows the inductively coupled antenna structure .The lumped element model is shown in Fig.10.

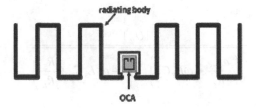

Fig. 9. Tag far-field inductively coupled antenna

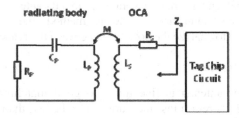

Fig. 10. Lumped element model of the antenna

The input impedance of the antenna can be calculated as

$$Z_a = R_S + j\omega Ls + \frac{(\omega M)^2}{Z_{rad}} \tag{4}$$

Z_{rad} is the individual impedance of the radiating body. At the resonate frequency ω_0 of the radiator, Z_a can be expressed as

$$Z_a = R_S + \frac{(\omega_0 M)^2}{R_R} + j\omega_0 Ls \tag{5}$$

The imaginary part of Z_a is dependent only upon L_s. A conjugate impedance matching between the antenna and the tag chip maximize the power transfer and minimize reflections .So the OCA must be designed to resonate with tag chip circuits, as the same design considerations in near field communication mentioned in section 2. The real part of Z_a is related to mutual inductance between the OCA and the off-chip radiator .Once the geometric parameter of the OCA is determined, we can change the structure of the radiator to impedance matching. The challenge of this inductively coupled antenna design is that it's hard to integrated high-Q inductor in silicon technology. If the quality factor of the chip impedance is much higher than the integrated inductor, it is not easy to give a good impedance matching according to the equation (5).

In this work, an off-chip meandered dipole, printed on a FR4 board, is used as the radiating body to verify the feasibility of the inductively coupled antenna with an on-chip feed loop. The antenna is designed at the operational frequency of 910-930MHZ. The chip die is fixed to the coupling feed port of the radiator on PCB, as shown in Fig.11. With an 8dBi reader antenna, the proposed RFID tag can operate at a distance of 3m when the reader transmits 1W RF power.

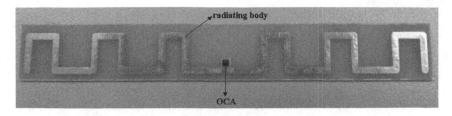

Fig. 11. The photo of the proposed antenna

4   Conclusions

A RFID tag with on-chip antenna has been fabricated in 0.18μm CMOS technology. A communication distance of 4 mm is achieved with 20dBm RF power generated by the reader. An off-chip radiator, inductively coupled to the integrated coil, is designed for far-field communication. This structure can be used as a contactless package method between RFID tag chips and off-chip antennas.

References

1. Finkenzeller, K.: RFID handbook: Fundamentals and Applications in Contactless Smart Cards and Identification, 2nd edn. Wiley (2003)
2. UHF Gen2 for Item-level Tagging,
   http://www.impinj.com/files/Impinj_ILT_RFID_World.pdf
3. Nikitin, P.V., Rao, K.V.S., Lazar, S.: An Overview of Near Field UHF RFID. In: IEEE International Conference on RFID, pp. 167–174 (2007)
4. Finocchiaro, A., Ferla, G., Girlando, G.: A 900-MHz RFID system with TAG-antenna magnetically-coupled to the die. In: Radio Frequency Integrated Circuits Symposium, RFIC 2008, pp. 281–284. IEEE (2008)
5. Kotani, K., Sasaki, A., Ito, T.: High-Efficiency Differential-Drive CMOS Rectifier for UHF RFIDs. IEEE Journal of Solid-State Circuits 44(11), 3011–3018 (2009)
6. Lin, J.-J., Wu, H.-T., et al.: Communication Using Antennas Fabricated in Silicon Integrated Circuits. IEEE Journal of Solid-State Circuits 42, 1678–1687 (2007)
7. Chen, X., Yeoh, W.G.: A 2.45-GHz near-Field RFID System with Passive On-Chip Antenna Tags. IEEE Transactions on Microwave Theory and Techniques 56(6) (2008)
8. Xi, J., Yan, N., Che, W., et al.: On-chip antenna design for UHF RFID. Electron Letter 45, 14 (2009)
9. Li, Y., Serkan Basat, S., Tentzeris, M.M.: Design and development of novel inductively coupled RFID antennas. In: Antennas and Propagation Society International Symposium, pp. 1035–1038. IEEE (2006)
10. Son, H.-W., Pyo, C.-S.: Design of RFID tag antennas using an inductively coupled feed. Electronics Letters 41, 994 (2005)

An Ultra Low Power RF Frontend of UHF RFID Transponder Using 65 nm CMOS Technology

Yongpan Wang, Chun Zhang, Ziqiang Wang, and Yongming Li

Institute of Microelectronics，Tsinghua University, Beijing, China
wangyp08@mails.tsinghua.edu.cn

Abstract. This paper presents the research about the RF frontend of a 0.6 V passive UHF RFID transponder, with Temperature Sensor (TPS) and Random Sequence Generator (RSG) as its digital load, using the TSMC 65 nm Mixed Signal RF SALICIDE Low-K IMD process. The sensitivity of the transponder is -19 dBm according to the simulated results. All of the blocks are based on 0.6 V DC supply voltage. The power consumption of the RF frontend of the transponder at 0.6 V is 656 nW.

Keywords: UHF RFID, low voltage, low power, 65 nm.

1   Introduction

RFID is a contactless automatic identification technology. Different from barcode identification technology, the RFID system can be used without human manipulating. Besides, RFID system can read in a high speed successfully and lots of tags can be read synchronously [3]. Today, RFID technology has been widely used in kinds of service industries, such as logistics, biomedical detection, traffic surveillance, and so on. Then it will be necessary to get knowledge about the RFID system. There are two basic parts in the RFID system, which are called interrogator and transponder. The communication between the two parts of the system depends on the inductively coupling in the near field and backscatter while the tag is in the far field. The research of this paper is about an ultra low power RF frontend of a passive transponder, integrated with TPS and RSG.

Two critical challenges that constrain the industrialization of passive RFID tags are power consumption and integrating with kinds of sensors. Confronting the fore challenge, this paper will present a design based on ultra low voltage implementing advanced process.

The TSMC 65 nm process offers a rich choice of supply voltages, e.g., 1.0 V, 2.5 V or 3.3 V. By choosing the appropriate devices, it can be seen that the whole transponder system will work properly at 0.6 V. This advantage of the 65 nm process has greatly reduced the supply voltage for UHF RFID transponders, which will reduce power consumption greatly, leading to longer reading distance.

The main interest of this project is the RF frontend. There are six blocks (Fig. 1):

Vddgen. Receiving RF carrier energy and giving DC output.

Bias. Supplying stable current bias for the Reset and Osc.

J. Luo (Ed.): Soft Computing in Information Communication Technology, AISC 158, pp. 85–92.
springerlink.com   © Springer-Verlag Berlin Heidelberg 2012

Osc. Supplying a clock signal for the digital part.

Demodu. Analyzing what has been received from the electromagnetic wave transmitted by the interrogator.

Reset. Supplying an initial reset signal for the digital part.

Mod. Sending back the responding signal to the interrogator in an "identified" mode.

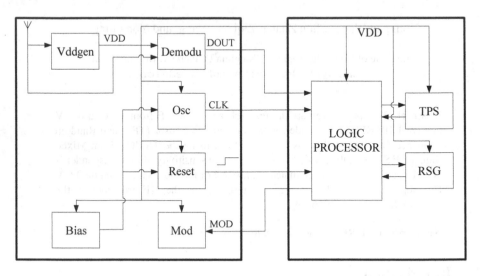

Fig. 1. Architecture of RF Frontend with Digital Load

2   Building Blocks

2.1   Vddgen

The energy received by the transponder should be converted by the AC-DC circuit. The DC output of this circuit will supply other blocks of the transponder. The performance of this circuit can be characterized by the parameter η:

$$\eta = \frac{P_o}{P_o + P_{loss}} \tag{1}$$

Where η is power efficiency, P_o is the power gained by the load of the AC-DC circuit, P_{loss} is the power consumed by the circuit.

Besides power efficiency, the DC output voltage must be big enough to supply the total transponder. Dickson topology (Fig. 2) meets the two demands [4]. ANT1 is connected to a 915 MHz RF signal: $V_{in} = V_a \cos(\omega t)$, while ANT2 is connected to ground. The 2N-stage topology will multiply the amplitude of the output to $2N$ ($V_a - V_{th,D}$), which will be high enough for the residual circuits (Where $V_{th,D}$ is the threshold voltage of the diode).

If the distance between transponder and interrogator is too close, the output voltage of the Dickson circuit will be too high, then a limiting circuit is needed (D00, D01, R0, M0 in Fig. 2).

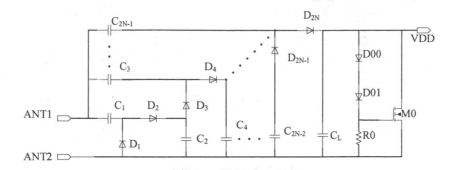

Fig. 2. 2 N-stage AC-DC circuit and Voltage Limiter

2.2   Bias

The fact that the RF carrier transmitted by the interrogator has been modulated makes the supply voltage suffer from considerable fluctuating, which will break the transponder's response. As a result, the bias circuit is required. As a matter of fact, the bias circuit will play an essential role in the RF frontend. It will determine both the Osc's frequency and the Reset's delay time.

Here is an example of bias circuit (Fig. 3). The circuit has introduced Wilson current mirror, the cost of which is lower than traditional band-gap current reference source. First of all, the theory of the circuit in the left part of Fig. 3 will be analyzed. The basic bias part is composed of devices M0~M3 and R1. The biasing current flows out from the drain of M4 by mirroring the current flowing into M0 and M1.The dimensions of M0 and M1 are set to be equal, then the currents flowing into R1 and M3 will be equal, assuming as I. Supposing M3 is working in the strong inversion field, there will be:

$$IR_1 = V_{T3} + \left(\frac{2I}{\mu_n C_{ox} (W/L)_3} \right)^{1/2}$$

(2)

$$I = \frac{V_{T3}}{R_1} + \frac{1}{\beta_3 R_1^2} + \frac{1}{R_1} \sqrt{\frac{2V_{T3}}{\beta_3 R_1} + \frac{1}{\beta_3^2 R_1^2}}$$

(3)

Where $\beta_3 = \mu_n C_{ox} (\frac{W}{L})_3$, V_{T3} is the threshold voltage of M3, $(W/L)_3$ is the dimension of M3. Not considering the second-order effects, the current will be independent of VDD.

While taking the second-order effects into account, cascode architecture should be used in the current mirror.(The right part of Fig. 3) For turning on the bias circuit, self start-up circuit is in need, too. (M0,M1,M2 in the right part of Fig. 3)

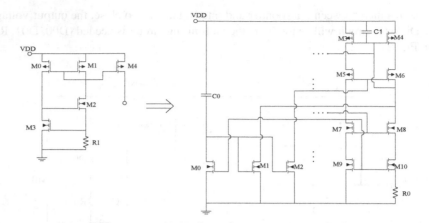

Fig. 3. Schematic of Bias

2.3   Osc

In order to decode the desired data transmitted from the interrogator, the decoder in the digital part of the transponder needs a reliable clock signal.

Traditionally, there are three kinds of oscillators: crystal oscillator, LC oscillator and ring oscillator. The crystal oscillator can give an accurate frequency; nevertheless, it can't be integrated on the chip. The LC oscillator will take up too much size of the chip if the frequency is in the magnitude of MHz. So, the ring oscillator is the only one we can choose for the passive UHF RFID transponder.

The difficulties of the design are frequency stability and power consumption.

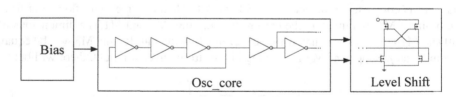

Fig. 4. Architecture of Osc

The total clock generator is composed of bias and core oscillator (Fig. 4). The core of the circuit is a three-stage ring oscillator with an inverter for each stage. While the inverter is working in the on-off status, it equals to a current switch, which can be seen as a resistor at some frequency. The core supply voltage will be produced by both the resistor and the mirroring bias current. The frequency, which is determined by this part, has nothing to do with the VDD. Voltage-shift block has been used to shift the output of the core oscillator to the whole system's supply voltage (VDD).

2.4   Demodu

In the field of short-distance wireless communication system, the ASK modulation mode is ubiquitous, because of its simple architecture and grown-up technology. One typical architecture of the ASK demodulator is based on the envelope detection, which performs low power characteristic. For the passive communication system, the forward link terminal not only receives data but also absorbs energy from the RF electromagnetic field. The energy of the carrier is in positive relation to the amplitude of itself, while in negative relation to both the modulation depth and the modulation pulse width [3]. This tells us that the demodulating circuit must be able to detect signal with very low depth (such as 30% in this project) in order to raise the sensitivity of the transponder successfully.

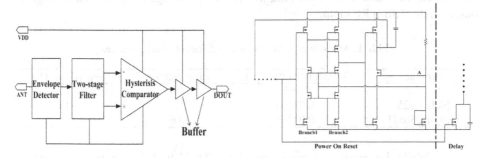

Fig. 5. Architecture of Demodu   **Fig. 6.** Schematic of Reset

The project has adopted the architecture based on envelope detection and two-stage filter net (Fig. 5).The architecture of the envelop detector is four-stage Dickson topology, with each stage selecting the smaller dimension devices than Vddgen. The junction of each stage of the filter net will be connected to the two inputs of the hysteresis comparator [2]. The hysteresis comparator and output-driving buffers are used to introduce digital signal that can be dealt with in the digital part.

2.5   Reset

The main function of the Reset circuit is to offer an initial reset signal for the digital part, forcing the state machine into the initial state. For the digital circuit can work normally at 0.6 V in the TSMC 65 nm process, the step voltage of the Reset should be set at 0.55 V~0.65 V best. Delay circuit is still needed because the stability of clock signal and the start-up of the bias will take some time.

There are two parts in the Reset block (Fig. 6): Power-On-Reset, Delay. Two voltage-detection branches are the core of the POR part: Branch1 and Branch2. The positive feedback of this circuit makes the Reset operate much more sensitively. The delay of the Reset is achieved by charging capacitor with bias current.

2.6   Mod

One way to design the Mod is to use a transistor M_{MOD} connected in series with a capacitor, the impedance of the transponder can be modulated by turning on or cutting off M_{MOD}, which will change the amplitude reflected by the transponder. The value of the modulating depth is determined by the capacitor as well as M_{MOD}.

3   Simulation and Verification

3.1   Simulation

Vddgen. Here are main simulation results of three kinds of devices that are suitable for AC-DC circuit (Table 1) for their ultra-low threshold voltages. The output voltage is based on the 0 dBm RF input.

Table 1. Analyzing of AC-DC using different Native devices

Attributes	Threshold voltage(V)	Efficient(%)	Output voltage(V)
Nch_na	0.095	33.8	2.31
Nch_na25	-0.03365	21	2.34
Nch_na25od33	-0.03365	23.6	2.12

Bias. While the supply voltage swings from 0.6 V to 1 V, the value of current reference is about 65 nA~68 nA .The power consumption of the Bias at 0.6 V is 103 nW. (Fig. 7.a)

Osc. The design specification of the Osc is: 2.4 MHz ± 0.4 MHz. While the supply voltage swings from 0.6 V to 1.0 V, the frequency range is 2.37 MHz~2.56 MHz, meeting the requirements. And the power consumption at 0.6 V is 216 nW. (Fig. 7.b)

Demodu. The minimal detecting amplitude for Demodu is below 300 mV, while the power consumption at 0.6 V is 123 nW. (Fig. 7.c)

Reset. The Power-On-Reset voltages at different corners are: 600 mV, 608 mV, 644 mV. The power consumption is 224 nW at 0.6 V. (Fig.7.d)

3.2   Verification

From 3.1, we can get that the total power consumption of the RF frontend at 0.6 V is 656 nW. Considering the power consumption of the digital load (TPS and RSG) is about 3 uW at 0.6 V and the AC-DC transfer efficient is about 30%, the sensitivity of the transponder will be -19 dBm while perfectly matching.

The layout of the project can be seen in Fig. 8.

The verification has not been finished yet. So far, we have got the sensitivity of -13 dBm while matching the transponder with Tuner instrument. Much more detailed testing will be carried out.

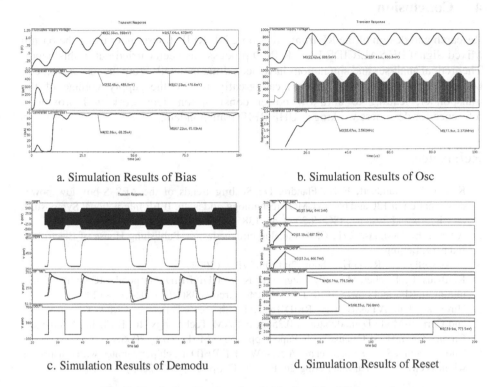

a. Simulation Results of Bias   b. Simulation Results of Osc

c. Simulation Results of Demodu   d. Simulation Results of Reset

Fig. 7. Simulation Results of main blocks of the RF frontend

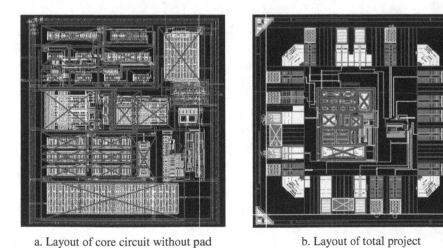

a. Layout of core circuit without pad   b. Layout of total project

Fig. 8. Layout of the Transponder

4 Conclusion

Research and design of the RF frontend of an UHF RFID transponder using the 65 nm Mixed Signal RF SALICIDE Low–K IMD process has been carried out in this paper. Theoretical analyses and simulations have been done in detail. From the research, it can be seen that the advanced process can greatly reduce the supply voltage of the transponder, leading to ultra-low power consumption. Our work will provide a reference for designers working at UHF RFID transponders.

References

1. Kamel, D., Standaert, F.-X., Flandre, D.: Scaling trends of the AES S-box low power consumption in 130 and 65 nm CMOS technology nodes. In: IEEE International Symposium on Circuits and Systems, ISCAS 2009, pp. 1385–1388 (2009)
2. Allen, P.E., Douglas, R.H.: CMOS analog circuit design. Publishing House of Electronics Industry, Beijing
3. Finkenzeller, K.: RFID Handbook, 2nd edn. Wiley (2004)
4. Karthaus, U., Fischer, M.: Fully integrated passive UHF RFID transponder IC with 16.7-uW minimum RF input power. IEEE J. Solid-State Circuits 38(10), 1602–1608 (2003)
5. Hiroyuki, et al.: A Passive UHF RFID Tag LSI with 36.6% Efficiency CMOS-Only Rectifier and Current-Mode Demodulator in 0.35um FeRAM Technology. In: IEEE Int. Solid-State Circuits Conf. (ISSCC) Dig. Tech. Papers, pp. 310–311 (February 2006)
6. Cho, N., Song, S.-J., Kim, S., et al.: A 5.1-μW UHF RFID tag chip integrated with sensors for wireless environmental monitoring. In: ESSCIRC, pp. 279–282 (2005)

Privacy Treat Factors for VANET in Network Layer

Hun-Jung Lim[1] and Tai-Myoung Chung[2]

[1] Dept. of Computer Engineering, Sungkyunkwan University
[2] School of Information Communication Engineering, Sungkyunkwan Universitu
`hjlim99@imtl.skku.ac.kr, tmchung@ece.skku.ac.kr`

Abstract. For a long term of vehicle communication research, now, VANET is in a stage of implementation. However, most of the VANET researches focus on message transmission, address allocation, and secure communication. Vehicle is extremely personal device; therefore personal information, so called privacy has to be protected. There are three kinds of privacy including identity privacy, location privacy, and data privacy. Data privacy is easily achieved by encryption method in application layer. In this paper, we analysis the identity and location privacy treat factor, problem, and solutions in network layer which is the most important layer for end-to-end data transmission. Our analysis includes four IP families: IPv4, IPv6, Mobile IPv6, and Proxy Mobile IPv6. The result of this paper could guide a way to design a privacy protection solution and present existing solution's trend.

Keywords: VANET, Identity Privacy, Location Privacy, Network Layer.

1 Introduction

VANET is developed to support Car-to-Car (C2C) and Car-to-Infra(C2I) communication. For many years, global researchers and projects have been investigating VANET's research issues: routing, security, address allocation, and etc. For an additional research, they focused on privacy issues on VANET. Since the vehicle is extremely personal device, its communication data should be secured and the user's privacy should be unrevealed. Generally, privacy means "Right of an individual to decide for himself/herself when and on what terms his or her attributes should be revealed"[1]. Without privacy protection, user's attributes such as 5W1H can be revealed and used by adversaries. Privacy in the context of VANET can be categorized into three parts [2].

— Data Privacy: Prevent others from obtaining communication data.
— Identity Privacy: Prevent others from identifying subject of communication.
— Location Privacy: Prevent others from learning one's current or past location

Usually, Data Privacy easily achieved through encryption method in an application layer. For that reason, Identity Privacy and Location Privacy are usually mentioned as privacy issues on VANET [16][3].

This paper is structured as follows. In Section 2, we describe the network layer privacy treating factors and its solutions. Section 3 concludes the paper.

J. Luo (Ed.): Soft Computing in Information Communication Technology, AISC 158, pp. 93–98.
springerlink.com © Springer-Verlag Berlin Heidelberg 2012

2   Privacy Treat and VANET

Network layer supports the Internetworking Protocol(IP). IP is the transmission mechanism used by the TCP/IP protocols. This layer is responsible for delivering a message from the source host to the destination host based on their addresses. In the scope of VANET environment, existing IPv4 and IPv6 protocol can be used. Additionally, Mobility support IP can be used for Vehicle's mobility. For the privacy viewpoint, a logical Address is a privacy treat factor. Because, it uniquely allocates to a node and distinguishes the node within global area.

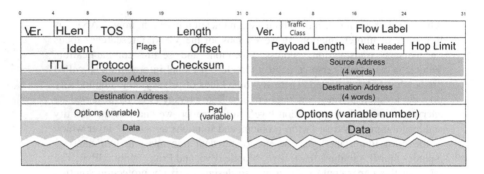

Fig. 1. IPv4 and IPv6 packet format and privacy treating fields

Both IPv4 and IPv6 packet include address(so called logical address or IP address) fields with other fields. All of IP family have privacy problems because transport and application layer information can be encrypted by means of IPSec, but network layer IP addresses still provide means to disclose end-point identities [4].

2.1   IPv4

IPv4 is the most widely deployed Internet Layer protocol. IPv4 uses 32-bit addresses, which limits the address space to 4.2 billion (2^{32}) possible unique addresses [5]. However, some addresses are reserved for special purposes and make it hard to allocate IP address uniquely to node.

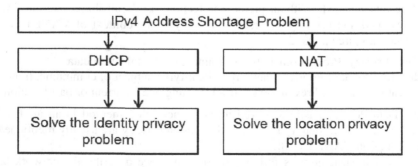

Fig. 2. IPv4 Problem and Privacy Solutions

To solve the address shortage problem, Dynamic Host Configuration Protocol [6], which generates host addresses based upon availability, and Network Address Translation [7], which maps the private address to the public address, are applied. These two techniques unintentional benefit of privacy protecting a host's address by hiding it within a private address space and by periodically change a host's address.

2.2   IPv6

IPv6 is a version of the Internet Protocol that is designed to succeed IPv4. To overcome the shortage of addresses in IPv4, IPv6 employs 128-bit addresses, which limits the address space to $5*1028$ (2^{128}) possible unique addresses. IPv6 support Stateless auto-configuration which is described in rfc4862. Stateless auto-configuration makes an administrator to configure the network of the address while each device automatically configures the interface identifier (IID), of the address. When IPv6 node attaches a network, it receives router advertisement message from router and retrieves network prefix information for the first 64bit of IPv6 address. Then, it extends the 48-bit MAC address to a 64-bit IID number for the last half of the IPv6 address.

Problems: From a privacy point of view, the IPv6 stateless auto-configuration scheme has two problems: Location privacy problem due to IPv6 network prefix information[8] and Identity privacy problem by IPv6 IID information[9]. IPv6 address uses a IID that remains static unless replacing the NIC. As a result, no matter what network the node accesses, the IID remains the same. Consequently, simple network tools such as ping and traceroute can be used to track a node's geographic location from any-where in the world [10]. Lindqvist, J. [11] insisted that the 64bit is enough to figure out the individuals and it is a major threat for user's privacy.

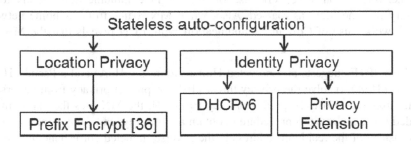

Fig. 3. IPv6 Privacy Solutions

Solutions: To solve the location privacy problem, Trostle, J [9] suggested to encrypt the parts of the prefix such that only appropriate routers in the network can decipher the prefix and obtain the topological information. To solve the identity privacy problem, IETF suggests a hash value of a nonce with the EUI-64 generated IID[12] and Cryptographically Generated Address as an IID [13]. Both [12] and [13] use a random number; therefore the address is dynamically obscured each time a node connects to a network. DHCPv6 also be used as a location and identity privacy solution[14].

2.3   Mobility Support IP

IPv4 and IPv6 usually allocate an address based on network domain and route a message according to the network domain. However, in the VANET environment, a vehicle moves across network domains and periodically causes network handover. When every handover occurs, its IP address has to be changed and for that the connection will be broken. To solve the problem, mobility support IP technologies, called Mobile IP, are required. Mobile IP allows user to move from one network to another while maintaining the connection. Mobile IP for IPv4 is described in IETF RFC 3344 and Mobile IPv6 is described in RFC 3775. In this paper, we focus on the Mobile IPv6 with the same reason as the Internet protocol case. MobileIPv6 is also divided into two categories based on subject of mobility support: Host based and Network based.

Host Based Mobility Support: MobileIPv6.
The core of MIPv6 is that the device maintains two kinds of address to support a mobility: Home-Agent-Address and Care-of-Address. The advantage of MIPv6 is that the MN handles all of the mobility signals, even the network does not fully support the mobility service. The disadvantage is that the MN requires additional mobility stack and battery consumption.

Problems: From a privacy point of view, the advantage of host based mobility threats the privacy. The end-point source and destination IP addresses are revealed to others because MN handles entire mobility signals. The privacy problems in the context of Mobile IPv6 are defined in rfc4882 [15]. The primary goal is to prevent adversary on the path between the MN and the CN from detecting roaming due to the disclosure of the HoA and exposure of a CoA to the CN. For example, when a MN roams from its home network to other network, use of a HoA in communication reveals to an adversary that the MN has roamed. Also, when MN roams from its home network other networks, use of CoA in communication with a CN reveals that the MN has roamed.

Solutions: IETF suggests an Encrypted Home Address (eHoA) and a Pseudo Home Address (pHoA) to solve the privacy threat [16]. To protect privacy from adversary, the MN uses the eHoA. To protect privacy from CN, the MN uses the pHoA in the extended home address test procedure to obtain a home keygen token; then, it uses the pHoA instead of the real home address in the reverse-tunneled correspondent binding update procedure.

Network Based Mobility Support : Proxy Mobile IPv6
PMIPv6 is network based mobility support protocol that is described in rfc5213. The core of PMIPv6 is that the network handles all of mobility support operations and MN does not participate in mobility. There are two methods to detect MN's movement. Beacon information in layer2 and network prefix information in layer3 router advertisement message. In MobileIPv6, MN uses layer3 information to detect its movement. When MN detects its movement by receiving a different network prefix in

RA message, it generates new CoA for communication. However in PMIPv6, Network uses layer2 information to detect MN's movement. When network detects MN's movement by receiving a beacon message, it sent RA message with MN's previously used network prefix(i.e. Home Network Prefix). Even if MN receives the layer3 RA message, it could not detect its movement because of the same network prefix information. Instead of MN in MobileIP, in PMIPv6, the LMA supports the mobility. Whenever CN sends message to MN, the message is firstly transmitted to LMA. Then, LMA checks MN's location and forwards the message.

Problems and Solutions: From a privacy point of view, network based mobility support mechanism's No IP change feature solves the privacy problem that existed in MIPv6. Even the PMIPv6 MN moves to other network, MN does not change its IP address. It maintains the same IP address in every network. i.e., network prefix of the IP address does not guarantee MN's Location. However, in Identity Privacy viewpoint, MN uses the IID based IP address and inherits IP identity privacy problem described in 2.2.

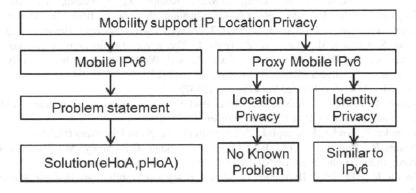

Fig. 4. Mobility Support IP Privacy Solutions

3   Conclusion

In our paper, we investigate IP families' privacy problems and solutions. IPv4, IPv6, and MIPv6 have each solution for identity and location privacy problem. Proxy Mobile IPv6 protects location privacy problem by itself. However, identification privacy problem is inherited due to the basic stateless address auto-configuration scheme and could solve by IPv6 solutions. In our future works, we plan to investigate all network layer's privacy treating factors and its solution.

Acknowledgments. This work (Grants No. 00044301) was supported by Business for Cooperative R&D between Industry, Academy, and Research Institute funded Korea Small and Medium Business Administration in 2010.

References

1. Kent, S.T., Millett, L.I.: IDs–not that easy: Questions about nationwide identity systems. Natl. Academy Pr. (2002)
2. Beresford, A.R., Stajano, F.: Location Privacy in Pervasive Computing. IEEE Pervasive Computing 2, 46–55 (2005)
3. Fuentes, J.M., González-Tablas, A.I., Ribagorda, A.: Overview of Security Issues in Vehicular Ad-Hoc Networks (2010)
4. Bagnulo, M., García-Martínez, A., Azcorra, A.: An Architecture for Network Layer Privacy. In: IEEE International Conference on Communications, ICC 2007, pp. 1509–1514. IEEE (2007)
5. Postel, J.: RFC 791: Internet Protocol (1981)
6. Droms, R.: Dynamic host configuration protocol (1997)
7. Srisuresh, P., Holdrege, M.: RFC 2663. IP Network Address Translator (NAT) Terminology and Considerations (1999)
8. Haddad, W., Nordmark, E., Dupontand, F., Bagnulo, M., Park, S., Patil, B.: Privacy for Mobile and Multi-homed Nodes: MoMiPriv Problem Statement (2005)
9. Trostle, J., Matsuoka, H., Tariq, M.M.B., Kempf, J., Kawahara, T., Jain, R.: Cryptographically Protected Prefixes for Location Privacy in IPv6. In: Martin, D., Serjantov, A. (eds.) PET 2004. LNCS, vol. 3424, pp. 142–166. Springer, Heidelberg (2005)
10. Groat, S., Dunlop, M., Marchany, R., Tront, J.: The privacy implications of stateless IPv6 addressing. In: Proceedings of the Sixth Annual Workshop on Cyber Security and Information Intelligence Research, pp. 1–4. ACM (2010)
11. Lindqvist, J.: IPv6 is Bad for Your Privacy (2007)
12. Narten, T., Draves, R., Krishnan, S.: RFC 4941-Privacy Extensions for Stateless Address Autoconfiguration in IPv6. IETF (September 2007)
13. Nikander, P., Arkko, J., Kempf, J., Zill, B.: SEcure Neighbor Discovery (SEND)
14. Droms, R., Bound, J., Volz, B., Lemon, T., Perkins, C., Carney, M.: Dynamic host sconfiguration protocol for IPv6, DHCPv6 (2003)
15. Koodli, R.: RFC 4882: IP Address Location Privacy and Mobile IPv6: Problem Statement (2007)
16. Qiu, Y.: RFC 5726: Mobile IPv6 Location Privacy Solutions (2010)

Study on the State-Owned Forest Resources Management System Based on Forestry Classified Management

Jie Wang[1,2], Yude Geng[1,*], and Kai Pan[2,*]

[1] College of Economics and Management, Northeast Forestry University
[2] Northeast Agricultural University
Harbin, China
wangjie-0825@163.com

Abstract. With the intensifying of forestry economic system reform in our country, the classified management issue on forestry resources management appears to be an ongoing issue. On the basis of focusing on the state-owned forestry resources management system reform based on classified management and carding the state-owned forestry resources management system, the thesis analyzes the theories on subject, object, route P of resources value realization, fund resources and management system of state-owned forest resources management, and gives a reference on the classified management of state-owned forest resources.

Keywords: classified management, state-owned forest resources, management system.

1 Introduction

The state-owned forestry resources contains all the forest, woods, woodland and all the wild animals, plants, germs which living on the forest and woodland in the country. The state-owned forestry resources play an important role in the ecological environment development, timber production and state-owned land security. The forestry resources mainly concentrates in the state-owned northeast and southwest forest zones, which are directly managed by 138 large and medium-sized forest industry enterprises of the state-owned northeast, Inner Mongolia forest land and southwest, northwest forest land, as well as 4,466 state-owned forest land around the country.

With the intensifying of forestry economic system reform and the increasing demand for forestry ecological development and substance diversification in the society, in 1995, an overall plan for forestry economic system reform issued jointly by the former Ministry of Forestry and State System Reform Committee. It proposed a forestry economic system reform plan focusing on classified management reform, which means the forestry resource is divided into commercial land and public welfare land, and they can be managed separately. In 1996, the State Ministry of Forestry selected 11 units for forestry classified management experiment among northeast, Inner Mongolia forest land, the two southwest provinces, south group forest land and

---

* Corresponding authors.

J. Luo (Ed.): Soft Computing in Information Communication Technology, AISC 158, pp. 99–105.
springerlink.com   © Springer-Verlag Berlin Heidelberg 2012

10 provinces (areas) in north China, and the provinces selected more than 60 counties for classified management testing as well. But in practice, there are still some problems in the forestry-classified management, such as low productivity of commercial land and public welfare land, failure to meet the requirement and so on.

2   The Classified Management Object of State-Owned Forestry Resource

Proceed from the social economic benefit for forest and the demand of ecological, social benefit, one part of forest land is used to produce commodities and service in terms of various function-oriented and developing methods, which can get the economic benefit on the basis of market mechanism; the other part of forest land is mainly used to develop the economic and social benefit for providing production of public welfare and developing the public-benefit function in the society. On this basis, timber stands, protection forests, economic forests, firewood forests, forests for special purpose designated in Forest Law, are classified as follows: timber stands, economic forests, firewood forests are placed in commercial woodland, protection forests and forests for special purpose are public-benefit woodland, the forests suited for both commercial and public welfare functions is called amphibious forest or compatible forest. The public welfare forest productions do not entry the market, which belongs to non-profit-making forestry; while commercial forest production can entry the market, which belongs to industrial forestry.

People classify the forestland on the basis of five big forests classification in our country. According to Forestry Industry Standard in the People's Republic of China, the five big forests can be summarized as two types: one type covers the ecological public-benefit forests for developing ecological benefit function (protection forests and forests for special purpose) and the commercial forests for mainly developing economic benefit (timber stands, economic forests, firewood forests); the other contains state-owned forests and private forests from a forest right perspective. The state-owned forestry resources are classified into two objects, state-owned ecological public-benefit forests and state-owned commercial forests.

The public-benefit forest is defined by two norms, ecological vulnerability and ecological importance. Besides the former two norms, the commercial forest should consider three more norms, internal rate of return, woodland productivity and land utilization index for plotting. According to the different leading function, the public welfare forest is divided into protection forest and forest for special purpose, and the protection forest includes soil and water conservation forest, water conserving forest, bank and road protective belts, wind protection and sand bind forest, farmland and grassland protection forest and other protection forests; forest for special purpose includes the national defense forest, scientific and educational experimental forest, the woods seed resource (the parent stand), the environmental protection forest, the scenic beauty forest, cultural forest (woodland places of historic interest and revolutionary memorial forest) and natural reservation forest (natural protection forest). The commercial forests contain timber stands, economic forests and firewood forests. The timber stands includes the common timber stands and industrial fiber forests; the economic forests includes fruit woodland, woody edible oil forest, chemical engineering materials forest and other economic forests.

3   The Forest Resource Value Realization Processing Based on Classification Management

Forestry industry can provide the intangible public-benefit production and tangible economic production such as wood production and traveling service. The labor in the public welfare production process does not have a concrete product carrier, so the manager cannot control the products in their own capability and method. Therefore, the benefit comes to the customers without a exchange, while the customers obtain the use value and need not pay for it, which means that the public-benefit products cannot get compensation through exchange. The market mechanism like exchange of equal values, impartial competition and selecting the superior and eliminating the inferior are ineffectual for the public-benefit products. Economic products (forest products and service) have a tangible product carrier or a controlled form, through which the manager can control it. The kind of products are exchanged in the market and according to the exchange of equal values the products can obtain the value or price when transferring the use value. The labor used in the products can be get the compensation. The law of value and market mechanism limits the production and circulation of the kind of products.

Therefore, the value of the ecological public-benefit forest is realized by the way of charging ecological benefit compensation from the direct or indirect beneficiaries. The commercial forest management benefit is the value coming from the production processing, which is determined by both seller and buyer, i.e. commercial forest producing enterprise (organization) and forest industry determine the value according to the quality and quantity of the trading commercial forest resource. The establishment of standing forest trading market is a beneficial exploring for profit realization, however, the specialty of the standing wood resource makes the trading limited in a binding legal agreement to realize the transfer of commercial forest proprietary right and woodland use right.

4   The State-Owned Forest Resource Management Pattern Based on Classification Management

Based on the forest-classified management, the two state-owned forest resource management are different in management system, source channel of fund and operation mechanism.

4.1   The Subject and Management System of State-Owned Forest Resource Management

The determination of management subject sets up standards for the division of duty, right and benefit in forest resource management.

The management subject mainly points to government, enterprise or individual. Government mainly manages the state-owned ecological public-benefit forest, and state-owned forestry enterprise, private forestry enterprise or individual manage the state-owned commercial forest.

According to the protection importance principle of ecological forest, the state-owned ecological public-benefit forest is divided into available and unavailable two types, i.e. key ecological public welfare forest and common ecological public welfare forest. The former should be in a strict government control, especially for natural protection forest and water conservation forest; we should try to develop the maximal economic benefit for the latter, the amphibious forest, on the premise of not reducing the ecological function. The management subject of state-owned ecological public-benefit forest is government, because the state-owned fund should be gathered in the crucial industry and key field related to the national economy and the people's livelihood, and the government should provide the public commodity. In the light of local conditions, the ecological public-benefit forest in state-owned woodland and collective woodland can rely mainly on government management while making private management subsidiary, which forms the institutional management pattern in the way of managing revenue and expenditure separately. Non-public sector of the economy take part in the ecological public-benefit forest construction, which can improve the public economic subject management for ecological public-benefit forest. At present, in the practice of public-benefit forest division, the private managed forest has been placed in ecological public-benefit forest, and the government cannot buyout for the government management completely. Therefore, the suitable management pattern of state investment and private management, private ecological public-benefit forest, should be considered, which means that the government purchases ecological products for ecological compensation and developing the advantage of private investment for public products.

The state-owned commercial forest resources are mainly distributed throughout the large-scale state-owned forest industry district and state-run forestland. The management subject of the state-owned commercial forest is the state-owned forestry enterprise, private forestry enterprise or individual. In order to ensure the national strategic goods reserve, the government should regulate the forest products price and the material requirement for national development construction on the premise of macro-control and guiding supervise. At the same time, the management subject of the state-owned commercial forest seeks the maximal economic benefit on the basis of market regulation mechanism. As for the state-owned commercial forestland, the non-public economic subject is required urgently to join in the commercial forestland management. In terms of the state-owned land reform principle and state-owned property management method, people should make full use of market operation mechanism to cultivate various management modes for property ownership, such as individual, cooperation, joint venture, joint management, joint stock and so on. As for the state-owned and private controlling management, we should set up Administration of State Forest Resource to manage the property right of state-owned forestland, forestland use right and the property right transfer for forestland, and preserve and increase the value of state property.

4.2   The Fund Source of State-Owned Forest Resource Management

The forest-classified management reform has been through the development of many years, and it forms the ecological public-benefit forest management mode that mainly relies on the state investment and the commercial forest management mode that mainly relies on market regulation.

1)   The fund source of state-owned ecological public-benefit forest resource management

The state-owned ecological public-benefit forests mainly rely on state management and implement the public welfare institutional management system. According to all the levels of government right division, governments at all levels solve their fund issue through fiscal measures and set up suitable fiscal compensation system to attract social power to construct together.

The fund investment for state-owned ecological public-benefit forest obeys the principle of "the beneficiaries are the investors, and the beneficiaries compensate for it". For the benefit externalization of public-benefit forest, the beneficiaries are the social public, and the government mainly invests the fund. In terms of the ecological area and beneficial scope of public-benefit forest, governments at all levels should take their duties separately. The other construction fund source of ecological public-benefit forest comes form ecological efficiency compensation fund, which mainly collects the ecological efficiency compensation charge from beneficiaries (related to industries relying on public-benefit forest resource and directly obtaining economic benefit), taking for the national fund reserved for ecological public-benefit forest management and development. Other fund sources of state-owned ecological public-benefit forest management and construction are from the government guiding multiple investments. The construction of infrastructure facilities takes franchising and bidding management modes, such as the usage of BOT financing pattern, making a contract with an individual or a company, allowing the construction, operation, management for the facilities during the contract term, transferring the facilities to the government gratis after it expires, withdrawing the management right by the government. This pattern makes full use of social fund and accelerates fund collection, which is an effective funding source off-fiscal budget.

2)   The fund source of the state-owned commercial forest resource management

The state-owned commercial forest mainly relies on enterprise (including state-owned forestland, forest products enterprise) management. Based on the basic industry management and market regulation mechanism, the government should provide necessary support to develop the commercial management of the forest resource and resources allocation efficiency. According to the market requirement, we should cultivate the forest resourced to realize the integration and industrialization of production and processing, to implement intensive management, and to take the responsibility of meeting demands on timber and forest products and solving wood shortage problem. The funds for commercial forest management mainly come from self-financing, social fund, financial organ credit and other channels.

4.3   The State-Owned Forest Resource Management System

The ecological public-benefit forest is a special forest for making use of the ecological forest function to provide ecological benefit. The management purpose is to develop various ecological benefits and obtain the maximal economic benefit on the premise of not doing harm to ecological function. According to Forest Law and its Enforcement Rules, people should strictly manage the protection forest for state land security, environment improvement, bio-diversity conservation and the forest for special purpose, and control forest harvesting. The key public-benefit forest is forbidden to cut and there is no cutting quota for it; while the economic loss of the forest cutting forbidden can be compensated through forest ecological benefit compensation fund. The other public-benefit forests should allow suitable forest usage, non-wooden products cultivation, forest traveling, sightseeing, visiting and many other activities. On the premise of protecting and not damaging public-benefit function, the suitable usage of the public-benefit forest can bring certain benefit for function compensation. The ecological compensation fund is the main approach to make up the manager's loss.

The main purpose of commercial forest is to develop the forest economic benefit. With the support and protection of the state industrial policy, the forest should rely on market regulation to improve development. People can make use of high-tech, high investment, implement setting cultivation, base practice and collective management, improve forestland productivity, and obtain the maximal benefit; in the view of management, people should make use of economic method and law to find a management method according with the market economic rules and organize production based on the market demand. The state-owned commercial forest management is mainly through the following patterns, such as asset like management, independent accounting, taking full responsibility for their own risks, profits and losses, development, and self-restraint. People can manage the forest by the way of transferring, renting, mortgaging, dispersing share and exchanging according to law and on a voluntary and compensatory basis so as to realize the efficient allocation of forest resources. On the way of the deep processing of forest products and the integration of forestry, industry and trading, people should develop economic forest and industrial raw material forest, abandon the single timber production and develop forest by-production and industrial material. The beautiful forestland and natural landscape, with a convenient transportation, can be developed into forest traveling industry. We should set up both state and local forestry industry association to improve the standardization development trend of forest market and develop the maximal economic benefit of the commercial forest.

In the view of the state-owned forest management mode of subject, operation system, investment channel, the management for the state-owned forest should ensure the management environment construction, such as to further improve the fiscal mechanism construction of forestry development, to put forest management in the list of central government budget, to provide fiscal subsidies for forest cultivation, improved varieties of forest, forestry machines, to make the local public-benefit forest compensation standard in line with local conditions, to provide subsidies for forestation; people should further improve forestry financial policy, take forest insurance, and take the forest use right as one item in the mortgage field. The central government should provide forestry for loan interest subsidizing; the government also should lighten the forestry tax burden and the government budget should support the local forestation fund; people can strengthen the supervision of state-owned forest asset management and improve the development of forestry social service organization.

Acknowledgment. Fund Project: Heilongjiang philosophy and social sciences planning program "research on economic transition of sustainable development forestry resources-oriented cities in Heilongjiang" and initial funding on postdoctoral scientific research in Heilongjiang.

References

1. Wang, F.: Study on Eco-forest Planting in Forestry Classified Management in China. State Academy of Forestry Administration Journal (1), 10–15 (2007)
2. Li, G.: Study on Forestry Classified Management Mode in China. Heilongjiang Science and Technology Information (5), 76 (2007)
3. Zhan, C., Zhong, S.: Exploring for Ecological Forest Management Model. Jiangxi Forestry Science and Technology (1), 10–12 (2002)

An Analysis of the Decision-Making Model about the Talents' Flow in China's University

Lin Hongquan[1], Zhang Jinzhu[2], and Shi Yuxia[3]

[1] Personnel Department Tianjin Polytechnic University, Tianjin, China
[2] School of Foreign Languages of Tianjin Polytechnic University
[3] Library of Ludong University, Tianjin and Yantai, China
hqlin@tjpu.edu.cn

Abstract. It is of great significance to carry on the research of the basic principles for talents' flow and build reasonable mechanisms for the reform and development in higher institutions. Based on the decision-making process of the talents' flow in higher institutions, the author proposes two kinds of models for the talents' flow. They are the partial balanced model and the general balanced model. Meanwhile, the author discusses their practical applications to talents' flow in higher institutions.

Keywords: Higher institutions, talents'flow, model of decision-making, application.

1. Introduction

A rational movement of talents is the basic conditions and fundamental ways for the human resources optimization. Higher institutions are the places abundant highly qualified human resources and relatively frequent mobility of them, so an unidentified and uncontrolled mobility will be far away from the purpose of human resources optimization.

Presently, many problems still exist in the current situation of talents flow in china's higher institutions, which has greatly impeded the fair, orderly and healthy talents' flow. Thus, it is of great significance to carry on the research of the basic principles for talents flow and build reasonable mechanisms for the reform and development in higher institutions. As for the decision –making of the talents' flow, we are trying to build the micro and macro models.

2 The Decision-Making Process of Talents' Flow in Higher Institutions

A number of factors may influence talent's flow. Generally they are the individual's subjective reasons as well as the objective environment of organization. The main decision-making processes are shown in figure 1.

J. Luo (Ed.): Soft Computing in Information Communication Technology, AISC 158, pp. 107–114.
springerlink.com   © Springer-Verlag Berlin Heidelberg 2012

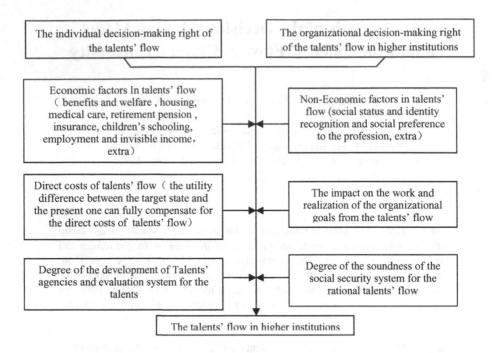

Fig. 1. The decision–making process of talents' flow in higher institutions

3   The Partial Balanced Model

3.1   The Positioning of the Roles

The concept that competition rather than the official administration can vitalize the higher institutions is contradictory to the traditional Chinese campus management system. Higher institutions, as non-profit organizations, must relocate their positions in the current market-driven society. The management of the higher institutions from government should be indirect guide rather than the direct intervention. As the combinations of typical stakeholders, if the higher institutions want to get the paid or unpaid supports from the society, an element to balance all the interest-holders must be found in autonomous institution management. So, a rational relationship in macro level among higher institutions, government and market will be the macro regulation from government, the main configuration from market, and the autonomous adjustment from the higher institutions.

It is a pre-task to conduct the appropriate relocation and arrangement of their roles for different forces in affairs of personnel mobility. We will use the following decision-making model to analyze the gaming among different forces. Our model will take the government as the internal variable, thus to illustrate the concept that macro regulation from government, the main configuration from market, and the autonomous adjustment from the higher institutions. Universities and the faculty, as the gaming subjects, are the mutual independent variables to construct the partial equilibrium model.

3.2   The Hypotheses Put Forward

We set the following hy potheses in the model so as to simplify the gaming process and related variables between schools and faculty.

1) Hypothesis 1:
The human capital flows in higher institutions is based on the idea that the market is the macro mechanism for the resources configuration;

2) Hypothesis 2:
Schools and teachers are the independent subjects in the decision-making, and both are the independent decision-makers of their own, this is not only the precondition that whether a teacher can achieve reasonable flow but also the basic premise that Talent Resources Market mechanism is perfectly taking its role.

3) Hypothesis 3:
Schools and university teachers, as the participants of the mobility, both parties have their rational and clear targets. The Decision to transfer is a careful and rational decision after a comprehensive and independent consideration.

4) Hypothesis 4:
Relevant property transactions of human capital between the school and university teachers are under the guidance of fair trade principles. Both parties should be bounded by the agreement signed within the duration of contract; and the independent decision made by both parties will not ask for the permission from each other when the contract expires.

3.3   The Contents of the Model

The analysis for talents flow can be conducted from many sides. But back to its very nature, it is always the joint decisions between the individuals and the organizations served.

For the talents themselves, two preconditions will decide their choices:

Firstly, the target state can provide greater utility than the present one; Secondly, The utility discrepancy between the target and present one can fully compensate for the direct cost of the talents flow. If we take the utility provided by the present state as the opportunity cost, we can use a cost-benefit model to show the individual decision of the talents.

F1 :   If $V_{FJ} - V_{NJ} - C_F \rangle 0$, then flow;

If $V_{FJ} - V_{NJ} - C_F \langle 0$, then stay.

V_{FJ} ——Value Future Job   (the utility offered by future jobs)

V_{NJ} ——Value Now Job   (the utility offered by present jobs)

C_F —— Cost Flow   (direct cost of flow)

When the current income of transfer are greater than current costs (opportunity cost and direct costs included), that is, $V_{FJ} - V_{NJ} - C_F \rangle 0$, the individuals will make the decision to flow; When the current incomes of transfer are fewer than current costs (opportunity cost and direct costs included), that is, $V_{FJ} - V_{NJ} - C_F \langle 0$, the individuals will make the decision to stay.

However, it is not just the individual decision of the talents that will determines whether he chooses to flow or stay. It also rests with decisions from the units and organizations where the talents serve.

In the market-driven economy, all units, including enterprises, public organizations or government agencies etc, decisions are all made based on their long term developments and costs analyses. Therefore, organizational decision making of talents flow also can be shown in the cost-benefit model, which is:

F2: If $I - C \rangle 0$, then they flow;

If $I - C \langle 0$, then they stay.

I —— Income (the benefits of talents' utility)

C —— Cost (the costs of talents' utility)

When income of talents outweighs the cost of the talents, that is, $I - C \rangle 0$, and the units will make the decision to flow; when Income of talents underlies the Cost of the talents , that is, $I - C \langle 0$, the organizations will make the decisions not to flow [1].

Due to the combination of the individual and organizational decision above mentioned, we can get a decision-making model of talents' flow.

That is, the possible implementation of talents' flow rests with the combination of individual and organizational decisions $F = (F1, F2)$.

$F1$, which represents the result of individual decision, is determined by the comparison between the current income and cost of talents' flow;

$F2$, which represents the result of organizational decision, is determined by the comparison between the benefits and use-costs of talents' flow.

Figure 2 shows the four combinational results of the individual and organizational decisions.

Com 1: flow	Com 2: stay
$V_{FJ} - V_{NJ} - C_F \rangle 0$ have the will to flow	$V_{FJ} - V_{NJ} - C_F \langle 0$ Have no will to flow
$I - C \rangle 0$ have the chance to flow	$I - C \rangle 0$ have no chance to flow
Com 3: stay	Com 4: stay
$V_{FJ} - V_{NJ} - C_F \rangle 0$ have the will to flow	$V_{FJ} - V_{NJ} - C_F \langle 0$ Have no will to flow
$I - C \langle 0$ have no chance to flow	$I - C \langle 0$ have no chance to flow

Fig. 2. The partial balanced model of talents' flow decision –making

From the model, we know that a normal flow could happen in combination 1st. As for this model, you should pay attention to the following points:

1) The composition of the utility function of the individual working state is very complicated, which may be influenced by numerous factors. As V_{FJ} or V_{NJ} is shown in the model. Although the current economic factors accounts for absolute significance in the utility function, some non-economic factors, such as occupational preference, and convenience degree of family life, interpersonal relationship etc, are all the elements that should not be neglected. And this also causes objective, accurate and detailed utility evaluation in different working states more difficult to be done, thus increasing the uncertainty of individual decision-making.

2) The direct costs of talents' flow are also decided by many factors. Such as the necessary financial costs and psychological burdens they have to take in dealing with some current institutional obstacles (such as domicile, archives) and costs of transferring and resettling, etc..

3) The benefits of talents' utility are not only related with individual factors, such as the professional ability, morality and the willingness of contribution, but also with the provisions of other facilities and resources, which has caused great difficulties in evaluating the benefits of talents' utility.

3.4   The Conclusion of the Model

Based on the above assumptions and the model analysis, we can know that under the current conditions that the systematic market mechanism functions well and the government serves as the macro regulators rather than direct administers, the free gaming between universities and teachers may result in four possible outcomes of talents' flow in universities:

Conclusion a : the flow of teachers happens when the teacher decides flow to the university, and the university has decided to bring in the teacher.

Conclusion b : the flow-out of teachers happens when the teacher decides to resign from the university and the university decides to fire the teacher.

Conclusion c : the flow-out of teachers happens when the teacher decides not to resign from the university, but the university has decided to fire the teacher.

Conclusion d : the flow-out of teachers happens when the teacher decides to resign from the university but the university decides not to fire the teacher.

4   A General Balanced Model

Under the traditional personnel system in higher institutions, university teachers are allocated by the official instruction. The negotiations between the teachers who are being partly balanced and the employers are prohibited. We have good reasons to believe that there is a great deal of personal preference distortions and inadequate incentives human capital. Now we must consider the general balance between all the college teachers (supply: S) and all the positions offered (demand: D) within the certain scope and period.

4.1   The Contents of the Model

The presupposition of the General balance model is also the premise of the partial equilibrium the model .it is still the free gaming between the teachers and university on the basis of the market allocation and government regulation, the only difference between lies in the extended larger range of analysis to all teachers and all universities in a country.

Let's assume that the total amount of college teachers is n; each individual college teacher is a_1 、 a_2 、 a_3 ...... a_n ; The total number of positions offered by universities is set as m; each position is b_1 、 b_2 、 b_3 ...... b_m . So respectively, teacher a_1 corresponds with b_1 , teacher a_2 corresponds with b_2 ,......, teacher a_n corresponds with b_m .

We call this stage before the flow stage I, the corresponding relationship can be listed as follows:

$$S: \quad a_1 、 \quad a_2 、 \quad a_3 \cdots\cdots a_n$$

$$D: \quad b_1 、 \quad b_2 、 \quad b_3 \cdots\cdots b_m$$

The total utility acquired by teachers : $A = \sum a_i \bullet \alpha_i$ (formula 1)

(α_i represents the utilization of human resources of a_i)

The total utility acquired by the units : $B = \sum b_i \bullet \beta_i$ (formula 2)

(β_i stands for the sparing degree of practical human resources capital of b_i)

Al stands for the total utility of teachers in the original stage;
B1 represents the total utility of the universities.

In the process of agreement-based free personnel flow, after numerous bargaining between employers and teachers, the corresponding relationship between the teachers and posts offered has changed dramatically.

We call the state after flow as the state II, and it is shown by the following formulation. Presently, the teacher a_i finds the post b_i based on his own preference and the information at hand. Let's suppose the state following happens:

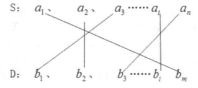

In state II, the total utility of teachers is A2, the total utility of universities is B2.

Under the assumption of a more transparent information publication, the changes of both parties takes place, the old balance of flow is continually broken and a new one is being formed. a dynamic balance has been achieved .in this balance, relationship between (A2, B2)and (A1, B1)is shown as follows:

1) A2>A1, B2>B1;
2) A2=A1, B2>B1;
3) A2>A1, B2=B1;
4) A2<A1, B2>B1;
5) A2<A1, B2<Bl.

4.2   The Conclusion of the Model

1) A2>A1 B2>B1. The total utility of both teachers and universities has been increased through flow, that's the famous Pareto improvement;

2) A2=A1, B2>B1. The total utility of teachers has not been increased through flow, but the total utility of universities has been up, that's also the Pareto improvement;

3) A2>A1, B2=B1. The total utility of universities has not been increased through the flow of faculty, but the total utility of teachers has been increased, that's also the Pareto improvement;

4) A2<A1, B2>B1. The total utility of teachers has been decreased through the flow of faculty, but the total utility of universities has been increased, except for the compulsory allocation, this will not happen in a contract-driven society.

5) A2<A1, B2<Bl. The total utility of both has been decreased through the flow of faculty, but this situation will not happen in any forms of society [2].

5   The Application of the Model

We can know that the realization of the talents' flow rests with the combination of results between individual and organizational decision; it also depends on the gaming of total utility between universities and teachers. The above discoveries have offered the foundation on the motivation to study the talents' flow in higher institutions.

The motivation of the talents' flow in China mainly covers the following aspects: Factor of Need, such as: the need for better material life, spouse arrangements, children's schooling, the individual development, work autonomy, the career achievements etc; Factor of environment. Such as: the macro social background, micro social structure and working environment etc; Factor of culture.

At the same time, the model has profoundly revealed the positive connection between the value realizations and the reasonable flow of the talents. Modern economic theory tells us that the decrease of marginal effectiveness of input is widespread in all aspects. Generally, at the initial stage, the benefits may be on the rise with the increase of the input, but when the input reaches a certain amount, and the marginal effectiveness rises to its utmost, and then gradually reduces even though the input is still being increased [3].

Moderation of teacher allocation is also a problem to be considered. No matter how great the benefits that input brings might be, it is sure to be down after reaching its utmost .so a full development of teachers' roles requires vested benefits at least to be stable, or else, a reasonable flow will be needed to change the form and structure of the input.

6   Conclusions

Theory and practice have proved that the optimal combination of human capital and natural resources can only be achieved in the capital flow. And greater advancement in benefits of talents' utility can only be achieved in the flow mechanism. Just like the Chinese saying goes that: running water is never stale and a door-hinge never gets worm-eaten.

The flow of talents in higher institutions is a complex systematic project. Firstly, we should try to change the current chaotic situations in higher institutions. Secondly, we should try to establish the scientific and rational talents' flow mechanism in higher institutions, Thus to facilitate the construction of a reasonable flow system from the introduction, internal management and exit inducement[4]. As shown in figure 3.

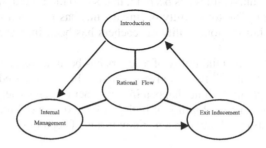

Fig. 3. The model of rational talents' flow

References

1. Cheng, Y., Sun, H.: The Economic Analysis of Talents' Flow in Higher Institutions. Economist, 7 (2003)
2. Liu, X.: Present Situation Analysis and Research of the Feasible Solutions on Human Resource Administration and Development in Universities and Colleges. Hefei Polytechnic University (2002)
3. Guest, D.: Human resource management and performance: a review and research agenda. The International Journal of Human Resource Management (August 1997)
4. Lin, H.: The research to the Current Conditions of Talents' Flow in China' Universities and Construction of the Reasonable Talents' Flow. Journal of Tianjin Polytechnic University 26 (2010)

A Review on Service Quality and Student Satisfaction of Higher Education

Xiaoyang Zhao

Department of Management and Economics
Tianjin University
Tianjin, China
zhaoxiaoyang_607@163.com

Abstract. This paper reviews the research on service quality of higher education and the successful experience of foreign student satisfaction measurement. Firstly, the author introduces the global trend and intensively competitive situation of higher education service. The important function of student satisfaction survey in higher education management is also mentioned. Secondly, the related notions of service quality and student satisfaction are discussed. Then, the qualitative and quantitative methods widely used in student satisfaction assessment are compared. Finally, the conclusions obtained by present research are summarized.

Keywords: Higher Education;Service Quality;Student Experience, Student Satisfaction.

1 Introduction

The foreign academic circles have begun to attach importance to the theoretical study of customer satisfaction since the beginning of 1970s. The customer-centered management concept of attempting to satisfy their requirements and expectations and pursuing customer satisfaction and customer loyalty was first formed in economically developed countries and has achieved rapid development since 1980s. The notion of customer satisfaction was first introduced to China in the late 1990s and then was extensively applied to retailing trade, catering industry, tourism and other service industries while introducing the satisfaction theory and service perspective to the higher education industry still lingers in the infant stage.

Educational quality is the core issue of higher education development and improving educational quality is the eternal theme of higher education. Quality was put forward as the central issue of higher education in a world conference of higher education thus making providing students with high-quality service an important subject of higher education (Chaffee, 1990; McMillen, 1991; Sines and Duckworth, 1994; Soutar and McNeil, 1996; Soutar et al., 1994). The application of service quality model to the field of education and training industry should be paid more attention (Harvey *et al.*, 1992).

The service industries are confronted with increasingly intense competition and higher education is no exception as a kind of service industry. The rapid expansion of

J. Luo (Ed.): Soft Computing in Information Communication Technology, AISC 158, pp. 115–122.
springerlink.com   © Springer-Verlag Berlin Heidelberg 2012

colleges and universities, the significant increase of educational cost and the change of humanistic characteristics of educatees force colleges and universities to perceive the function of student satisfaction from different angles in order to survive (Kotler and Fox, 1995). Many scholars have begun to take cognizance of globalization trend of higher education (Wilson *et al.* 1997; Chalmers 2007; Richardson 1994; Rhodes and Nevill 2004; DiBiase 2004; Umbach and Wawrzynski 2004). The intense competition of educational industry impels colleges and universities to adopt a market-oriented strategy to make a distinction between the service of their own and that of the competitors, that is, to understand the target markets including students, stakeholders and their demands, to adjust their own services to meet the market demands and then to improve the customer satisfaction degree through providing excellent services (Keegan and Davidson, 2004). Since the development of colleges and universities is oriented toward the development of students and other stakeholders, it is bound to adopt the suggestions of students. Regarding student satisfaction as the bridge of communication, students and colleges and universities can contribute to improving higher education according to student experience and assisting colleges and universities in judging and adapting to the constantly changing economic society thus grasping opportunities and meeting challenge in a better way.

This article intends to make an overall review of the research into the service quality of foreign higher education and the successful experience of student satisfaction measurement.

2   Related Notions

2.1   Higher Education Service and Service Quality

The foreign scholars studying the service quality of higher education maintain that higher education service is one of the basic outputs of higher education. Shank, Walker and Hayes (1995) hold the opinion that higher education service possesses the properties of service industries, such as, intangibility, heterogeneity, inseparability from the service transmission process, variability, volatility, and students participating in service process. In fact, various educational activities are implemented in accordance with laws of the market. Education, as a management action, is not only a service provided by schools or teachers as the selling party unilaterally but also a process of educational and learning activity process conducted by teachers and students together. Each school doesn't provide ready-made commodities by quality and the connotation of school is determined by teachers and students jointly. Even so, the service property of education industry is still very obvious. According to the definition of PZB (1988), the service expectation of students refers to 'the desire for service or desirable things of students under certain circumstances' and the performance perception of students refers to 'the practical experience of service provided by students'. PZB (1988) considers that service quality is the comparative result between performance perception and service expectation of customers for practical performance of services. This relationship can be summarized as the following formula:

Service Quality (SQ) =Perception (P)-Expectation (E)

PZB indicated that consumers' quality perceptions are influenced by a series of four distinct gaps occurring in organizations. These gaps on the service provider's side, which can impede delivery of services that consumers perceive to be of high quality, are:

Gap 1: Difference between consumer expectations and management perceptions of consumer expectations; Gap 2: Difference between management perceptions of consumer expectations and service quality specifications; Gap 3: Difference between service quality specifications and the service actually delivered; Gap 4: Difference between service delivery and what is communicated about the service to consumers. Perceived service quality is defined in the model as the difference between consumer expectations and perceptions (Gap 5), which in turn depends on the size and direction of the four gaps associated with the delivery of service quality on the marketer's side.

Higher education pertains to the service industry and one of its products is higher education service. From the perspective of higher education service quality, students are the direct customers of higher education service hence the primary task of colleges and universities is to satisfy the need of students as the direct customers. It is necessary to have a general knowledge about the characteristics of higher education service so as to offer satisfactory service for students.

2.2   Student Satisfaction

Students are the principal and most primary customers of higher education. Hill (1995), Wright (1996), Cook (1997), Emanuele (2006) have pointed out students as the principal and pivotal customers confirms students to be the principal part to evaluate education. Kerlin (2000), Soekisno Hadikoemoro (2001), Carrie Leugenia Ham (2003) have indicated that students are the principal part for assessing the quality of higher education service.

As for the definition of student satisfaction, there are different voices but the universally acknowledged definition is that student satisfaction refers to the attitude toward certain study and living environment of undergraduate thus being provided with the characteristic of general attitude.

Danielson (1998), Fitzgerald (1992) and other scholars maintain the opinion that student satisfaction degree refers to the appealing, proud and positive emotion of students for their schools. Bryant (2001), Schreiner and Juillerat (1993) consider that student satisfaction is the satisfactory feeling for their university experience reported by college students when their expectation gets met or the satisfactory state is exceeded. Aldemir and Gulcan (2000) point out that student satisfaction is the positive or negative attitude of students toward their universities. Although the notions of student satisfaction are not consistent, its multidimensional property has achieved general consent (Hartman & Schmidt 1995).

3   Research Methods

3.1   Student Interview and Focus Group

Qualitative research method can obtain dynamic, experiential and interactive data thus enabling the discovery of deeper and richer conclusions (Gilmore & Carson, 1996;

Swan & Bowers, 1998). While qualitative research mainly deliberates on the representativeness of sample data, the qualitative research mainly considers the depth and breadth of obtaining information (Hussey & Hussey, 1997; Denzin & Lincoln, 1998). In regards to the research into student satisfaction, the frequently used qualitative method is to make sampling interview of the research population, namely, the method of Focus Group. In the process of interview, it is common that the organizers raise question and guide the interviewees to put forward their own viewpoints thus obtaining relevant information.

Brenders (1999) investigated into the perception of university service of students in a university in Brisbane of Australia in a period of six months through the Student Focus Project. The research has established 24 Focus Groups composed by undergraduates and has evaluated the perception and satisfaction degree of university service of students through problems such as a series of positive or negative influencing factors of college students, success or failure of student service, solutions for deficiencies in university life and the relation with university expected by students, etc.

In the related subjects put forward by student Focus Groups, malignant bureaucracy and balkanization of Information are the major systematic factors to engender negative influence on student service perception. Moreover, it is discovered in the research that student service recovery strategy can contribute to improving the overall confusion state of student service and the poor performance of university service.

Clewes (2003) has made a longitudinal analysis of three years of study experience of part-time postgraduates in a British university by means of interviewing 10 samples. With the qualitative research method, namely, the method of in-depth interview, this research has made five times of interviews of ten interviewees within three years.

There were ten students receiving three times of interviews before the first academic year began, ten weeks after the first academic year began and after the first academic year ended respectively; eight students continuing their studies received the fourth time of interview at the end of the second academic year; two students further continuing their studies received the fifth time of interviews in the third academic year. The interview contents include the definition of service quality, service expectation, the importance of various factors influencing student satisfaction, etc. Each interview lasts 36-95 minutes and the interview contents are recorded in case of summary and analysis; the research results divide students' educational service experience into three different phases: before-class phase, during-class phase and after-class phase. The focus of the first phase is the input of student service expectation and service process; the focus of the second phase is to point out the aspects of satisfaction and dissatisfaction in student service experience; the focus of the third phase is the output of the service process and the evaluation on the service value of students.

3.2   Questionnaire Survey and Quantitative Analysis

Compared with qualitative analysis, quantitative analysis can obtain a large amount of accurate data and then make analysis by statistical method and statistical software thus arriving at more convincing conclusions. Quantitative analysis often requires the representativeness of samples, that is, the characteristics of samples can reflect the general more adequately. Questionnaire survey and quantitative analysis are the main approaches to carry out student satisfaction survey and assessment.

The first student satisfaction survey throughout the country was conducted by the way of questionnaire in America in 1994; the Student Satisfaction Inventory used in the survey was designed by Doctor Laurie Schreiner and Doctor Stephanie Juillerat in 1993 and was published by Noel-Levitz Incorporation. The objective of the survey was to assess students' opinion on the importance of various experiences at university and satisfaction degree of all kinds of expectations thus discovering the key factors which are practically influencing the academic achievements and concerned by students.

With the development of the Internet and electronic technology, on-line satisfaction questionnaire gradually rises which is convenient and not limited by space distance but the response rate may be relatively low sometimes. Therefore, it is necessary to choose appropriate time for survey and combine with some incentive measures in order to get satisfactory response rate (Aldridge, 1998).

4   Research Achievements and Conclusions

4.1   Assessment Model on Service Quality of Higher Education

The thoughts of Gronroos (1988) and SERVQUAL (e.g. Parasurraman *et al*,1988) support that service quality stems from the comparison between expectation and perception made by the customers. After developed, SERVQUAL has been used to measure the service quality of all trades and professions. In recent years, many foreign scholars have made efforts to apply SERVQUAL model to the field of higher education, and the main emphasis has been put on the empirical study to test the service quality of higher education and student satisfaction.

Theoretical basis for service quality research of higher education is the SERVQUAL model designed by PZB group. Superiority or inferiority of service quality of higher education depends on the discrepancies considered by the students between expectation and perception of service quality of higher education. Fewer discrepancies, higher quality on service quality of higher education, students will find greater satisfaction. On the contrary, they will be less content with it. There are relevant relations among the service quality, student satisfaction, and performance tendency in higher educational institutions. Performance tendency of students is based on satisfaction of students and directly leads to the complaint or loyalty of students. Devinel (1995) utilized SERVQUAL method and he found that except the visible dimension of student group, all students and teachers in every kind of dimensions of service quality of higher education had higher expectation than perception. Shank, Walker and Hayes (1995) analyzed the service quality of higher education. They designed a research related to 686 business major students and 14 professors respectively. This research took the student expectation for potential satisfaction and long-term measure of college service quality perception, which aimed to assess the service expectation of higher education in the view of the service providers (professors) and the customers (students). Vieiva (1996) found that positive interaction from students with staff, especially with teachers had an influence on student maintain rate and satisfaction of institution. Schwantz (1996) utilized the SERVQUAL model designed by PZB to make a research which involved 92 traditional students and 116 non-traditional students and he found that there were no ordinary differences between expectation and perception of service quality; there were no ordinary differences between staff and teachers on the service expectation either; however, there were remarkable differences between staff and teachers on perception of service quality considered by students. Every dimension of staff score in SERVQUAL was lower than the teacher's score. Napapom (2001) utilized EDSERVQA scales, a kind of tool which combines SERVQUAL with QUALED, to measure the service quality of higher education, and he made a research about the differences between the student service quality expectation and perception of partial time MBA students in public and private colleges of Thailand; meanwhile, Napaporn (2000) inspected the different conditions about the expectation and perception service quality set off by student demographic statistics features (gender, income level and state of employment). Research indicated that there were ordinary differences between expectation and perception of service quality for MBA part time students of Thailand in the five dimensions; meanwhile, it found that there are still differences between expectation and perception of service quality, due to the different gender and income level. Soekisno (2001) studied two public and private colleges of Jakarta, Indonesia with his self-designedly revised SERVQUAL tool, and he measured the college students' expectation and perception of the provided service quality. Researchers found that there were no remarkable differences on expectation of service quality for the college students in the public and private college of Indonesia; however, there were differences on perception of service quality. Assessments about service quality from private college students were higher than that from the public college students. Researchers even found that differences of service quality depended on students' gender, age, parents' income and parents' vocational. But, only involving parents' education level, differences of service quality are different. Carrie Ham (2003) found that service quality of higher education and customer satisfaction (students) made some impacts on the customer (students) behavioral tendency. He considered that it was universal that there were differences between student expectation and perception in higher education service. Research confirmed that service quality and improvement were of importance for college students in Southern Wesleyan University and Western Michigan University. This research indicated that the service quality in the academic field and customer satisfaction had the influence on customer behavior tendency. Li-Wei Mai (2005) applied the conception of service quality and variable quantity to check the difference on educational quality and mainly influential factors of Britain and American students. The questionnaire set up a frame with SERVQUAL and related to 20 variable of service quality. Through the investigation and research of satisfaction for 322 Britain and American Business graduates, the result found that there was a larger difference on student service perception in Britain than American universities: the educational quality in American is better than that in Britain. Besides, that research also found that the two aspects, "the whole impression on university" and "entire impression to educational service", are the most important indexes influencing the educational satisfaction.

Due to the different particularities between higher education and the other service industries, there was still a debate over whether it is appropriate for SERVQUAL method to test and appraise the service quality of higher education. Generally speaking, when testing and appraising to the service quality of higher education field, it should take the proper revision to SERVQUAL scales and should make conversion, modification, addition or cut to dimension and variables of SERVQUAL in the field of higher education. At the same time, it needed to retest the credit and validity of measuring scales.

4.2   Influential Factors of Student Satisfaction

Student satisfaction has the strong relation with higher education quality perception made by students, and which will be influenced by many kinds of educational service factors, such as, demographic statistics features, service expectation, institution image, intellectual environment and so on. Dozark (1999) made a discovery that some demographic statistics features had an impact on the satisfaction of students in his research about quality practice and students satisfaction. Kerlin (2000) considered that there were tremendous differences between male and female students. Bean and Vesper (1994) studied on sex difference of higher education students and they found that social relation factors were more important to female than to male. Attractive courses and faculty contact for all students' experience and satisfaction would have a great effect. Research from Debnath (2005) found that getting a good job or not after graduation was the most significant influence factor to satisfaction rates. Research conclusion from Li-Wei Mai (2005) was that the whole image of college and educational quality presented by student affected the satisfaction rates greatest. Adela (2009) investigated the satisfaction rates of educational experience among the higher education graduates in Europe. Research put emphasis on several key factors affecting students satisfaction, such as, environment factors, field of study, usefulness of study, individual-specific characteristics. The result indicated that those most satisfied students usually made very high assessment to course content and social aspects; however, opportunity to participate in research projects and poor supply of teaching materials were the main reasons why college students dissatisfied with their study experience. Yen-Ku kuo (2009)proposed a model based on student overall experience to test the influential factors over student satisfaction and loyalty, this model found that service quality and institution image were the main factors for improving student satisfaction level, eventually would affect the loyal performance. Research analyzed 321 samples from high vocational education of Taiwan with structure equation model, and the result showed service quality and institution image were the key factors to maintain student satisfaction rates, and indirectly affected the student degree of loyalty. Brown (2009) utilized least square structure equation model to test the driving factors of customer satisfaction and loyalty under the higher education. This model, based on the five dimensions of SERQUAL, separated perception quality into human ware and hardware. The research took four kinds of college students as the sample and got the conclusion: student loyalty directly depended on student satisfaction and student satisfaction depended on perception of institution image proposed by college students. Jen-HerWu (2010) proposed a research model based on social cognitive theory to test the decisive factors of student satisfaction under the Blended Electronic Learning System (BELS). We could drew the conclusion from 212 samples that computer self-efficacy, performance expectations, system functionality, content feature, interaction and learning climate were the main factors influencing student learning satisfaction under the condition of Blended Electronic Learning System (BELS). Moro-Egido (2010) studied how employment status (full-time or part-time student) affects student satisfaction, and the respondents were 116 graduates who were Bachelor Program in Computing from Autonomous University of Barcelona. The result showed that part-time students had lower level satisfaction on their educational experience; students tended to learn more specialized rather than diversified acknowledge; higher GPA and shorter degree completion time had the positive influence on overall satisfaction of students.

References

1. Shank, M.D., Walker, M., Hayes, T.: Understanding Professional Service Expectation: Do We Know What Our Students Expect in a Quality Education. Journal of Professional Services Marketing 13(1), 71–89 (1995)
2. Aldridge, S., Rowley, J.: Measuring customer satisfaction in higher education. Quality Assurance in Education 6(4), 197–204 (1998)
3. Clewes, B.: A Student-centred Conceptual Model of Service Quality in Higher Education. Quality in Higher Education 9(1), 69–85 (2003)
4. Brenders, D.A., Hope, P., Ninnan, A.: A systemic student-centred study of university service. Research in Higher Education 40(6), 665–685 (1999)
5. Moro-Egido, A.I., Panades, J.: An Analysis of Student Satisfaction: Full-Time vs. Part-Time Students. Social Indicators Research 96, 363–378 (2010)

Effects of Path Length on the Similarity of Network Nodes Based on RTT*

Nafei Zhu[1] and Jingsha He[2]

[1] College of Computer Science and Technology
Beijing University of Technology, Beijing 100124, China
zyy@emails.bjut.edu.cn
[2] School of Software Engineering,
Beijing University of Technology, Beijing 100124, China
jhe@bjut.edu.cn

Abstract. Correlation of network nodes in terms of round-trip-time (RTT) can be quantified by the similarity metric which is defined as the extent to which two source nodes have the same changing trends in terms of RTT to the same destination node. With this definition, the effects of path length, more specifically, the path length ratio that is defined as the ratio of the length of the common path to that of the longer private path, on such correlation is analyzed. Our analysis shows that path length ratios from 1:1 to 8:1 have obviously more positive effects on the upward trend of similarity when the common path load increases or the private path loads decrease and that similarity will get better when path length ratio gets higher while path load combinations remain the same.

Keywords: similarity, RTT, load, path ratio.

1 Introduction

Round-trip-time (RTT) is a basic metric for network measurement and understanding the relationship between the static RTTs of all the nodes in the Internet is important to network measurement. A lot of work has been done in this area. An architecture called IDMaps was proposed to estimate the latency of arbitrary network paths from some known nodes [1]. Netvigator is a tool for network proximity estimation based on landmark clustering technique [2]. Agarwal and Lorch designed a latency prediction system for game matchmaking scenarios [3]. Internet Iso-bar was proposed as an overlay distance monitor system [4]. However, the focus of the above work has been primarily on the measurement of delay and the estimation of delay by clustering nearby nodes to a given node. None of them has thoroughly studied the similarity of network nodes in terms of delays. Studying the correlation or the similarity among network nodes can give us a more precise understanding of the architecture of the Internet based on delays. It can also help us to determine where to place monitors for better performance of network measurement.

* The work in this paper has been supported by funding from Beijing Education Commission.

J. Luo (Ed.): Soft Computing in Information Communication Technology, AISC 158, pp. 123–128.
springerlink.com   © Springer-Verlag Berlin Heidelberg 2012

In this paper, we study the effects of path length, more specifically, the path length ratio that is defined as the ratio of the length of the common path to that of the longer private path, on the correlation or the similarity of two network nodes to a common destination node in terms of RTT in which we use the simulation tool OPNET to perform the experiments. Our main findings are as follows. Firstly, path length ratios from 1:1 to 8:1 have obviously more positive effects on the upward trends of similarity when the common path load increases or the private path loads decrease. Secondly, similarities will get closer when private path load combinations are simply exchanged while the path length ratio gets higher, especially when the ratio is higher than 1:1. Thirdly, when load combinations remain the same, similarities will be less different when the path length ratio gets higher.

The rest of the paper is organized as follows. In the next section, we state the research target and present experimental methodology. In Section 3, we describe the formula for measuring similarity as well as some notations used in our discussion. In Section 4, we study the effects of path length ratios under different path load combinations on similarity. Finally, we conclude this paper in Section 5.

2   Problem Statement and Experimental Strategy

2.1   Problem Statement

We are interested in the correlation or the similarity between two network nodes to a common destination node in terms of RTT. In this paper, we study the effects of path length on the similarity, which describes to what extent the two sequences of RTT values have the same changing trend.

2.2   Experimental Method and Parameters Configuration

In our study, we first use PING to get the RTT values. Then, we do the experiment using OPNET to get RTT values with defined hops and loads for the paths. Our simulation topology include at least three subnets, two of which containing the two source nodes, respectively, that send PING packets and the third containing the destination node for the PING packets. Of course, there may be one or more other subnet between the source and the destination subnets.

Three parameters need to be carefully determined in the experiment. The first is path length. Since the average number of hops in the Internet is 15~19 and the default maximum path length is 15 hops in OPNET [6], in our experiment, the maximum path length for PING is chosen to be 15 hops, which makes our experiment more practical while still meaningful. The second is path load. We classify path loads in 3 tiers: light with an average utility of 20% in the interval [15%-25%], medium with an average utility of 50% in the interval [40%-60%] and high with an average utility of 80% in the interval [65%-95%]. The third is traffic pattern. We use the Raw Packet Generator (RPG) mode in OPNET to generate self-similar traffic [7] with the typical Hurst value of about 0.7 [8]. We have performed 11,259 experiments for all the possible path loads and path lengths and got 2,748,816 data for our similarity analysis.

3   Formula and Denotations

3.1   Formula of Similarity

We use rank-based Kendall's method to calculate similarity in which the parameter τ is defined in Equation (1) [9].

$$\tau = 1 - 4s(\pi,\sigma)/N(N-1).$$ (1)

Let $Y = y_1...y_n$ be a set of items to be ranked and let π and σ be two distinct orderings of Y. $s(\pi,\sigma)$ denotes the minimum number of adjacent transpositions needed to bring π to σ. N is the number of objects (i.e., items) to be ranked.

3.2   Path Length and Path Load

Let A and B be the two source nodes and T be the common destination node. Paths AT and BT share a common path TM whose length is H_{com} and the load on the common path is U_{com}. Similarly, the length of the private path AM is H_{pri_a} and the load on the path is U_{pri_a}. Same can be said for node B. Note that $H_{com}+\max(H_{pri\_a},H_{pri\_b}) \leq 15$ and $0 \leq U_{com},U_{pri\_a},U_{pri\_b} \leq 1$. We call "$H_{com}$, H_{pri_a}, H_{pri_b}" a length combination and "U_{com}, U_{pri_a}, U_{pri_b}" a load combination.

We assume in our study that H_{com} is at least 1 and the shorter private path length is at least 1. The total number of length combinations is 417 with 71 path length ratios [5]. We will mainly study the 27 path length ratios which include 14 integer path length ratios and 13 reciprocal values of the integral ones. For the path load, we use the three tier values which result in a total number of 27 path load combinations.

4   Effects of Path Length Patios on the Similarity

There are 3 characteristics regarding the effects of path load on the similarity [5]. We now perform further analysis regarding the effects of path length ratios on the similarity with different path load characteristics.

4.1   Effects with the "Common+Private Load" Characteristic

"Common+Private Load" characteristic refers to the situation in which similarity will likely get better as the common path load gets heavier and get worse as the private path load gets heavier [5]. We examine the effects of every path length ratio with this characteristic to find out which path length ratios are more sensitive for this characteristic. We first standardize the data and then use Equation (2) which is the slope of the linear regression to calculate different effects of the path length ratios.

$$\hat{\beta}_1 = (\sum_{i=1}^{n}(x_i - \bar{x})(y_i - \bar{y}))/(\sum_{i=1}^{n}(x_i - \bar{x})^2).$$ (2)

There are 3 common path loads and 9 private path load combinations. First, for each of the 9 private path load combinations, we calculate the slope values for the similarities when the common path load changes from 20% to 80% and then the average of the 9 slopes. Second, we calculate the average slope of the slopes for the similarities when one private path load changes from 20% to 80% while the common path load and the other private path load remain the same. Third, we give every path ratio a value from 1 to 27 based on its place in the ordering of the average slopes for the common path load. Fourth, we do the same based on the reverse ordering of the average slopes for the private path load. Both orderings are from small to large. Lastly, we add these two values for every path ratio and the results are shown in Table 1.

Table 1. Sum of the Values for Different Path Ratios

Ratio	1:14	1:13	1:12	1:11	1:10	1:9	1:8	1:7	1:6
Value	7	7	7	10	11	18	12	23	24
Ratio	1:5	1:4	1:3	1:2	1:1	2:1	3:1	4:1	5:1
Value	20	26	28	37	45	54	50	51	41
Ratio	6:1	7:1	8:1	9:1	10:1	11:1	12:1	13:1	14:1
Value	45	41	43	34	25	36	27	19	15

From Table 1, we can see that the values get larger when path ratios change from 1:1 to 8:1 and are thus more sensitive to this characteristic. Meanwhile, path ratios between 1:14 and 1:10 are insensitive to this characteristic.

4.2   Effects with the "Exchanged Private Load" Characteristic

We examine the effects of path ratios for the "Exchanged Private Load" characteristic, meaning that similarity will mostly remain the same if loads on the two private paths are exchanged [5]. We use change in terms of percentage to measure the effects of path ratios with this characteristic with Equation (3).

$$\text{Percentage change} = \Delta V / V_1 = (V_2 - V_1)/V_1. \tag{3}$$

For each of the 3 common path loads, there are 3 pairs of load combinations for the exchanged private path loads. For each common load, we calculate the percentage of change for similarity by exchanging the loads on the private paths for each path ratio and can thus get 3 sets of values. Consequently, we get 9 sets of values for each path ratio and the results are shown in Fig. 1 by using boxplot.

We can see that for path ratio 1:1, the changes are close to 0, which means that path ratios higher than 1:1 will result in better performance for the "exchanged private load" characteristic. For path ratios that are lower than 1:1, although the percentage of change could be higher, it still does not violate the characteristic.

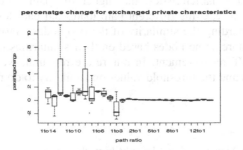

Fig. 1. Percentage of change by exchanging loads on the private paths

4.3   Effects with the "Com+HeavyPri Load" Characteristic

The "Com+HeavyPri load" characteristic refers to the situation in which two load combinations have the same common and heavy private loads and the similarities for the same length combinations will have no significant difference to each other [5]. We have 5 groups of such load combinations in the study on the effects of path ratio for this characteristic which is shown in Fig. 2 in which "20com50pri" means that the common path load is 20% and the heavier private path load is 50%.

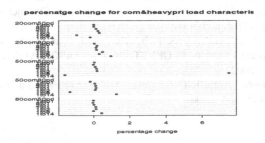

Fig. 2. Percentage of change for "Com+HeavyPri Load" characteristic

We again use percentage of change to express the performance of this characteristic for every path ratio. In the figure, although we only show 7 path ratios, we can clearly see the trends in which from 1:14 to 14:1, especially from 1:1, the percentage of change is close to 0. Therefore, the higher the path ratio is, the more obvious the performance of this characteristic will be.

5   Conclusion

In this paper, we studied the effects of path length, more specifically, the path ratio, on the correlation or similarity of two network nodes to a common destination node in terms of RTT for different path load combinations. Our main findings are as follows. First, for the "Common+Private Load" characteristic, path ratios from 1:1 to 8:1 have better performance than others. Second, for the "Exchanged Private Loads" and the

"Com+HeavyPri load" characteristics, the higher the path ratio is, the better the performance is. Therefore, using different path loads and path lengths, we can get a lot of information regarding the similarity of the two nodes. Based on the results on similarity, we can ignore some nodes based on their similarities to some other nodes when performing RTT measurement. In our future research, we will study how to determine such nodes and the threshold values on similarity to do the elimination.

References

1. Francis, P., et al.: IDMaps: A global Internet host distance estimation service. IEEE/ACM Trans. Networking 9(5), 525–540 (2001)
2. Sharma, P., Xu, Z.C., Banerjee, S., Lee, S.J.: Estimating network proximity and latency. In: Proc. ACM SIGCOMM 2006, Pisa, Italy (2006)
3. Agarwal, S., Lorch, J.R.: Matchmaking for online games and other latency-sensitive P2P systems. In: Proc. ACM SIGCOMM 2009, Barcelona, Spain (August 2009)
4. Chen, Y., Lim, K.H., Katz, R.H., Overton, C.: On the stability of network distance estimation. ACM SIGMETRICS Performance Evaluation Review 30(2), 21–30 (2002)
5. Zhu, N., He, J.: Experimental study of the similarity of network nodes based on RTT. In: Proc. 3rd International Workshop on Computer Science and Engineering 2010, Ganzhou, China (December 2010)
6. Fei, A., Pei, G., Liu, R., Zhang, L.: Measurements on delay and hop count of the Internet. In: Proc. IEEE GLOBECOM 1998, Sydney, Australia (November 1998)
7. Legend, W.E., Taqqu, M.S.: On the self-similar nature of Ethernet traffic. IEEE/ACM Trans. Networking 2(1), 1–5 (1994)
8. Leys, P., Potemans, J., Van den Broeck, B., Theunis, J., Van Lil, E., Van de Capelle, A.: Use of the raw packet generator in OPNET. In: Proc. OPNETWORK 2002, Washington D.C., USA (August 2002)
9. Lapata, M.: Automatic evaluation of information ordering: Kendall's tau. Computational Linguistics 32(4), 471–484 (2006)

Efficient CT Metal Artifacts Reduction Based on Improved Conductivity Coefficient

Yi Zhang, Yifei Pu, and Jiliu Zhou

College of Computer Science, Sichuan University,
South 1st part of 1st ring road, 24. 610065 Chengdu, China
maybe198376@gmail.com,
{puyifei, zhoujl}scu.edu.cn

Abstract. In this paper, we propose a efficient metal artifacts reduction method based on improved conductivity coefficient for computed tomography (CT) and the numerical implementation of our method is also given. The experiment results show that on both visual effect and peak signal to noise ratio (PSNR), the method we propose is superior to conditional interpolation methods and classic total variation model.

Keywords: Image inpainting, metal artifact reduction, computed tomography (CT).

1 Introduction

Metal artifacts reduction(MAR) is still an important research in CT imaging. As a high-density object, when X-ray goes through the human body, the metal has a much higher attenuation factor than organisms and projection data distortion appears. Consequently there will be star-burst streak artifacts in the images after reconstruction and misdiagnosis may be caused[1]. Now, there are many MAR methods used in different clinic applications[2].

In CT image processing field, there appears two main classes of algorithms: projectioncorrection methods[3][4][5][6][7] and iterative reconstruction methods[8][9]. After theoretical analysis, MAR methods based on iterative reconstruction have better performance than projection correction methods, but they have much more computational consumption and higher cost to implement. At here, we focus on the projection correction methods for MAR. The idea we firstly got was interpolation. Lewitt and Bates's used Chebyshev polynomial to implement the interpolation[3]. Kalender et al employed the linear interpolation[4] and Lonn and Crawford added some assistant process based on it[5] Zhao et al proposed to interpolate the wavelet coefficients of the projection data[6] To obtain better visual effect, Duan et al brought in the classical image inpainting method based on total variation (TV) for MAR[7]. However, the existing PDE sinogram inpainting methods can not connect wide inpainting region, so when wide regions exist, it can not get satisfactory results.

J. Luo (Ed.): Soft Computing in Information Communication Technology, AISC 158, pp. 129–134.
springerlink.com   © Springer-Verlag Berlin Heidelberg 2012

In this paper, we applied an improved conductivity coefficient instead of the classic TV inpainting model to deal with the damaged sinogram with wide data gap. Firstly, we will introduce the curvature driven diffusions image inpainting model (CDD) based on the new improved conductivity coefficient[10]. Then, we present the numerical implementation for our method. After the experiments section, the conclusion will follow.

2   Our Method

The main difference between our method and the conventional sinogram inpainting methods lies in the projection data correction with CDD inpainting model.. When the metal projection data contaminated the original projection data as a form of a gap. Conventional methods applied linear or polynomial interpolation and PDE inpainting algorithms to correct the data gap. But the conventional methods have some drawbacks. Firstly, the gap boundaries after inpainting are lack of smoothness. Secondly, when the gap is wide, the inpainting result can not get satisfied visual effect. Here we apply CDD image inpainting model to CT metal artifacts reduction. It can get smoother results and meanwhile, cope with the data gap.

The original TV inpainting model, the conductivity coefficient of the diffusion strength which only depends on the numerical value of the isophotes is:

$$F = |\nabla u|^{-1}. \tag{1}$$

The geometric information of the isophotes is not be considered. That is why the wide gap can not be restored perfectly. To recover from this situation, we use a new conductivity coefficient instead of the old one to :

$$F = f(|\kappa|)|\nabla u|^{-1}, \tag{2}$$

where f is a function satisfied the following property:

$$f(s) = \begin{cases} 0, & s = 0 \\ \infty, & s = \infty. \\ \text{between 0 and } \infty, 0 < s < \infty \end{cases} \tag{3}$$

When the isophote is having large curvature, the diffusion strength will be large too. Thus, the Euler-Lagrange equation of CDD model is

$$\frac{\partial u}{\partial t} = \nabla \cdot \left[f(|\kappa|) |\nabla u|^{-1} \nabla u \right], \quad \text{in } D$$

$$u = u^0, \qquad\qquad\qquad \text{in } D^c \tag{4}$$

Here the inpainting domain D is an open set mathematically, D^c denotes the outer of D and u^0 is the available part of the image. The curvature κ is defined as

$$\kappa = \nabla \cdot \left[\nabla u |\nabla u|^{-1} \right]. \tag{5}$$

3   Numerical Implementation

In this section, we use the time marching scheme to discretise our model. Assuming a time step size of Δt and a space grid size of h, we let:

$$x_i = ih, \, y_j = jh, i, j = 0,1,...,N, \text{with } Nh = 1,$$

$$t_n = n\Delta t, n = 0,1... \tag{6}$$

To avoid a numerical risk, we choose

$$f(s) = s, s > 0. \tag{7}$$

The explicit scheme iterates as

$$u^{n+1} = u^n + \Delta t (sign(\Delta u) |\Delta u|^2 |\nabla u|^{-2})$$

$$= u^n + \Delta t (sign(u_x + u_y)(u_x + u_y)^2 (u_x^2 + u_y^2 + \varepsilon)^{-1}), \tag{8}$$

where ε denotes a small positive number to prevent from dividing by 0 and $sign(s)$ is the sign function.

To detail the spatial discretization, we use upwind finite difference scheme given by Osher and Sethian[11].

$$u_x = \min \bmod(u_x^c(i, j), \min \bmod(2u_x^b(i, j), 2u_x^f(i, j))),$$

$$u_y = \min \bmod(u_y^c(i, j), \min \bmod(2u_y^b(i, j), 2u_y^f(i, j))),$$

$$u_{xx} = u(i+1, j) + u(i-1, j) - 2u(i, j),$$

$$u_{yy} = u(i, j+1) + u(i, j-1) - 2u(i, j), \tag{9}$$

where the supindexes c, b and f denote central, backward and forward differences respectively and the minmod function satisfies

$$\min \operatorname{mod}(x, y) = sign(a) \cdot \max(0, \min(|a|, b \cdot sign(a))). \tag{10}$$

4   Experiments

In this section, we present the experiment results of our algorithm. Meanwhile, we compare our CDD sinogram inpainting model with linear interpolation (LI), cubic spline interpolation (CSI) and conditional total variation inpainting (TV). To test the methods, a region with much higher attenuation is added into the Shepp-Logan(S-L) phantom (256×256) to simulate the metal artifact. As there is no quantitative method to measure the CT metal artifacts reduction performance, we apply PSNR which is commonly used in image inpainting[12] to be an available criterion:

$$PSNR = 10 \times \log_{10} \left(\frac{255^2}{\|u - u_0\|_2^2} \right). \tag{11}$$

u is the image after inpainting and u_0 is the original image. Bigger the value of PSNR is, better the performance is. PSNR is usually used to measure the similarity between inpainted image with real image and it is a suitable standard for test.

The concrete processing steps are following. Firstly, because in this paper, we focus on the inpainting algorithm, so for simplicity, only use the threshold method to extract the metal region. It is easy to say that, a more accuracy segmentation algorithm will enhance the performance of metal artifact reduction. Secondly, we locate the corresponding metal part in the projection data set. Thirdly, we employ the CDD algorithm to inpaint the metal region. At last, we reconstruct the image from the inpainted sinogram and insert the metal region.

Fig. 1 shows the results of different algorithms. Fig. 1a is the original phantom with metal region. Fig. 1b is the reconstructed phantom from projection data and we can see there is obvious metal artifacts here. Fig. 1c to Fig. 1f in sequence are the results of LI, CSI, TV and CDD. In Fig. 1c to Fig. 1f, we can see that metal artifacts in all of them are significantly reduced. Compared with other methods, our method get the best visual effect. In Fig. 1c and Fig. 1d, the shapes of the organs in the middle of image is obviously unnaturally. TV recovered better than LI and CSI in the same region, but near the metal, the artifacts were still remarkable. At both locations mentioned above, CDD achieved best effect. On other side, the PSNR value of CDD is also the biggest one.

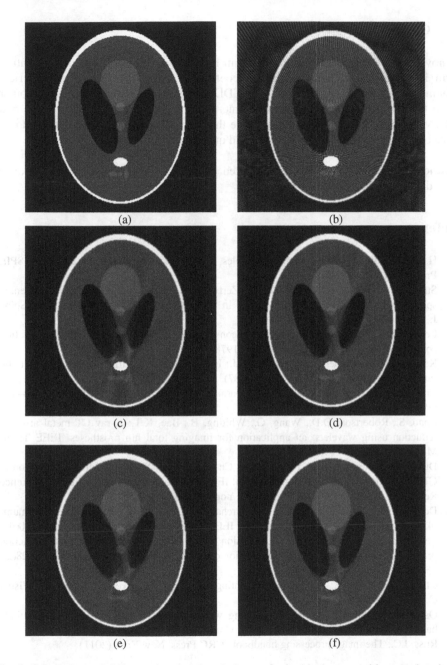

Fig. 1. Comparison of different sinogram inpainting methods. (a) phantom with a metal region, (b) filtered backprojection with metal artifacts, (c) filtered backprojection after LI(PSNR=26.9005), (d) filtered backprojection after CSI(PSNR=27.1819), (e) filtered backprojection after TV(PSNR=27.0987), and (f) filtered backprojection after CDD(PSNR=27.2985).

5   Conclusion

A novel metal artifacts reduction algorithm is proposed by employing the curvature term into the conditional conductivity coefficient. After analyzing the numerical scheme of our model, we compare our CDD sinogram inpainting model with other three typical methods, including linear interpolation, cubic spline interpolation and total variation, and the results demonstrate that our CDD sinogram inpainting model have better performance in both quality and quantity.

Acknowledgments. This work is supported in part by the National Natural Science Foundation of China (60972131).

References

1. Hsieh, J.: Computed tomography: principles, design, artifacts, and recent advances. SPIE Press, New York (2003)
2. Stradiotti, P., Curti, A., Castellazzi, G., Zerbi, A.: Metal-related atifacts in instrumented spine. Techniques for reducing artifacts in CT and MRI: state of the art. Eur. Spine J. 18(suppl. 1), S102–S108 (2009)
3. Lewitt, R.M., Bates, R.H.: Image reconstruction from projections(III): Projection completion methods. Optik 50, 189–204 (1978)
4. Kalender, W.A., Hebel, R., Ebersberger, J.: Reduction of CT artifacts caused by metallic implants. Radiology 164(2), 576–577 (1987)
5. Crawford, C.R.: Reprojection using a parallel backprojector. Med. Phys. 13, 480–483 (1986)
6. Zhao, S., Robertson, D.D., Wang, G., Whiting, B., Bae, K.T.: X-ray CT metal artifacts reduction using wavelets: an application for imaging total hip prostheses. IEEE Trans. Med. Imag. 19(12), 1238–1247 (2000)
7. Duan, X., Zhang, L., Xiao, Y., Cheng, J., Chen, Z., Xing, Y.: Metal artifacts reduction in CT images by sinogram TV inpainting. In: IEEE Nuclear Science Symposium Conference Record, pp. 4175–4177. IEEE Press, New York (2008)
8. De Man, B., Nuyts, J., Dupont, P., Marchal, G., Suetens, P.: An iterative maximum-likelihood polychromatic algorithm for CT. IEEE Trans. Med. Imaging 20, 999–1008 (2001)
9. Lemmens, C., Faul, D., Nuyts, J.: Suppression of metal artifacts in CT using a reconstruction procedure that combines MAP and projection completion. IEEE Trans. Med. Imaging 28(2), 250–260 (2009)
10. Chan, T.F., Shen, J.: Nontexture inpainting by curvature-driven diffusions. J. Visual Comm. Image Rep. 12, 436–449 (2001)
11. Osher, S., Sethian, J.: Fronts propagating with curvature dependent speed: algorithms based on Hamilton-Jacobi formulations. J. Computer Phys. 79, 12–49 (1988)
12. Russ, J.C.: The image processing handbook. CRC Press, New York (2011)

Research and Design of Teaching Strategies and Rules Base on Fuzzy Comprehensive Evaluation

Fengshe Yin[1] and Lei Jiao[2]

Shaanxi Polytechnic Institute Shanxi Xiaanyang 712000, China
[1] Vice Professor. Research Interests: Intelligent Software Design
[2] Teaching assistant. Research Interests: Intelligent Software Design
stgi001@163.com

Abstract. Through learning ability and effects analysis, research the evaluation index system theory and bring in fuzzy comprehensive evaluation system. Then propose and design out the teaching strategies and rules based on fuzzy comprehensive evaluation, in order to realize personalized teaching and fully mobilize the learner's initiative.

Keywords: fuzzy comprehensive evaluation, teaching strategy, rules, personalization.

1 Introduction

Fuzzy comprehensive evaluation is to analysis in the fuzzy environment, consider of the multiple effects, works out the comprehensive evaluation and decision base on the definite target or standard. According to the complexity of research object and involved judgment factors, the fuzzy comprehensive evaluation [1] should divided into single stage fuzzy comprehensive evaluation and multistage fuzzy comprehensive evaluation. Extremely complicated judgment system considers more factors, so we need to apply multistage fuzzy comprehensive evaluation to evaluate.

2 Learning Ability and Effect Analysis

Students' cognitive ability and students interested in this course is the main consideration on the student's learning level and effect. Comprehensive evaluate the student's learning level and effect by using multistage fuzzy comprehensive evaluation to get the evaluate results.

2.1 Establishment of Evaluation Index System [2]

Evaluation index set is U and it is involved two parts: cognitive competence (U1), learning interest (U2), marked as U={U1, U2}.

From the fuzzy comprehensive evaluation, the evaluation index set is U1 and students' cognitive competence is definite into U1 = {memory ability, comprehensive ability, application ability, analysis ability, integrative competence}. The corresponding

J. Luo (Ed.): Soft Computing in Information Communication Technology, AISC 158, pp. 135–141.

factor is U1={u11, u12, u13, u14, u15}. Moreover, the relevant comment index is V and can be definite as V= {excellence, good, middling, pass, worse, and failure}. The corresponding factor is V={v1, v2, v3, v4, v5, v6} and they are distinguish express the score of student as $90 \leq v1 \leq 100$; $80 \leq v2 < 90$; $70 \leq v3 < 80$; $60 \leq v4 < 70$; $40 \leq v5 < 60$; $0 \leq v6 < 40$.

2.2   Construct Membership Function

The construction each element in the evaluation index U about the membership function of comment set is:.

to v1:
$$\mu_{v1} = \begin{cases} \frac{1}{10}(x-90) & 90 \leq x \leq 100 \\ 0 & x < 90 \end{cases}$$

to v2
$$\mu_{v2} = \begin{cases} \frac{1}{10}(100-x) & 90 \leq x \leq 100 \\ \frac{1}{10}(x-80) & 80 \leq x < 90 \\ 0 & x < 80 \end{cases}$$

to v3:
$$\mu_{v3} = \begin{cases} 0 & x \geq 90 \\ \frac{1}{10}(90-x) & 80 \leq x < 90 \\ \frac{1}{10}(x-70) & 70 \leq x < 80 \\ 0 & x < 70 \end{cases}$$

to v4:
$$\mu_{v4} = \begin{cases} 0 & x \geq 80 \\ \frac{1}{10}(80-x) & 70 \leq x < 80 \\ \frac{1}{10}(x-60) & 60 \leq x < 70 \\ 0 & x < 60 \end{cases}$$

......

In the formula, X is the result score of each element in U.

2.3   Weight Coefficient Vector Determination of Evaluation Index

Evaluation protocol of two-stage factor set and each level factors' weight distribution (from domain expert):

U={U1 ,   U2}={cognitive   ability,   learning   interest   },   weight coefficient $A = \{0.6\ 0.4\}$

In this part:

U1={u11, u12, u13, u14, u15}={memory ability, comprehensive ability, application ability, analysis ability, integrative competence}

Weight coefficient $A_1 = \{0.1\ 0.25\ 0.35\ 0.2\ 0.1\}$ (weight coefficient of each unit can be different)

2.4   First Grade Evaluate the Integrative Ability Value of Student[3,4]

After learning one unit (suppose it is unit 2), the system will provide one testing paper to test the students' learning ability and effect. First, test the memory ability. Suppose the full score of this test is 100. Compare with the correct answer, the test score of the student is 76. Fuzzy evaluate on the basic of our definition, this student is belongs to V3.

From the memory ability score we can get the single element of evaluate vector is R11= (0, 0, 0.6, 0.4, 0, 0) . In addition, suppose the comprehensive ability score of this student 72 is, application ability score is 71, analysis ability score is 68 and integrative competence score is 64. The relevant evaluate vector is R12= (0, 0, 0.2, 0.8, 0, 0) ; R13= (0, 0, 0.1, 0.9, 0, 0) ; R14= (0, 0, 0, 0.8, 0.2, 0) ; R15= (0, 0, 0, 0.4, 0.6, 0) . So we will get

$$
R_1 = \begin{bmatrix} 0 & 0 & 0.6 & 0.4 & 0 & 0 \\ 0 & 0 & 0.2 & 0.8 & 0 & 0 \\ 0 & 0 & 0.1 & 0.9 & 0 & 0 \\ 0 & 0 & 0 & 0.8 & 0.2 & 0 \\ 0 & 0 & 0 & 0.4 & 0.6 & 0 \end{bmatrix}
$$

After normalization we will get B_1 = (0 0 0.145 0.755 0.1 0) . That is the comprehensive evaluation result of this student's cognize ability. The result shows in the integrative ability of this student, the middling element is 14.5%, the pass element is 75.5% and the worse element is 10%. We suppose there is 1000 person to evaluate this student. The 145 people think the integrative competence of this student is medium, 755 people think it is in the pass level and 100 people think it is worse. Now set the representative of "excellence", "good", "middling", "pass", "worse", and "failure" as 95、85、75、65、50、20. They constitute one rating fraction matrix, so the test score of this student is:

$$
S_1 = (0 \quad 0 \quad 0.145 \quad 0.755 \quad 0.1 \quad 0) \begin{bmatrix} 95 \\ 85 \\ 75 \\ 65 \\ 50 \\ 20 \end{bmatrix} = 64.95
$$

2.5   Learning Ability Score from Secondary Evaluation

Suppose the interest score of this student is 75, from the formula we can get the single factor evaluation vector is R21= (0, 0, 0.5, 0.5, 0, 0) , so there has:

$$R_2 = (0, 0, 0.5, 0.5, 0, 0)$$

$$B_2 = (0, 0, 0.5, 0.5, 0, 0)$$

Then get the fuzzy relation matrix

$$R = \begin{bmatrix} B_1 \\ B_2 \end{bmatrix} = \begin{bmatrix} 0 & 0 & 0.145 & 0.755 & 0.1 & 0 \\ 0 & 0 & 0.5 & 0.5 & 0 & 0 \end{bmatrix}$$

From formula （2.1）, we get

$$B = A \circ R = (0.6 \quad 0.4) \circ \begin{bmatrix} 0 & 0 & 0.145 & 0.755 & 0.1 & 0 \\ 0 & 0 & 0.5 & 0.5 & 0 & 0 \end{bmatrix} = (0 \quad 0 \quad 0.287 \quad 0.653 \quad 0.06 \quad 0)$$

After normalization will get $B = (0 \quad 0 \quad 0.287 \quad 0.653 \quad 0.06 \quad 0)$, set the representative of "excellence", "good", "middling", "pass", "worse", and "failure" as 95、85、75、65、50、20, they constitute one rating fraction matrix, so the integrative test score of this student is:

$$S_{1z} = (0 \quad 0 \quad 0.287 \quad 0.653 \quad 0.06 \quad 0) \begin{bmatrix} 95 \\ 85 \\ 75 \\ 65 \\ 50 \\ 20 \end{bmatrix} = 66.97$$

3   Teaching Strategy and Rule Design

Teaching strategy research is the key point of personalized network teaching system research. Teaching strategy is the driving mechanism to control the teaching content and present to the students. It can be construe from teaching target, design and adjust the teaching sequence on the basic of student condition then to reach the heuristic teaching.

3.1   Teaching Strategy Design

Design the teaching strategy by using the linkage of teaching knowledge will adjust the students' requirement. This system is to build the perfect teaching sequence under the linkage of teaching knowledge and student model.

   a)   *Confirm the type of student and knowledge point [5, 6], as well as different knowledge points of different students.*

   We use 0-100 to perform the cognitive ability of student. It can also divide into several grades. Take feeble, common and strong for instance. The membership function is μ (x)cog={0.3/1, 0.6/2, 1/3}. The interest of this course we also use 0-100 to perform and divide into several grades as well. Take low, common and high for instance. The membership function is μ(x) inte ={0.3/1, 0.6/2, 1/3}.

Therefore, the score of learning ability can be getting by fuzzy comprehensive evaluation on the basic of filling-in cognitive ability score and course interest core after students' registration. Then divide students into 9-student grade. As table 1 shows.

Table 1. Student Grade

Learning ability score	Student grade
90-100	A
80-90	B
70-80	C
60-70	D
50-60	E
40-50	F
30-40	G
20-30	H
≤20	I

Then, the types of knowledge points are different from different students. As table 2 shows.

Table 2. Knowledge point of different student

Student type	Knowledge point
A	1-15
B	1-14
C	1-13
D	1-12
E	1-11
F	1-10
G	1-8
H	1-6
I	1-5

b)   Acquire knowledge sequence according to the student type and corresponding knowledge points.[7]

We have already known some student need to learn what kind of knowledge. Then we get the knowledge sequence through prior search algorithm to abstract the correct knowledge point.

c)   Acquire the perfect teaching sequence according to the hierarchical relationship of knowledge points.

We get the better knowledge points from previous sift. However, the knowledge point should be sequencing. That means some knowledge points need after the guide knowledge point. These antecedence conditions define the support relation among knowledge points. Moreover, the support relation should show as figure 2.

3.2   Teaching Regulation Design

During the studying and ceaseless thorough, the learning interest and cognitive ability is changing. We need to take consider while chose the learning content and method and adjust the teaching strategy on time. For this reason, we build teaching regulation database to restore the regulations of teaching strategy then convince to the teaching adjustment [8].

During the application, the database of each course is provides from expert [9, 10] research. The part regulation of this system is as follows:

R1   If first time study ‖ ability score change // judge student grade, confirm study content
R11   If ability score > =90 Then student type="A"
If student type ="A" Then study all the knowledge points
R12   If ability score > =80 && ability score <90 Then student type ="B"
If student type ="B" Then delete the 15th type of knowledge point
If ability score > =70 && ability score <80 Then student type ="C"
If student type ="C" Then delete the 14th and 15th type of knowledge point
……
R18   If ability score > =20 && ability score <30 Then student type ="H"
If student type ="H" Then delete 7th～15th type of knowledge point
R19 If ability score <20 Then student type ="I"
If student type ="I" Then delete 6th～15th type of knowledge point
……
If integrative test score >=80 And integrative test score <90
Then ability score = ability score +1.5
If integrative test score >=70 And integrative test score <80
Then ability score = ability score +1
If integrative test score >=60 And integrative test score <70
Then ability score= ability score+0.5
If integrative test score >=50 And integrative test score <60
Then ability score = ability score +0
If integrative test score >=40 And integrative test score <50
Then ability score = ability score -0.5
If integrative test score >=30 And integrative test score <40
Then ability score = ability score -1
If integrative test score >=20 And integrative test score <30
Then ability score = ability score -1.5
If integrative test score <20 Then ability score = ability score -2
R42   If ability score change, Then turn to R1
R43   If ability scores no change, Then continue to next unit
……

4   Summary

During the teaching strategy and rules research, detailed how to use the multistage fuzzy comprehensive evaluation to analysis the students' learning ability and effect then to obtain the next step personalization. In the teaching strategy and inference design, detailed how to create the only perfect teaching sequence according the relation among knowledge points.

References

1. Shen, J.: Personalization Strategy Research of Network Teaching. Computer Research and Development 40(4), 589–595 (2003)
2. Yin, F.: Personalized Computer Assisted Instruction System Research and Application Base on Inference Engine. Master's Thesis. Xi'an Jiaotong University, Xi'an (2008)
3. Yang, K., Teng, Z.: Network Base Courseware Design Base on Client/Sever, vol. 2, pp. 35–38. Modern Education Press (2001)
4. Song, J.S., Hahn, S.H., Tak, K.Y., et al.: An Intelligent Tutoring Systems for Introductory C Language Course. Computer & Education 28(2), 93–102 (1997)
5. Cao, X.: Computer Assisted Instruction System Improvement and Application Base on Personalization and Intelligent of Web. Master's Thesis. South China University of Technology, Guangzhou (2002)
6. Shi, Y., Zhang, S., Xiang, C., et al.: Knowledge Point Performance and Relevance Technical Research of Network Course. Zhejiang University Journal (Engineering Edition) 37(5), 508–511 (2003)
7. Wang, X.: Knowledge Database System Modeling and Application Research Base on Ontology. Ph.D. and other doctoral theses. Information Technology Department of East China Normal University, Shanghai (2007)
8. Thibodeau, M.-A., Bélanger, S., Frasson, C.: WHITE RABBIT- Matchmaking of User Profiles Based on Discussion Analysis Using Intelligent Agents. In: Gauthier, G., VanLehn, K., Frasson, C. (eds.) ITS 2000. LNCS, vol. 1839, pp. 113–122. Springer, Heidelberg (2000)
9. Wooldridge, M.: Agent-based software engineering. In: IEEE Proceedings on Software Engineering (February 1997)
10. Frankin, S., Grasser, A.: Is it an Agent, or Just a Program?: A Taxonomy for Autonomous Agents. In: Tambe, M., Müller, J., Wooldridge, M.J. (eds.) IJCAI-WS 1995 and ATAL 1995. LNCS, vol. 1037, Springer, Heidelberg (1996)

References

A Novel Fraction-Based Hopfield Neural Networks

Jinrong Hu[1], Jiliu Zhou[1], Yifei Pu[1], Yan Liu[2], and Yi Zhang[1]

[1] College of Computer Science, Sichuan University, Chengdu, China
[2] College of Electronic and Information Engineering, Sichuan University, Chengdu, China
dewhjr@hotmail.com

Abstract. In this paper, we propose a novel fractional-based Hopfield Neural Network (FHNN). The capacitors in the standard Hopfield Neural Network (HNN) with traditional integer order derivatives are replaced by fractance components with fractional order derivatives. From this, continues Hopfield net is extended to the fractional-based net in which fractional order equations describe its dynamical structure. We also prove the stability of FHNN through the Lyapunov energy function. In addition, we analyze the performance of FHNN by performing printed number recognition experiments. The simulation results in comparison with the standard HNN, showed some salient advantages in the fractional-based Hopfield Neural Network containing the higher capacity.

Keywords: Fractional Calculus, Hopfield Neural Networks, Lyapunov Energy Function, Number Recognition.

1 Introduction

Fractional calculus has long history as same as integral calculus and was presented 300 years ago relative to traditional integral calculus. Fractional calculus has been credited as being the natural mathematical model for power-law relations. These relations are often observed as accurate descriptors for natural phenomena. The real objects are generally fractional [1], so fractional calculus allows describing a real object more accurate than the classical "integer" methods. A typical example of a non-integer (fractional) order system is the voltage-current relation of a semi-infinite lossy RC transmission line [2], [3] or diffusion of the heat into a semi-infinite solid, where heat flow is equal to the half-derivative of the temperature [4].

In the fractional calculus, differential equations have non-integer order. The engineers could understand the importance of the fractional-order equations only during last 30 years, especially when they observed that the description of some systems are more accurate, when the fractional derivative is used [5]. Now, it has been successfully applied in many research fields, such as diffusion process, viscoelastic theory, stochastic fractal dynamic and modern signal analysis and processing. In the fractional calculus, the main operator is the fractance. The fractance is a generalized capacitor.

It is well-known that HNN are based on integral order differential and integral order circuit, such as the capacitors based on integral order derivative. Actually it is

J. Luo (Ed.): Soft Computing in Information Communication Technology, AISC 158, pp. 143–150.

an electrical circuit in which its voltage and current are related by the fractional order differential equation, and the fractional calculus result of recurrent signal is quite similar with biology neural impulse signal. The main reason for using the integer-order models was the absence of solution methods for fractional differential equations. We have to identify and describe the real object by the fractional order models. The first advantage is that we have more degrees of freedom in the model. The second advantage is that we have a "memory" in model. Fractional order systems have an unlimited memory, being integer-order systems cases in which the memory is limited. Therefore we need a memory term (e.g. fractional integral or derivative) in the fractional order model. This memory term insure the history and its impact to present and future. This is a very important thing especially for fractional order model of the Hopfield neural networks.

In recent years, some results of the research issue about fractional calculus based neural network have been obtained. For example, Arena et al introduced a new class of the Cellular Neural Networks (CNN) with fractional order cells, they replaced first order cells with m-th order ones, where m being a non-integer quantity [6]; Yi-fei PU et al designed a any ractional-based multilayer dynamics associative neural network and implemented it with analog fractance circuit [7]; Arefeh Boroomand proposed a fractional-based Hopfield Neural networks and used it to solve parameter identification problem[8]. However, deep-going and concrete study about this research issue is still required. Thus, in this paper, on the bases of the existed results, in the first, we introduce the fractional order model based Hopfield neural networks. Then we prove the stability of FHNN through by the Lyapunov energy function. In the last, we analyze the performance of FHNN by performing recognition experiments with the handwritten numeric. The theoretical deduction confirms that the standard Hopfield neural network is a special example of fractional calculus based Hopfield neural networks. And the simulative results promise some salient advantages of the fractional calculus based Hopfield neural network.

The rest of the paper is organized as follows. In section 2, we briefly introduce the fractional calculus and its implementation algorithms as well. Fractional-based Hopfield neural network is introduced in section 3. In section 4, performance comparison between standard Hopfield neural network and fractional-based Hopfield neural network for printed number recognition is described. Finally, section 5 concludes the paper.

2   Definitions of Fractional Derivatives

The idea of fractional calculus has been known since the development of the regular calculus, with the first reference probably being associated with letter between Leibniz and L'Hospital in 1695. Fractional calculus is a generalization of integration and differentiation to non-integer order fundamental operator $_a D_t^\alpha$, where a and t are the limits of the operation. The continuous integro-differential operator is defined as

$$_aD_t^\alpha = \begin{cases} \dfrac{d^\alpha}{dt^\alpha} & \Re(\alpha) > 0, \\ 1 & \Re(\alpha) = 0 \\ \int_a (d\tau)^{-\alpha} & \Re(\alpha) < 0 \end{cases} \tag{1}$$

The two definitions used for the general fractional differintegral are the Grunwald-Letnikov (GL) definition and the Riemann-Liouville (RL) definition [9], [10]. The GL is given here

$$_aD_t^\alpha f(t) = \lim_{h \to 0} h^{-\alpha} \sum_{j=0}^{\left[\frac{t-a}{h}\right]} (-1)^j \binom{\alpha}{j} f(t - jh), \tag{2}$$

where [.] means the integer part. The RL definition is given as

$$_aD_t^\alpha f(t) = \frac{1}{\Gamma(n-\alpha)} \frac{d^n}{dt^n} \int_a \frac{f(\tau)}{(t-\tau)^{\alpha-n+1}} d\tau, \tag{3}$$

for $(n-1 < \alpha < n)$ and where $\Gamma(\cdot)$ is the Gamma function.

Some others important properties of the fractional derivatives and integrals we can find out in several works (e.g.: [9], [10], etc.). We can compactly model many of physical phenomena, material properties and processes, using fractional-order differential equations (FODEs). In some researches [11,12,13], it is noted that many dynamical systems such as electrochemical processes, membranes of cells of biological organism, certain types of electrical noise, and chaos are more adequately described by FODEs.

3   Fractional-Based Hopfield Neural Networks

The continuous Hopfield model of size N is a fully interconnected neural network with N continuous valued units. It contains linear and non-linear circuit elements, which typically are capacitor, resistor, op-amp, and sources. The topological structure of Hopfield net is shown in Figure 1. The input and output of the net are analog signals. The resistance Ri0 and capacitor Ci are parallel to simulate the time-delay characteristics of biologic neurons. The resistance Rij(i,j=1,2,...,N) and the op-amps are used to simulate the synapse and the non-linear characteristic of biologic neurons, respectively.

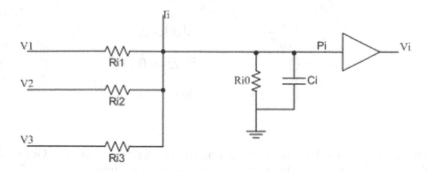

Fig. 1. The dynamic neuron of continuous Hopfield net

The state and output equations of continuous Hopfield with N neurons are given as follows:

$$C_i \frac{dP_i}{dt} = \sum_{j=1}^{N} W_{ij} V_j - \frac{P_i}{R_i} + I_i, \quad P_i = \left(\frac{1}{\lambda}\right)^{-1}(V_i) \tag{6}$$

where Pi(t) and Vi(t) are the input and output of op-amp, respectively, for the i-th neuron at time t, λis the learning rate and Wij is the conductance between the i-th and j-th neuron, and the following equations hold.

$$W_{ij} = \frac{1}{R_{ij}}, \quad \frac{1}{R_{ij}} = \frac{1}{R_{i0}} + \sum_{j=1}^{N} W_{ij}. \tag{7}$$

Consider the following energy function as a Lyapunov function

$$E = -\left(\frac{1}{2}\right) \sum_i \sum_j W_{ij} V_j V_i - \sum_i I_i V_i + \frac{1}{\lambda} \sum_i \left(\frac{1}{R_i}\right) \int_0^{V_i} {}^{-1}(v) dv \tag{8}$$

Hopfield showed that, if the weights are symmetric $W_{ij}=W_{ji}$, then this energy function has a negative time gradient. This means that the evolution of dynamic system (5) in state space always seek the minima of the energy surface E. Because of easy structure of Continuous Hopfield neural networks for implementation, they often used for solving optimization problems and associative memory.

In 1994, Westerlund proposed a new linear capacitor model based on Curie's empirical law that for a general input voltage u(t), the current is shown as formula (9) [15].So the capacitor based on integral order differential can be generalized to fractance by fractional order relationship between the voltage's terminal and the current passing through, and the fractance can be described as follows:

$$i(t) = F \frac{d^\alpha u(t)}{dt^\alpha} = F \cdot D^\alpha u(t), \tag{9}$$

From this idea, some researchers have proposed several kinds fractional based neural networks [6,7,8]. In this section, we use this generalized capacitor in the continuous Hopfield neural networks instead of common capacitor and obtained a new continuous neural network, which can be described by fractional order differential equations, and named it with FHNN. The topological structure of fractional-based Hopfield net is shown in Figure 2. And the state and output equations of FHNNs are as follows:

$$F_i \cdot D^\alpha P_i = F_i P_i^{(\alpha)} = \sum_{j=1}^{N} W_{ij} V_j - \frac{P_i}{R_i} + I_i, \quad P_i = \left(\frac{1}{\lambda}\right)^{-1}(V_i). \tag{10}$$

where $0 < \alpha < 1$ and $D^\alpha(\cdot)$ is the defined in (1).

We now analyze the stability of fractional-based Hopfield neural networks through Lyapunov energy function of neural networks. From the FHNNs structure in (9), the formula of Lyapunov energy function for this network is as follows:

$$E = -\left(\frac{1}{2}\right)\sum_i \sum_j W_{ij} V_j V_i - \sum_i I_i V_i + \frac{1}{\lambda}\sum_i \left(\frac{1}{R_i}\right)\int_0^{V_i} \varphi^{-1}(v)dv. \tag{11}$$

Calculating the derivative of E:

$$\begin{aligned}
\frac{dE}{dt} &= -\sum_{i=1}^{N}\frac{dV_i}{dt}\sum_{j=1}^{N}W_{ij}V_j - \sum_{i=1}^{N}\frac{dV_i}{dt}I_i + \frac{1}{\lambda}\sum_{i=1}^{N}\left(\frac{1}{R_i}\right)\frac{d\int_0^{V_i}\varphi^{-1}(v)dv}{dt}\cdot\frac{dV_i}{dt} \\
&= -\sum_{i=1}^{N}\frac{dV_i}{dt}\sum_{j=1}^{N}W_{ij}V_j - \sum_{i=1}^{N}\frac{dV_i}{dt}I_i + \frac{1}{\lambda}\sum_{i=1}^{N}\left(\frac{1}{R_i}\right)\varphi_i^{-1}(V_i)\cdot\frac{dV_i}{dt} \\
&= -\sum_{i=1}^{N}\frac{dV_i}{dt}\cdot\left(\sum_{j=1}^{N}W_{ij}V_j + I_i - \frac{1}{\lambda}\left(\frac{\varphi_i^{-1}(V_i)}{R_i}\right)\right) \\
&= -\sum_{i=1}^{N}\frac{dV_i}{dt}\cdot\left(\sum_{j=1}^{N}W_{ij}V_j + I_i - \frac{P_i}{R_i}\right) \\
&= -\sum_{i=1}^{N}\frac{dV_i}{dt}\cdot C_i \cdot \frac{dP_i}{dt} \\
&= -\sum_{i=1}^{N}\left(\frac{dV_i}{dt}\right)^2\cdot C_i \cdot \frac{d\varphi_i^{-1}(V_i)}{dV_i}
\end{aligned} \tag{12}$$

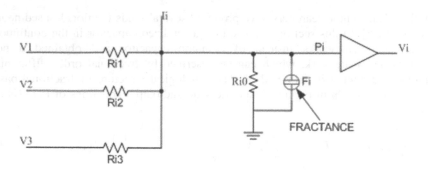

Fig. 2. The dynamic neuron of fractional-based Hopfield net

4   Experimental Results

In this section, we present experimental results and analyze the performance of the fractional-based Hopfield neural network. In order to verify the performance of the fractional-based Hopfield neural network, recognition experiments using the number signs: 0~9. The training data include 1 group of printed number sign; the testing data include 12 groups of printed number sign, each group has 10 number signs: 0~9, which are corrupted by random noise of various levels. Each printed number image are composed of black and white pixels, and they are all 60×36 pixels format. Fig. 3 shows some representative samples taken from the training data and the testing data used in this paper.

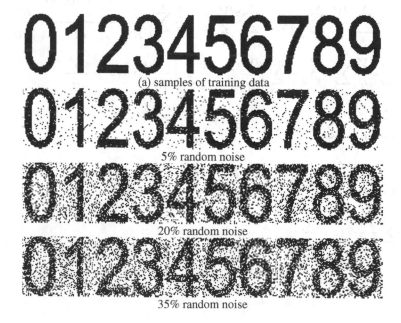

(a) samples of training data

5% random noise

20% random noise

35% random noise

Fig. 3. Partial samples used in experiment

50% random noise
(b)samples of testing data : image corrupted with various levels random noise

Fig. 3. *(continued)*

In our experiment, we compared the performance between the fractional-based Hopfield neural networks and the standard Hopfield neural networks through noise corrupted printed numbers. The topology structure, neuron number and learning algorithm are all the same for fractional-based Hopfield neural networks and the standard Hopfield net. At the same time, the Hebbian learning method is used and the number of neuron is 60.

In our experiment, we compared the performance between the fractional-based Hopfield neural networks and the standard Hopfield neural networks through noise corrupted printed numbers. The topology structure, neuron number and learning algorithm are all the same for fractional-based Hopfield neural networks and the standard Hopfield net. At the same time, the Hebbian learning method is used and the number of neuron is 60.

Table 1. Results of standard HNN and FHNN

Samples	Recognized correctly by HNN	Accurate rate of HNN	Recognized correctly by FHNN	Accurate rate of FHNN
Training samples (10)	9	90.00%	10	100.00%
Test samples (100)	78	78.00%	93	93.00%
All samples (110)	87	79.09%	103	93.63%

5   Conclusion

In this paper, we chose Hopfield net and extended its dynamical equations to the fractional differential equations using fractional order operators and introduced the fractional-based Hopfield neural networks (FHNN). Then, we proved the stability of FHNN through Lyapunov energy function. At last, we analyzed the performance of the proposed FHNN by performing printed number recognition experiments. The simulation results in comparison with the standard HNN, showed some salient advantages in the fractional-based model containing the higher capacity.

References

1. Westerlund, S.: Dead Matter Has Memory! Causal Consulting, Kalmar, Sweden (2002)
2. Wang, J.C.: Realizations of generalized warburg impedance with RC ladder networks and transmission lines. J. of Electrochem. Soc. 134(8), 1915–1920 (1987)
3. Nakagava, M., Sorimachi, K.: Basic characteristics of a fractance device. IEICE Trans. Fundamentals E75-A(12), 1814–1818 (1992)
4. Podlubny, I.: Fractional Differential Equations. Academic Press, San Diego (1999)
5. Jenson, V.G., Jeffreys, G.V.: Mathematical Method in Chemical Engineering, 2nd edn. Academic Press, New York (1977)
6. Arena, P., Caponetto, R., Fortuna, L., Porto, D.: Bifurcation and chaos in noninteger order cellular neural networks. International Journal of Bifurcation and Chaos 8(7), 1527–1539 (1998)
7. Pu, Y.-F.: Implement Any Fractional Order Multilayer Dynamics Associative Neural Network. In: 6th International Conference on ASIC, ASICON 2005, vol. 2, pp. 638–641 (2005)
8. Boroomand, A., Menhaj, M.B.: Fractional-order Hopfield Neural Networks. In: 15th International Conference on Neural Information Processing of the Asia-Pacific Neural Network Assembly, Auckland, New Zealand (November 2008)
9. Oldham, K.B., Spanie, J.: The Fractional Calculus. Academic Press, New York (1974)
10. Podlubny, I.: Fractional Differential Equations. Academic Press, San Diego (1999)
11. Li, C., Chen, G.: Chaos in fractional order Chen system and its control. Chaos, Solutions & Fractals, 305–311 (2004)
12. Cole, K.S.: Electric Conductance of Biological System. In: Proc. Cold Spring Harbor Symp. Quant. Biol., New York, pp. 107–116 (1993)
13. Mandelbrot, B.B.: Some Noises with 1/f Spectrumm, a Bridge Between Direct Current and White Noise. IEEE Tran. on Info. Theory IT-13(2) (1967)
14. Charef, A., Sun, H.M., Tsao, Y.Y., Onaral, B.: Fractional Systems as Represented by Singularity Function. IEEE Trans. on Automatic Control 37(9), 1465–1470 (1992)
15. Westerland, S.: Capacitor Theory. IEEE Tran. on Dielectrics and Electrical Insulation 1(5), 826–839 (1994)

Design of Optimized Fuzzy PID-Smith Controller on Flow Control in Real-Time Multimedia Communication

Hong-yan Qi and Man Li

College of electrical and information engineering
Heilongjiang Institute of Science and Technology
Haerbin, China
qihongyan@126.com, lm980525@126.com

Abstract. The application of VOD has been expanded rapidly with the population of broadband integrated services digital network. But the playback of VOD is not guaranteed because of the jitter of propagation delay. The feedback control mechanism for the real-time multimedia traffic control is discussed. Due to the particularity of real-time multimedia flow model and performance demand, it is difficult for adopting conventional feedback control technology to overcome it. To these questions, on the basis of the control theory scheme of rate-based traffic control, Optimized Fuzzy PID-Smith controller is designed to overcome the adverse effect caused by propagation delay and jitter. The play quality is improved.

Keywords: real-time multimedia, delay and jitter, fuzzy PID.

1 Introduction

With the population of broadband network, VOD service based on internet is provided in the almost all the broadband ISP and the website. Multimedia terminal in the interactive video and audio business is Set-Top Box, but because buffer of the Set-Top Box is smaller, when unpredictable jitter occurs in the network, overflow and underflow happen in the buffer of Set-Top Box, ever if transmission errors have not happen, the playback quality will be seriously affected. Therefore, new rate-based control strategy is introduced[1] .

Yeali S.Sun puts forward the flow control based on the client queue length, but the control algorithm is too complicated, because control period is based on round trip time, the more of control period , the greater of possibility of divergence in the buffer, the system's robust performance is poor. Reference [2] and [3] show that control theory has been successfully applied to the rate-based video coding and high-speed network facilities. In particular, when load information is uncertain, feedback mechanisms acquire better system performance. Wang Chia-Hui[4] puts forward the PD control strategy, however, because design of control parameters is conservative, which cause the lower gain and different dynamic response performance, in addition, when network jitters, the system's robust performance is poor. In order to solve these problems, the paper designs the optimized fuzzy PID-smith controller.

J. Luo (Ed.): Soft Computing in Information Communication Technology, AISC 158, pp. 151–158.
springerlink.com   © Springer-Verlag Berlin Heidelberg 2012

2   Feedback Control Structure

Fig. 1 shows feedback control system structure. Data packet is transmitted from the server to the client server through the network, furthermore periodically the buffer length of current Set-Top Box is returned to the rate regulator, rate regulator calculates the next period sending rate in accordance with the control strategy and regulates the sending rate.

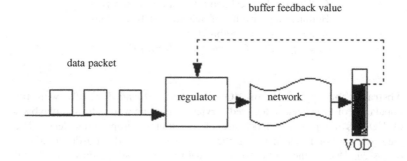

Fig. 1. Rate feedback control structure graph

A.   Relevant variable selection

According to fig. 1,   creation the relevant control variables:

- Intermediate variable(unknown for feedback controller): receiving rate $\lambda(k)$, is the sum of the sending rate and network jitter rate, that is $\lambda(k) = u(k) + q(k)$, $u(k)$ is the sending rate of the server, $q(k)$ is rate of network jitter that is caused by the network transmission delay and data retransmission; playback rate $\mu(k)$, which apply MPEG-1 File Format as Video data background, average playback rate is 172kB/s. $r(k) = \lambda(k) - \mu(k)$, is difference of receiving rate and playback rate.
- Controlled variable: buffer length $b(k)$.
- Set value or expected value (representative the performance of the controlled variable): buffer length set value $B_m = B^0 / 2 = 100kB$; the biggest value of buffer length and the smallest value of buffer length is respectively $B_h = 150kB$ and $B_l = 50kB$; when beyond its range, which easily causes the overflow or underflow phenomenon and affects broadcast quality.
- Control variable: sending rate of server $u(k)$.

B.   rate control model

Due to restriction of the buffer capacity and buffer dynamics is provided with saturation nonlinearity, if T_s is control period, the discrete equations of buffer as following:

$$b(k) = Sat_{B^0}\{b(k\quad 1) + T_S(\lambda(k)\quad \mu(k))\} \tag{1}$$

Where:

$$Sat_z = \begin{cases} 0, & z < 0 \\ a, & z > a \\ z, & other \end{cases}$$

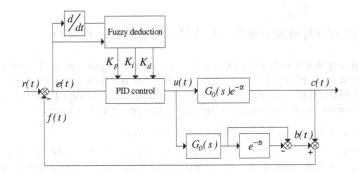

Fig. 2. Fuzzy PID-Smith control system structure graph

The paper analyses anti-disturbance performance of control system. In order to convenient analysis, variables are executed incremental treatment in the near steady-state. Assuming initial steady-state value are $\lambda_s(0) = \mu_s(0) = u_s(0) = 172kB/s$, $b_s(0) = 100kB$, $q_s(0) = 0kB/s$, because $\mu(k)$ and $q(k)$ act on the front of the object, and they are unpredictable, assumptions $d(k) = \Delta\mu(k) + \Delta q(k)$, which express unpredictable disturbance, communication network disturbance is caused by jitter of delay transmission and jitter of playback rate, therefore $\Delta r(k) = \Delta u(k) - d(k)$, thus sending rate control may take place rate control. Do not point out in particular, variables in the paper are executed incremental treatment on the basis of steady-state, $u(k)$ replaces $\Delta u(k)$, $b(k)$ replaces $\Delta b(k)$. In the buffer range, nonlinear factor in the formula 1 is removed, system dynamic equation is:

$$b(k) = b(k-1) + T_s\{u(k-1) - d(k-1)\} \tag{2}$$

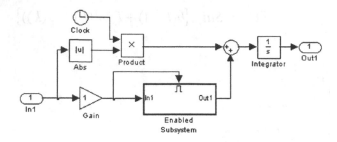

Fig. 3. Simulation model of objective function

3   Design of Optimized Fuzzy PID-Smith Controller

The system applies fuzzy-PID controller as master regulator and utilizes fuzzy reasoning achieving PID parameters of online self-tuning, thus achieves quick response and good stability. The system applies Smith predictor as vice-regulator, which is used to eliminate inertia and delay of the system [5]. System structure graph is fig. 2.

C.   design of optimized fuzzy PID-Smith controller

The parameters of the controller need to optimize tuning to enable achieve optimal control performance after determining the control strategy. There are optimization problems for fuzzy PID controller or other controller. Optimization of fuzzy controller is to optimized tune the relevant rules and parameters, Such as the division of membership functions, selection of fuzzy rules, adjustment of scale factor. In the Industrial production process, the quality of control system is improved by optimizing the controller parameters, which brings significant economic benefits, tuning problems of controller will become more prominent and important.

The fuzzy rules and parameter is optimized using multivariable constrained minimization function of the optimization toolbox on the based of conventional fuzzy PID, thereby, optimized fuzzy PID controller is designed. Design target is to make the system step response to "fast - no overshoot". ITAE criterion is used as the objective function[6].

$$J = \int_0^\infty t|e(t)|dt = \min \tag{3}$$

And constraints on the overshoot are added. Therefore, the objective function of the system is:

$$OBJ = \int_0^{t_s} t|e(t)|dt + k\int_0^{t_s} e(t)dt = \int_0^{t_s} [t|e(t)| + ke(t)]dt \tag{4}$$

Optimized system model is shown in fig. 3:   Absolute control, Multiplier, Integrator of the simulation model is used to implement ITAE criteria, Proportion part, enabled sub-system, Integrator is used to implement overshoot constraint.

In accordance with the characteristics of VOD system, two-dimensional fuzzy controller is established. Input variables are the absolute value of error e and the absolute value of error rate ec and output variables are k_p、 k_i、 k_d of PID parameters. Language value e and ec is NB, NM, NS, ZO, PS, PM, PB. Fuzzy domain is $\{0,10\}$. In order to conveniently realize, trimf is selected as membership function of input variables and output variables.

Based on the above analysis and settings of language variables, Optimized fuzzy control rules of k_p、 k_i、 k_d are summarized in the following table, which are respectively listed in table 1, table 2 and table 3.

Table 1. k_p control rule table

$\|e\|$ $\|ec\|$	NB	NM	NS	ZO	PS	PM	PB
NB	PS	PB	PM	PM	PS	ZO	ZO
NM	PB	PM	PM	PS	PS	ZO	NS
NS	PM	PM	PS	PS	ZO	NS	NS
ZO	PM	PM	PS	ZO	NS	NM	NM
PS	PS	PS	ZO	NS	NS	NM	NM
PM	PS	ZO	NS	NM	NM	NB	NB
PB	ZO	ZO	NS	NM	NM	NB	NB

Table 2. k_i control rule table

$\|e\|$ $\|ec\|$	NB	NM	NS	ZO	PS	PM	PB
NB	NB	NB	NM	NM	NS	ZO	ZO
NM	NB	NB	NM	NS	NS	ZO	ZO
NS	NB	NM	NS	ZO	ZO	PS	PS
ZO	NM	NM	NS	ZO	PS	PM	PM
PS	NM	NS	ZO	PS	PM	PM	PB
PM	ZO	ZO	PS	PS	PM	PB	PB
PB	ZO	ZO	PS	PM	PM	PM	PB

Table 3. k_d control rule table

$\|e\|$ $\|ec\|$	NB	NM	NS	ZO	PS	PM	PB
NB	PS	NS	NB	NB	NB	NM	PS
NM	PS	NS	NB	NM	NM	NS	ZO

Table 3. *(continued)*

NS	ZO	NS	NM	NM	NS	NS	ZO
ZO	ZO	NS	NS	NS	NM	ZO	ZO
PS	ZO	ZO	ZO	ZO	ZO	ZO	ZO
PM	PB	NS	PS	PS	PS	PS	PB
PB	PB	PM	PM	PM	PS	PS	PB

Based on the quantitative value and fuzzy relation, through fuzzy deduction aggregation rule computing, fuzzy sets of the corresponding control variable changes is obtained, which is fuzzy sub set and reflect a combination of the different values on control language. In the paper, fuzzy deduction method is Mamdani, method of and is min, method of Or is max, method of Implication is min, method of Aggregation is max, method of Defuzzification is centroid.

Online self-tuning algorithm is established between parameters k_p、k_i、k_d and input variables (error and error rate) on the fuzzy PID controller to meet the different needs of control parameters at the different error and error rate, so can be superior to conventional PID controller results.

D.   design of Smith predictor

It is difficult to eliminate radically the effects of dead time for PID control, so smith predictor control is introduced. Principle of Smith predictor estimates in advance the dynamic characteristics in the basic disturbance. Thereby, equivalent transfer function of compensation no longer contains pure lag, which is moved ahead of time, thus reduces the overshoot and speeds up the adjustment process[7].

Smith predictor control system structure graph is fig. 4.

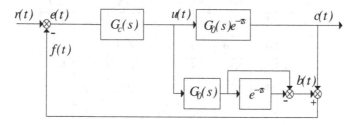

Fig. 4. Structure graph of Smith predictor system

Where, $G_0(s)$ is linear portion, $e^{-\tau s}$ is pure lag portion. If transfer function of PID controller is $G_c(s)$, feedback transfer function of system is :

$$\Phi(s) = \frac{C(s)}{R(s)} = \frac{G_c(s)G_0(s)}{1+G_c(s)G_0(s)} e^{-\tau s}$$ Equivalent structure graph is fig. 5.

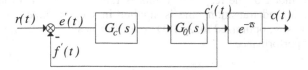

Fig. 5. Equivalent structure graph of Smith predictor system

It is obvious that the pure lag portion $e^{-\tau s}$ and linear portion $G_0(s)$ is disjoined after adding the Smith predictor. The pure lag portion $e^{-\tau s}$ is removed to outside the feedback system. Control of the output $C(s)$ is converted to the control of output $C'(s)$. Effect of pure lag is eliminated radically and performance is enhanced.

4   Analysis of Simulation Result

Taking into account data transmission rate in the network is affected by many factors, so that it can not be observed, and can not be forecast and uncertainty, which bring certain difficulties for the simulation of network traffic. In order to comprehensively assess he performance of feedback control, disturbance loading is given as follows:

Step disturbance: A step disturbance is commonly used signal in the system performance tests, if the control system has a better response to step disturbance, then it is able to easily overcome to other disturbance. So a step disturbance is introduced.

Providing that a step disturbance is $6kB/S$, simulation construction is showed under SIMULINK and the parameters of each module are adjusted[8][9]. Step response curve of buffer length is fig. 6. Comparison between Predictive fuzzy-PID controller and Smith predictor is showed in fig. 6.

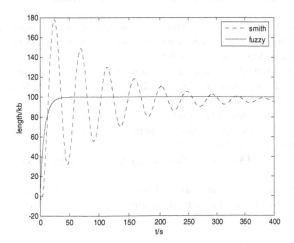

Fig. 6. The step response of the buffer length

Performance index of the response results is listed in table 4. From the response results, overshoot and transition time is smaller in the optimized fuzzy PID-Smith controller. Therefore, optimized fuzzy-PID controller is superior to Smith predictor on general control performance.

Table 4. Performance index of response results

disturbance	Control algorithm	$\sigma\%$	$t_s(s)$
step	Smith	77.8	400
step	fuzzy	0	30

5   Conclusion

Optimized fuzzy PID-smith control that is designed in the paper is provided with favorable response stability and definition, which is combination of conventional PID, optimized fuzzy control and Smith predictor, which determine intelligently PID parameters, based on fuzzy control rules, in accordance with bias and bias rate shift. The simulation results show that optimized fuzzy PID-smith control is an effective control scheme, and has better feasibility and practical application value.

References

1. Fan, W.-X., Sheng, S.-J., Quan, W.-Z.: Application of control theory to Internet congestion control. Control and Decision 17, 129–134 (2002)
2. Kolarov, A., Ramamurthy, G.: A control-theoretic approach to the design of an explicit rate controller for ABR service. IEEE/ACM Trans. on Networking 17, 741–753 (1999)
3. Liew, S.C., Tse, D.C.: A control-theoretic approach to adapting VBR compressed video for transport over a CBR communication channel. IEEE/ACM Trans. on Networking 16, 42–45 (1998)
4. Wang, C.-H., Ho, J.-M., Chang, R.-I., Hsu, S.-C.: A control-theoretic method for rate-based flow control of multimedia communication. Technical Report TR-IIS-01-007, Institute of Information Science, Academia Sinica (2001)
5. Lei, X., Li, X.-G., Yin, Z.-H.: Simulation study on adaptive fuzzy smith-PID controller based on sugeno inference. Journal of System Simulation 20, 4952–4955 (2008)
6. Wang, Y.-F., Yang, G.-X.: Design and research of time optimal fuzzy-PID control algorithm. Electric Machines and Control 8, 366–372 (2004)
7. Tan, W.: Tuning of a modified Smith predictor for processes with time delay. Control Theory and Application 34, 87–92 (2003)
8. Li, W.-X., Hui, L.-Z.: MATLAB auxiliary fuzzy system design. Xidian University Press, Beijing (2002)
9. Fong, D.-Q., Zhang, X.-P.: Optimization Design integrative Method of Fuzzy Controller Based on MATLAB. Journal of System Simulation 16, 849–852 (2004)

Lecture Notes in Computer Science:
Local Trinary Patterns Algorithm for Moving Target Detection

Xuan Zhan and Xiang Li

Department of software engineering, East China Institute of Technology,
Nanchang, Jiangxi Province, China

Abstract. In this paper, we present a novel moving target detection called Local Trinary Patterns which is based on Local Binary Patterns algorithm, The standard LBP mainly captures the texture information, and in some circumstances it results in misidentification. The proposed LTP feature, in contrast, captures the gradient information and some texture information. Moreover, the proposed LTP are easy to implement and computationally efficient, which is desirable for real-time applications. Experiments show that this algorithm can significantly improve the detection performance and produce state of the art performance.

Keywords: moving target detection, local binary patterns, local trinary patterns, texture feature.

1   Introduction

The ability to detect moving target in images has a major impact on applications such as video surveillance, smart vehicles, robotics. Changing variations in moving target such as clothing, combined with varying cluttered backgrounds and environmental conditions, make this problem far from being solved.

The proposed method has a simple flow: every pixel at every frame is encoded as a short string of ternary digits(trits) by a process which compares this frame to the previous and to the next frame. The encoding process itself is based on comparing nearby patches, in a manner inspired by the self-similarity approach. For every pixel of every frame, a small patch centered at this pixel is compared to shifted patches in the previous and in the next frame. In a manner pertaining to the Local Binary Pattern approach, one trit of information is used to describe the relative similarity of the two patches to the patch in the central frame: the shifted patch in the previous frame is more similar to the central one, the patch in the next frame shifted by the same amount is more similar, or both are approximately comparable in their similarity.

2   The LBP Feature

LBP is a texture descriptor that codifies local primitives(such as curved edges, spots, flat areas)into a feature histogram. LBP and its extensions outperform existing texture descriptors both with respect to performance and to computational efficiency[1].

J. Luo (Ed.): Soft Computing in Information Communication Technology, AISC 158, pp. 159–165.
springerlink.com

The standard version of the LBP feature of a pixel is formed by thresholding the 3×3 neighborhood of each pixel with the center pixel's value. Let g_c be the center pixel graylevel and g_i (i=0,1,...7) be the graylevel of each surrounding pixel. If g_i is smaller than g_c, the binary result of the pixel is set to 0, otherwise to 1. All the results are combined to a 8-bit binary value. The decimal value of the binary is the LBP feature. See Fig.1 for an illustration of computing the basic LBP feature.

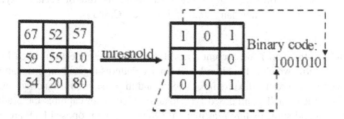

Fig. 1. Illustration of the basic LBP operator

In order to be able to cope with textures at different scales, the original LBP has been extended to arbitrary circular neighborhoods by defining the neighborhood as a set of sampling points evenly spaced on a circle centered at a pixel to be labeled. It allows any radius and number of sampling points. Bilinear interpolation is used when a sampling point does not fall in the center of a pixel. Let $LBP_{p,r}$ denote the LBP feature of a pixel's circular neighborhoods, where r is the radius of the circle and p is the number of sampling points on the circle. The $LBP_{p,r}$ can be computed as follows:

$$LBP_{p,r} = \sum_{i=0}^{p-1} S(g_i - g_c)2^i, S(x) = \begin{cases} 1 & \text{if } x \geq 0 \\ \\ 0 & \text{otherwise.} \end{cases}$$

Here g_c is the center pixel's graylevel and g_i (i=0,1,....,7) is the graylevel of each sampling pixel on the circle. See Fig.2 for an illustration of computing the LBP feature of a pixel's circular neighborhoods with $r=1$ and $p=8$. Ojala et al. proposed the concept of "uniform patterns" to reduce the number of possible LBP patterns while keeping its discrimination power. An LBP pattern is called uniform if the binary pattern contains at most two bitwise transitions from 0 to 1 or vice versa when the bit pattern is considered circular. For example, the bit pattern 11111111(no transition),00001100(two transitions) are uniform whereas the pattern 01010000(four transitions) is not. The uniform pattern constraint reduces the number of LBP patterns from 256 to 58 and is successfully applied to face detection in [3].

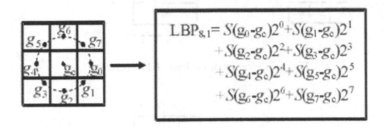

Fig. 2. The LBP operator of a pixel's circular neighborhoods with r=1, p=8

3   The LTP Feature

Although the local texture character can be described efficiently and the whole image character description can be easily extended. Due to the single transform and mapping, namely in the calculation of mapping around the neighborhood point and center pixel size relations, only consider threshold value 0, when the value of surrounding pixel points minus the center point pixel is greater or equal to 0, denoted by 1, or 0 vice versa. Thus the local texture character can be described efficiently, but focus only on texture background environment detail varies, in some dramatic changes will brought by mistake, and these details identify a plethora of information for classification may lead to a information redundancy or fitting. Thus in order to make LBP less sensitive to noise, particularly in near-uniform image regions, Tan and Triggs[3] extended LBP to 3 valued codes, called local trinary patterns(LTP). If each surrounding graylevel g_i is in a zone of width $\pm t$ around the center graylevel g_c, the result value is quantized to 0. The value is quantized to $+1$ if g_i is above this and is quantized to -1 if g_i is below this. The $LTP_{p,r}$ can be computed as:

$$LTP_{p,r} = \sum_{i=0}^{p-1} S(g_i - g_c)3^i, S(x) = \begin{cases} 1 & \text{if } x \geq t \\ 0 & \text{if } |x| < t \\ -1 & \text{if } x \leq t \end{cases}$$

Here t is a user-specified threshold. Fig.3. shows the encoding procedure of LTP. For simplicity, Tan and Triggs[27]used a coding scheme that splits each ternary pattern into its positive and negative halves as illustrated in Fig.4,treating these as two separate channels of LBP codings for which separate histograms are computed, combining the results only at the end of the computation.

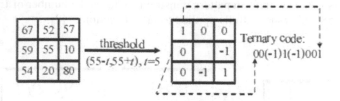

Fig. 3. Illustration of the basic LTP operator

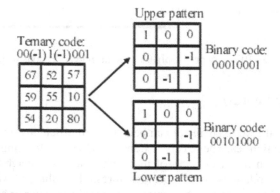

Fig. 4. Splitting the LTP code into positive and negative LBP codes

4 Encoding Motion in the Frame

The various flavors of Local Binary Patterns use short binary strings to encode simple properties of the local microtexture around each pixel. Here we propose an LBP like descriptor which captures the effect of motion on the local structure of self-similarities. Consider a small image patch moving from left to right. During its motion it will pass through a certain image location $(x - \Delta x, y)$ at time $t - \Delta t$, and continue to location (x, y) to the right at time t. This motion is probably going to induce image similarity between a patch of appropriate dimensions centered at location $(x - \Delta x, y)$ at time $t - \Delta t$ and the patch with the image center (x, y) at time t.

By itself, the increase of image similarity caused by the motion depends on the intensities of the moving patch and the appearance of the rest of the image. It may be difficult to distinguish between similarity caused by motion and similarity caused by similar static textures, without incorporating further statistics. Here we suggest to examine the similarity between a patch centered at (x, y) at time t and the patch around $(x - \Delta x, y)$ at time $t + \Delta t$ as the background statistic. One trit is used to encode whether one of the two similarities is significantly higher than the other or whether the two similarities are approximately the same. If the previous frame patch is more similar to the central patch -a value of -1 is assigned, if the patch in the next frame is more similar -a value of +1 is assigned. If both similarities are within a predefined threshold from each other, a value of 0 is assigned.

Note that in the absence of significant image motion the similarities of the patch at center location (x, y) at time t to the patches at location $(x - \Delta x)$ at times $t - \Delta t$ and $t + \Delta t$ are about equal, and the value of the encoding trit is zero. This implies that no appearance information is encoded in the absence of motion.

The full 8 trit encoding is described in Figure 5. Patches at eight shifted locations at times $t - \Delta t$ and $t + \Delta t$ are compared to a central patch at time t to produce 16 similarities. Due to its computational simplicity the SSD(sum of square differences)score as the basic distance between the patches. The lower the SSD score, the larger the similarity.

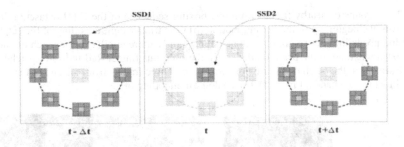

Fig. 5. An illustration of the encoding process

For each of 8 different locations at time $t - \Delta t$ and the same locations at time $t + \Delta t$ SSD distances of 3×3 patches to a central patch at time t are computed. SSD1 and SSD2 are computed patch distances at one of the eight locations. One trinary bit is used to encode if $SSD1 < SSD2 - TH$ (value of -1), $|SSD1 - SSD2| < TH$ (value of 0), or $SSD2 < SSD1 - TH$ (value of +1). We define gray values are between 0 and 255, and TH is set to 0.097. Also Δt is set to 3 frames, and the patches are spread around as close as possible using integer values to distance of 4 pixels from the center of the central patch.

5 Analyse the Simulation Result

We perform the experiments under Matlab platform. The first group image is from one boxing sequence of the KTH dataset, the rest are from two typical AVI file: intelligent_room.avi, highway_raw.avi, the detection result are as follows:

(a)   (b)

(c)   (d)

Fig. 6. Two groups of nearby frames from one boxing sequence of the KTH dataset.(a) Three frames from the beginning of the boxing motion.(b) One trinary digit encoding of the sequence in (a). Blue pixels indicate patches which are significantly more similar to the patch on the left in the next frame than to the patch on the left in the previous frame. Red indicates patches that are more similar to the patch of the previous frame. (c) Three frames from the end of the boxing motion, in which the hand returns. (d) The analog trit encoding of (c).

Fig. 7. Intelligent_room   **Fig. 8.** Highway_raw

From the result, we can see this method aims to extract features of a target for moving targets detection, and have a good combination of local texture information and edge mutations gradient information so that improve the measuring accuracy.

6   Conclusion

We present a novel moving target detection called Local Trinary Patterns which is based on Local Binary Patterns algorithm, the detection result shows that it can combine local texture information and edge mutations gradient information, and can significantly improve the detection performance and produce state of the art performance.This work was supported by the grant from Scientific research plan projects of JixiangXi Education Department(No. GJJ11495) and Key Laboratory of Nuclear Resources and Environment (East China Institute of Technology) of Ministry of Education of China(No. 101112).

References

1. Velisavljević, V., Beferull-Lozano, B., Vetterli, M., Dragotti, P.L.: Directionlets: anisotropic multi-directional representation with separable filtering. IEEE Transactions on Image Processing 15(7), 1916–1933 (2006)
2. Velisavljević, V.: Directionlets: anisotropic multi-directional representation with separable filtering. Ph.D. Thesis no. 3358, LCAV, School of Computer and Communication Sciences, EPFL, Lausanne, Switzerland (October 2005)
3. Velisavljević, V., Beferull-Lozano, B., Vetterli, M., Dragotti, P.L.: Approximation power of directionlets. In: Proceedings of IEEE International Conference on Image Processing (ICIP 2005), Genova, Italy, vol. 1, pp. I-741–I-744 (September 2005)
4. Zhang, L., Hai, T., Zhang, Y., Luo, C.G.: An infrared and visible image fusion algorithm based on image features. Journal of Rockets and Missiles 29(1), 245–246 (2009)
5. Jiao, L.C., Hou, B., Wang, S., Liu, F.: The theory and application of multi-scale image analyse, pp. 459–464. Publishing House of Xi'an Electronic and Technolony (2008)
6. Yan, J.W., Qu, X.B.: Analyse and application of super wavelet, pp. 46–60. Publishing House of Defense Industry (2008)
7. Bai, J., Hou, B., Wang, S., Jiao, L.C.: Noise suppression of SAR image with Gause regional mixed scale model based on lifting Directionlet
8. Chen, H., Liu, Y.Y.: The research on infrared image fusion based on wavelet transform. Infrared and Laser 1(39), 97–100 (2009)
9. Xydeas, C.S., Petrovic, V.: Objective image fusion performance measure. Electronics Letters 694, 308–309 (2000)

Intellectual Traffic System and Application in the Urban Road Safety

Xiaowei Wei[1,*] and Yongfeng Ju[2]

[1] Shaanxi College of Communication & Technology,
Chang'an University College of Electronics and Control Ph.D., X'an 710000 China
[2] Chang'an University, Professor, College of Electronics and Control
X'an 710000 China
stgi001@163.com

Abstract. To China's transport system is the forefront of transportation research. The core research is aimed at increasing traffic demand and transport resources, the use of information technology, communications technology, computer technology, management level. the composition of the framework plan, the analysis of urban road traffic safety based on the feasibility of information systems, based on road safety in cities unusual event detection system and its principle,that the Intelligent Transportation Systems information collection,road intelligent traffic information processing, application and performance improvement.

Keywords: Road safety, Intelligent Transportation Systems information collection unusual event.

1 Introduction

According to the present transportation in China, this paper puts forward the necessity, the feasibility and function of the intelligent transportation system. Its good functions include the smooth efficiency, security effectiveness, environmental effectiveness and economic efficiency. Intelligent transportation system concerns wide and comprehensive variety of high-tech fields of study. The following focuses on intelligent transportation system components and their application in road safety.

2 China's Current Traffic Situation[1]

With the sustainable economic development in China, the urbanization and motorization have increased very rapidly since implementing the policy of reform and opening up. Before the reform, the level of urbanization [2] was less than 19%. But now it has grown to more than 30%, in 2010 to nearly 50%. Vehicle ownership has now reached 60 million. With an annual growth rate of more than 10%, it is expected in 2010 to more than 130 million vehicles. Urban transport is characterized by mixed

---

* Male,1975.10~, Ph.D. Research Interests: Intelligent Traffic Control.

J. Luo (Ed.): Soft Computing in Information Communication Technology, AISC 158, pp. 167–174.
springerlink.com
© Springer-Verlag Berlin Heidelberg 2012

traffic because bicycle ownership now over 180 million. If the level of public transport services won't be raised and urban transport structure won't be improved, bicycle ownership will continue to increase. Though there has made great achievements after years of construction of transport infrastructure, traffic is still not optimistic with the increase in car ownership. To solve the traffic problems, it must rely on intelligent transport systems to improve traffic safety and convenience. The development of high-tech offers the possibility of implementation for intelligent transportation systems (ITS), including modern transportation technology, electronic technology, communication technology, computer technology and so on.

3   The Need for the Development of Intelligent Transportation Systems

Today there are traffic congestion problems among all major cities in the world. In the United States, in 1976 and 1997, annual vehicle miles increased by 77% while the increasing number of miles of road construction was only 2% at the same time. At the peak traffic in the city, 54% of the cars were congested. Due to heavy traffic, the time consumption in commute is 1.5 hours more than in usual. Moreover, the cost of transportation is increased because commercial vehicles led to the delays in transportation. Therefore, it is necessary to develop intelligent transport systems with the integrated use of modern information and communication technology and other means.

4   Intelligent Transportation Systems Functions as Followers

Smoothness: to increase traffic mobility, improve operational efficiency, the road network capacity and infrastructure efficiency, control traffic demand.

Safety: to improve traffic safety and reduce the likelihood of accidents; reduce accident damage and prevent the expansion of the disaster after the accident.

Road condition: to reduce congestion, pollution and the impact of motor transport.

Economies: the construction of intelligent transportation system as a new economic growth point.

5   The Elements of Intelligent Transportation Systems

5.1   ITS Structural System

ITS consists of four modular subsystems, namely, sub-centers, road subsystem, vehicle subsystems and remote access subsystems, which share a common communications equipment (Figure1). By the specific functional requirements ITS can be assigned to the four subsystems of the physical subsystem (a total of 19 physical subsystems). Data and information is exchanged between each subsystem and user.

Between the various subsystems are relatively independent and interrelated. As a single entity, each subsystem is able to complete a reorganization of organic ITS strategy.

5.2   ITS Related Technologies [3]

Technologies include advanced information technology, data communication transmission technology, electronic control technology, computer processing technology, sensor technology and so on.

a)   ITS information collection technology module.

The sensor technology provides ITS traffic information, road information, weather information around the clock

Hardware devices are including image sensors, infrared sensors, laser sensors, sound sensors, wireless sensors and so on

b)   ITS information processing technology modules .

The large amounts of data from the sensor has to be fast processed and converted to useful information technology.

Hardware devices are including mainframe computers, the main computer, workstation, mobile terminal, simulation, optimization algorithms, graphics, vehicle MPUs (for engine control, navigation, etc.), vehicle LAN's.

c) ITS information transfer technology modules.

A lot of information is transferred to the vehicle, mobile terminals and fixed equipment wired and wireless technologies.

5.3   Data Collected

Data acquisition is a prerequisite for system operation. the data source is based on a bus running data collection capability and means of Intelligent scheduling system, The basic data include: the passenger flow for bus stations, vehicle running speed, stop interval, the vehicle operation, vehicle positioning and so on. The data collection mainly is completed by public transport vehicles.

Vehicle traveling data recorder. Large capacity memory is built in to record real-time movement of vehicles during the operation, including: the date and time of vehicle start and stop; vehicle speed and the entire process of testing mileage, vehicle brake changing data, the time and place of traffic incidents, the safety data of 30 seconds prior to the accident and so on. In addition, the instantaneous real-time display speed, current road code, date and time; indicator with opening and closing light, speed indicator combined with the alarm buzzer to open the door and speed alarm; IC card management and the data acquisition of vehicles operation; support wireless data collection, GPS satellite positioning reserved interface. Extended functionality can be realized real-time vehicle tracking, vehicle scheduling, flow detection interface reserved and so on.

Short-range wireless communication device. Traffic recorder is designed for supporting wireless data acquisition devices, with long distance communication (50 ~ 80m), reliable communication (BER <0.2%), high speed (115KBPS) and so on. When the vehicle reached the terminal, the intelligent scheduling system can automatically identify the vehicle identification system stop time, road grades, staff numbers and driver's license number, etc., and use of wireless automatic acquisition software is used to automatically transmit data collected by the traffic recorder data to the terminal of intelligent scheduling systems without human intervention.

5.4   The Data Is Refined into Knowledge after Processing and Organization

Intelligent scheduling system has an effective data management and analysis capabilities, including operational data management and analytical data management to protect scientific, decision analysis and efficient management of daily operations planning and scheduling.

Through wireless data collection methods, collecting traffic data recorder, a specific algorithm running on the vehicle and the driver's driving behavior for processing analysis. The computer graphically display the location and operating status of vehicles and graphical analysis methods provide safety information at the same time. The recorder can also display traffic conditions and traffic anomalies associated equipment violation records. The problem of vehicles through the computer screen flashes red and give an alarm as phonetic indication.

On-site operation management. Dispatcher use intelligent scheduling system to easily run the line of vehicles for monitoring and scheduling. Entire site scheduling management modules compose of the front control program and intelligent scheduling module. Front control program is used to display the line running to complete the interaction between the system control and dispatch personnel.

5.5   The Timely Transmission of Information to Decision Makers

Intelligent Transportation Systems will record a variety of operational data and traffic reports, single-and other information, the transmission by setting the feedback to the bus company. Decision-making groups will access a variety of reports, charts and analysis of relevant data and information materials to master the vehicle operating conditions and the utilization of resources related to public transportation. Therefore, the bus resources are fully utilized after to making appropriate judgments and planning.

6   Its Road Safety in the City of [4]

Traffic is generated mainly due to traffic jams. It must be taken seriously because traffic congestion will result in significant property damage. To solve the problems of the traffic congestion and vehicle delay is to reduce all kinds of traffic impact on traffic flow. In fact, the traffic incident is inevitable, and therefore traffic congestion and vehicle delay is normal road traffic phenomenon. ITS incident management systems use the existing technology and effectively coordinate with relevant units to effectively reduce traffic delays and traffic congestion.

6.1   The Goal of Incident Management [5]

Effective event management system can generate huge economic benefits. To achieve this goal, it should have the following functions:

- to improve road safety (such as the reduction of secondary accidents);
- to improve the efficiency of the various implementing agencies;
- more effective use of existing human and material resources;
- to increase the scope and channels for the release of information;
- to reduce delays;
- to improve the mobility of goods vehicles;
- to reduce incident response time;
- reduce the impact on the environment;
- reduce operating costs;
- to expedite the processing time of the accident;
- to clear roads to speed up the time;

1) Traffic abnormal incident detection system[6]

a) The traffic incident detection system anomalies are the traffic conditions. test data processing, signal control and information displays and other accessories. When using magnetic or ultrasonic detector, the system consists of detector, data processing, the accident determining, the accident confirmed, the signal controller, traffic lights, variable traffic information displays and other components of hardware and software facilities; When testing facilities with a video camera, system consists of digital cameras, consoles, image processing apparatus, information control machine, lights, traffic information display facilities and so on.

b) The detection of traffic anomalies. The basic principle is: the vehicle detector running in traffic under normal conditions, the measured traffic parameters should conform to certain rules. Once the traffic does not match this rule, it means that a traffic unusual event possible. As the road traffic conditions are different, the detection of traffic anomalies is vastly different in the highway or long distance highway and urban roads.

c) Urban road traffic anomaly detection event [7].Vehicles on urban roads and highways are not the same as running; traffic flow is interrupted for normal stop on the road due to the impact of the intersection. Therefore, testing on the road in the city must first identify a normal stop or non-normal stop, and then press the principle of detection, so as to properly determine whether the unusual event occurred in traffic. So the difficulty of detection in the road is higher than in the highway. Besides, it requires determining the location of abnormal events.

I: a variety of different sources of data is to establish the appropriate traffic anomaly detection algorithm, such as detector data, probe vehicle observations, passing drivers, patrol reports and other data. The final results are based on the results of the various detection algorithms.

II: a historical data database is created to store the normal traffic state of each signal period detector. The deviation value between the current detection time real-time data is regarded as the basis for distinguishing abnormal events.

d) urban road traffic unusual event location identification [8].The layout of the detector is set to the location of traffic discrimination, with ① ② detector abnormal events that occur on roads, with ② ③ detector exception occurred in the imports of trail events, with ③ ④ ⑤ crossing detector in the middle of the anomalies.

e) Performance of detection of traffic anomalies and its improvement .The performance of abnormal traffic incident detection system is general evaluated by detection rate, false positive rate and an average of detection time. The basic principles of abnormal event detection algorithm are constantly being studied to improve the performance, such as fault tree analysis, analysis method, and traffic simulation technology and so on.

f) The application of image recognition to detect traffic anomalies.With the continual emergence of high-tech achievements, video camera and image processing technology continues to improve, which will soon be applied to the detection of traffic anomalies.

Ttraffic conditions up taken by the digital camera on the way is first sent to the emergency tests equipment, and then test the image is transmitted through the information transmission line to the control center. Image device in the control center records image, and turns into a control command by processing unit. The information transmission system sends control instruction information to the emergency site to display control instructions.

6.2   Site Management [9]

The main task of standardized on-site traffic management is to accurately evaluate the severity of the incident, determine the appropriate priority, coordinate the use of resources, ensure a clear and smooth communication, and ensure the event handlers effectively, fast and efficiently and so on. But event parties, and other vehicle drivers and passenger safety is the primary purpose of the event site management. Efficient scene of the incident management program must include:

- to identify a site command office
- to appoint a commander;
- to mobilize all personnel relevant to the event;
- to call the emergency vehicles and equipment in phases.

6.3   Event Cleaning

Event clean-up is to move the wrecks to clear debris on the road and other road traffic flow, which affect the normal operation of the barrier to return to normal levels before the incident process.

7   Achievements and Prospects of Urban Intersections Its Research

7.1   Recent Advanced Results [10]

Not long ago, German scientists have developed a preliminary intelligent traffic control system which can mimic biological nervous system to process operational analysis of traffic crossing information to help coordination of road traffic management. For statistics of traffic management, to install sensors in the pavement is now a feasible to calculate street traffic. But sensors built in a street not only cost a lot but can only measure the number of motor vehicles rather than speed information. It is intelligent traffic control system to overcome these shortcomings, which consists of digital cameras, computer and Image processing board. Digital cameras can capture traffic intersections. After computer analysis and evaluation of image information, the results can be directly sent to the Traffic Control Centre; one can also spot treatment, or send direct command of the traffic lights by computer conversion. In software, the German scientists learn from biological screening to determine the visual information to develop a program algorithm. In a certain period of time, the street congestion is calculate according to the number of cars and trucks through the intersection, the speed and the waiting time at a red light.

7.2   The Development of Urban Road ITS in China [11]

The study of advanced traffic management system aims for the development of urban road ITS in China, including the development of urban traffic signal control system, the modern command system , quick incident handling system, the regulations and law enforcement, etc. To achieve this goal, research and development for our country should consider the optimal signal control software systems and the modernization of traffic management control system concerning the mixed traffic, bicycles and pedestrians. Advanced traffic information system should focus on the following points, such as traffic congestion forecast, traffic information service system development, existing traffic units' supplement and improvement, real-time traffic information on travel and short-term forecasts and so on. To achieve this goal, the study should be strengthen, such as the analysis of dynamic traffic models and methods, pattern recognition, processing technology, communication technology between mobile body, the road and other vehicles and so on.

8   Conclusion

Intelligent Transportation Systems is a matter of wide and comprehensive variety of high-tech fields of study. This paper discusses safety measures only from the perspective of a number of road and urban road rather than passengers. To improve the overall level of Intelligent Transportation Systems need coordinated development of various industries to jointly promote the improvement of urban traffic.

References

1. Construction of Digital City Guide. Beijing Hope Electronic Press (2007)
2. National Bureau of Statistics. China Statistical Yearbook. China Statistics Press, Beijing (2008)
3. Yang, D., Wu, J., Zhang, Q.: Intelligent Transportation System and Its Information Model. Journal of Beijing University of Aeronautics and Astronautics 16(3) (June 2006)
4. Yue, X., Ding, Y., Huang, X., Li, X.: Traffic Safety Analysis and Evaluation in Fuzhou. Journal of Fujian Agriculture and Forestry University (7), 435–439 (2006)
5. Ding, W., Liu, K., Yang, W., Wu, X.: Intelligent Transportation and Traffic Safety in China. China Safety Science Report (6), 6–9 (2007)

6. Kim, K.: A Transportation Planning Model for State Highway management: A Decision Support system Methodology to Achieve Sustainable Development. Dissertation for the degree Doctor (2006)
7. Hecht, H.R.: Development of Major Transportation Projects in California. Dissertation for the degree Doctor (2007)
8. Zhao, Y., Da, Q., Yang, Q., Liu, Z., Zhang, W.: Study on Intelligent Traffic Safety System. China Safety Science Journal 11(3) (June 2007)
9. Yang, X., Zeng, S., Hang, M.: China's Real-time Adaptive Urban Traffic Control and Management System. Traffic and Transportation Engineering 1(2) (June 2006)
10. Gao, X.: Cross Century Transportation: Its Goals, Trends and Solutions. Journal of Wuhan Transportation University 23(2) (April 2007)
11. Zhang, Y., Li, X.: Intelligent Transportation System as a Key to Sustainable Development of Urban Traffic. Communication Science and Technology 26(1) (March 2007)

The Research of Network Transmission Error Correction Based on Reed-Solomon Codes

Shiying Xia and Minsheng Tan

* School of Computer Science & Technology, University of South China,
HengYang, Hunan 421001 P.R. China
10826778@qq.com

Abstract. This article introduced a new application of Reed-Solomon code on network transmission. Reed-Solomon code is a kind of Maximum Distance Separable code, it works in Galois Field. For a Galois Field $GF(2^3)$,it can correct 1 block error. The paper gives the example of $GF(2^3)$, from the analysis, C language code is also given. At last, real application scene is also presented. From the analysis, it can be seen that Reed-Solomon code can be well work in network communication.

Keywords: Reed-Solomon, Network communication, Galois Field, correction code.

1 Foreword

With the development of computer network, the network has covered each city, as communication subnet, physical layer and data link layer, the main function is to ensure that the data can be transmitted to the destination address accurately. No matter how reliable the communication system is, it can't be perfect. So it must be considered how to discover and correct the error in signal transmission.

There are two kinds of mechanism, error detecting code and error correcting code in the process of network transmission generally[1].The function of error detecting code is to detect whether a data stream transmission errors, and the error correcting code can not only detect data stream errors, but also correct transmission errors[2].

The common detection methods during parity, the principle is to increase one bit after seven bits' ASCII code,to make the code word "1"into an odd number (odd parity) or even (even parity). After transmission,if one of them (even the odd bit) error, the receiver can detect the error by the same rules.This method is simple and practical, but can only deal with a small amount of random error [3]. In order to detect unexpected bit string error,it can be used by the method of checksum. It takes each byte of data block as a binary integer, modulo 256 addition in the process of sending it. After sending the data block,take the result as a checking byte and send it out. The

---

* Shiying Xia(1982-),female, assistant, master,research interests include computer network and information security, 10826778@qq.com;Min-sheng Tan(1965-),male, Professor of Computer Science and Technology at University of South China, Master Instructor, research interests include computer network and information security.

receiver uses the same addition method during the process of receiving,after the data block plus,comparing the own checksum with received checksum,to discover whether the error is.

In 1950,Hamming studied the theory and method used by redundant data bits to detect and correct error code. According to his theory,it can be added a number of redundant bits to make up code word on the data code; error correction code is to take all legitimate code word in the n-dimensional hypercube vertices as far as possible, making sure the distance between any pair of code words as large as possible.If the Hamming distance between any two code words is d,the error which is less than or equal d-1 bits can be checked out,the error can be corrected which is less than d/2 bits. A native corollary is,for a length of error string, to correct it would be more than double the number of redundant bits to detect it [4].

The two methods of error detection and correction, which one is better, there are no certain standard,the detection mechanism is used in the network usually,such as parity and CRC-32[5].

2   Reed-Solomon Code

Reed-Solomon Code is a kind of Maximum Distance Separable (MDS) code; it is an important subclass of BCH code based on q numeric system,it has powerful error correction and especially in outburst errors. Reed-Solomon Code used in many engineering fields such as data storage system(hard disk drivers、CD、DVD et al), consumptive electronic system(Digital TV、Digital Audio System and Digital Video System), digital communication system (satellite communication、outer space detection、ATSC、DAB、DVB)[6].

Reed-Solomon Code is calculated in Galois Field, we introduce Galois Field as following.

2.1   Galois Field

Galois Field is a close field, the calculate results falls in the field, for example, $GF(2^8)$ means there are 256 elements in the field, two of them are 0 and 1, the others generated by a fontal polynomial P(x). A P(x) in $GF(2^8)$ is:

$$P(x) = x^8 + x^4 + x^3 + x^2 + 1 \tag{2-1}$$

Fontal vector of $GF(2^8)$ is:

$$\alpha = \begin{pmatrix} 0 & 0 & 0 & 0 & 0 & 0 & 1 & 0 \end{pmatrix}$$

Now constructing $GF(2^3)$, lets the fontal polynomial is:

$$P(x) = x^3 + x + 1 \tag{2-2}$$

Suppose α is the root of P(x)=0, that is $\alpha^3+\alpha+1=0$, and then $\alpha^3=\alpha+1$. This is XOR or modulo operation. The elements of GF(2^3) can be calculated as Table 2.1.

Table 2.1. Elements of GF(2^3)

Field elements	Results of modulo operation	Binary value
0	mod($\alpha^{3}+\alpha+1$) =0	000
α^0	mod($\alpha^{3}+\alpha+1$) =1	001
α^1	mod($\alpha^{3}+\alpha+1$) =α	010
α^2	mod($\alpha^{3}+\alpha+1$) =α^2	100
α^3	mod($\alpha^{3}+\alpha+1$) =$\alpha+1$	011
α^4	mod($\alpha^{3}+\alpha+1$) =$\alpha^2+\alpha$	110
α^5	mod($\alpha^{3}+\alpha+1$) =$\alpha^2+\alpha+1$	111
α^6	mod($\alpha^{3}+\alpha+1$) =α^2+1	101
α^7	mod($\alpha^{3}+\alpha+1$) =1	001

From Table 1, we can see α^7 is the same with α^0, and next elements will repeat the elements from α^0 to α^6.

2.2 RS Coding Algorithm

The theory of RS Coding is for a message M,choosing proper error correction code R, the length is t, then satisfy M×2^t+R can be divided exactly by Generation Polynomial G(x). If the receiver found the message can be divided exactly by G(x), and the quotient falls in expectation field, it can be consider that transmission is correct. RS Coding is a process of getting R.

In the field GF(2^m), symbol (n, k)RS described as follows:

m   size of symbol block, m=8 means the symbol is formed with 8 bit
n   length of symbol block
k   message length in symbol block
K=n-k=2t   symbol numbers of error correction code
t   error numbers of it can correct

Generation polynomial of RS commonly is:

$$G(x) = \prod_{i=0}^{K-1} (x - \alpha^{K_0+i})$$ (2-3)

Usually, K_0 can be chosen as 0 or 1.

3   Algorithm Realization

3.1   Coding

In the real engineering, GF(2^8) is chosen, but the algorithm using GF(2^3) will be took as an example to achieve the process of encoding and decoding here.Suppose (6,4)RS has 4 symbols: m_3, m_2, m_1 and m_0, symbol polynomial is $M(x)$, correction code Generation Polynomial is $G(x)$, residual polynomial is $R(x)$, correction code symbol is Q_1 and Q_0.

$$M(x) = m_3 x^3 + m_2 x^2 + m_1 x + m_0 \tag{3-1}$$

For (6,4)RS, K is 2, t=1, let $K_0=1$, and $G(x)$ is:

$$G(x) = \prod_{i=0}^{K-1} (x - \alpha^{K_0+i}) = (x - \alpha)(x - \alpha^2) \tag{3-2}$$

From the Equation 3-1 and 3-2, we can get Equation 3-3:

$$m_3 x^5 + m_2 x^4 + m_1 x^3 + m_0 x^2 + Q_1 x + Q_0 = (x - \alpha)(x - \alpha^2) \tag{3-3}$$

Let x= α and x= α^2, from the Equation 3-3, we can get equations:

$$\begin{cases} m_3 \alpha^5 + m_2 \alpha^4 + m_1 \alpha^3 + m_0 \alpha^2 + Q_1 \alpha + Q_0 = 0 \\ m_3 (\alpha^2)^5 + m_2 (\alpha^2)^4 + m_1 (\alpha^2)^3 + m_0 (\alpha^2)^2 + Q_1 \alpha^2 + Q_0 = 0 \end{cases} \tag{3-4}$$

Solute Q_1 and Q_0 from the Equation 3-4, we can get correction symbol.

$$\begin{cases} Q_1 = m_3 \alpha^5 + m_2 \alpha^5 + m_1 \alpha^0 + m_0 \alpha^4 \\ Q_0 = m_3 \alpha + m_2 \alpha^3 + m_1 \alpha^0 + m_0 \alpha^3 \end{cases} \tag{3-5}$$

3.2   Correction Process

Correction algorithm has three parts: (1) calculating syndrome,(2) calculating the positions of error and (3) calculating error values. We an use such syndromes based on $GF(2^3)$ above:

$$\begin{cases} s_0 = m_3 \alpha^5 + m_2 \alpha^4 + m_1 \alpha^3 + m_0 \alpha^2 + Q_1 \alpha + Q_0 \\ s_1 = m_3 (\alpha^2)^5 + m_2 (\alpha^2)^4 + m_1 (\alpha^2)^3 + m_0 (\alpha^2)^2 + Q_1 \alpha^2 + Q_0 \end{cases} \tag{3-6}$$

Two correction symbols can only correct one symbol error, use the values of Equation 3-5 to displace Q_1 and Q_0 in Equation 3-6, if s_0 and s_1 are both zero, it can be considered that there are no error or more than one error. Regarding α^i as the i-th block, if i-th block is error, suppose that m_i become m_i+m in transmission process, then displace m_i with m_i+m in Equation 3-6, we can get:

$$\begin{cases} s_0 = m \alpha^i \\ s_1 = m \alpha^{2i} \end{cases} \tag{3-7}$$

s_0 and s_1 are not zeros, so $\alpha^i = s_1/s_0$, we can get the value of i. Replacing i with the value in Equation 3-7,the value of m can be calculated. The goal of correction is to get correct the error symbol, so after receive the error symbols, we can add the value of m to i-th block to get correct value m_i.

3.3   Code Realization

This is a realization of $GF(2^3)$ Reed-Solomon by C Language.

For getting Table 2.1, we can create Discrete Logarithm array gflog[] and Discrete Inverse Logarithm array gfilog[] first[7].

Logarithm array of $GF(2^3)$ illustrated as Table 3.1.

Table 3.1. Logarithm array in $GF(2^3)$

i	0	1	2	3	4	5	6	7
gflog[i]	—	0	1	3	2	6	4	5
gfilog[i]	1	2	4	3	6	7	5	—

Obviously, gflog[gfilog[i]]=i and gfilog[gflog[i]]=i.

Multiplication and division operations can be done closely in Galois Field from the two arrays, addition and subtraction operations can be done by XOR operation.

This is the examples:

$3 \times 7 = $ *gfi* log[*gf* log[3]+ *gf* log[7]] = *gfi* log[3 + 5] = *gfi* log[8 mod 7] = 2

$3 \div 7 = $ *gfi* log[*gf* log[3]− *gf* log[7]] = *gfi* log[3 − 5] = *gfi* log[5] = 7

C Language code of multiplication and division is illustrated as List 3.1.

```
int Gfmul(int m, int exp)
{    int biexp, result;
        biexp = gfilog[exp mod 7];
 result = gfilog[(gflog[m]+gflog[biexp]) mod 7];
        return result;}
int Gfdiv(int dividend, int divisor)
{   return gfilog[(gflog[dividend]-gflog[divisor]) mod
7];}
```

List 3.1 Multiplication and Division in $GF(2^3)$

The code of RS Coding is illustrated in List 3.2.

```
int Q1(int m3, int m2, int m1, int m0)
{ return Gfmul(m3, 5)+ Gfmul(m2, 5)+ Gfmul(m1, 0)+
Gfmul(m0, 4);    }

int Q0(int m3, int m2, int m1, int m0)
{    return Gfmul(m3, 1)+ Gfmul(m2, 3)+ Gfmul(m1, 0)+
Gfmul(m0, 3);    }
```

List 3.2 Coding in GF(2^3)

To get syndromes, we use the code like List 3.3.

```
int S0(int m3, int m2, int m1, int m0)
{    return Gfmul(m3, 5)+ Gfmul(m2, 4)+ Gfmul(m1, 3)+
Gfmul(m0, 2)+Gfmul(Q1,1)+Q0;    }

int S1(int m3, int m2, int m1, int m0)
{    return Gfmul(m3, 10)+ Gfmul(m2, 8)+ Gfmul(m1, 6)+
Gfmul(m0, 4)+Gfmul(Q1,2)+Q0;    }
```

List 3.3 code of syndromes in GF(2^3)

When correcting,,the value of i can be get by the code as List 3.4 illustrated, if no error detected, the value is 100, this is the infinite value in GF(2^3) theoretical.

```
int geti(int s0, int s1)
{     if(s0!=0 &&s1!=0)
          {    return gflog[Gfdiv[s1,s0] mod 7];    }
      else
          {    return 100;    }
}
```

List 3.4 code of calculating the position i
It is easy to correct the error by i, the code is not illustrated here.

4   Analysis and Conclusions

Reed-Solomon coding which uses the matrix methods at fault-tolerant and error correction,is a wide range of technology in respect of RAID in current days .In the application of network, the data of network transmission is byte stream or bit stream, so it is not suitable at matrix form, it can be the basic RS algorithm. In real application, RS(255,223) can be used with 8-bit block size. Table 4.1 shows some applications of different RS algorithm at different rates [8].

Table 4.1. Some typical applications for RS

Code	Data rate
RS(255,251)	12Mbps
RS(255,239)	2.7Mbps
RS(255,223)	1.1Mbps

References

1. Wang, D.: Basic Network Engineering, pp. 31–95. Electronic Industry Press (2006)
2. http://www.51kaifa.com/html/zxyd/200511/read_z-75-441.htm
3. Tanenbaum, A.S.: Computer Network, pp. 46–205. Tsinghua University Press (2004) Translated by Pan, A.M.:
4. http://baike.baidu.com/view/2094490.htm
5. Lei, Z.: Computer Network, pp. 115–180. Machinery Industry Press (2010)
6. McEliece, R.J.: Information Theory and Coding Theory, pp. 73–95. Electronic Industry Press (2004) Translated by Li, D., Yin, Y., Luo, Y.:
7. Plank, J.S.: A Tutorial on Reed-Solomon Coding for Fault-Tolerance in RAID-like System. Software-practice and Experience 27(9), 995–1012 (1997)
8. http://www.cs.cmu.edu/afs/cs/project/pscicoguyb/realworld/www/reedsolomon/reed_solomon_codes.html

References

Technology of Pipa Panel Texture and Contour Extraction Based on Improved GVF Snake Algorithm

Ying-lai Huang[1,*], Yi-xing Liu[2], Zhen-bo Liu[2], Jjian-min Su[1], and Xiao-li Wang[1]

[1] Information and Computer Engineering College,
Northeast Forestry University, Harbin, 150040, China
[2] Material Science and Engineering College,
Northeast Forestry University, Harbin, 150040, China
nefuhyl@163.com

Abstract. Panel is an important component of pipa and it has an extremely relationship with the voice actor of musical instruments. The warts of textures, width of annual rings, presence of scars and cracks are all part of the main parameters when selecting panels. This paper uses a GVF Snake model based on the ant colony algorithm to improve image analysis of pipa panel, overcomes the external force field faults in the original GVF Snake model and finally obtains determined texture contours which is closer to the target object , the algorithm not only reduces the iterations, it also improves the calculating speed and expands the capturing ranges at the same time. It provides accurate data for realizing the scientific and automatic selection of pipa panels.

Keywords: Texture identification, Ant Colony Algorithm, GVF Snake model, Greedy algorithm, Contour extraction.

1 Introduction

Panel which has an intimate relationship with the voice quality is the important component of Pipa[2], and it is general requirement that panel should be made of Paulownia which has straight veins, proper width of annual ring, no scar and crack[3]. The image analyses of Pipa's panel plays an important role in automated production for national musical instruments. However, paulownia has a soft texture, low dense and little texture chromatism, which makes the image segmentation become a difficulty. Image segmentation is that first step of image analysis, the accuracy and precision of segmentation will directly affect later analysis results, and a good segmentation algorithm is the important guarantee of accurate analysis.

The late eighties, Kass proposed the Sanke model which was widely applied in image segmentation and object recognition fields [4]. Snake model have the rapid convergence and the accuracy which can achieve the sub-pixel advantages. But it has many shortcomings such as the dependent on the selection of the initial curve, easy to

* National natural science foundation of China (30871974) 、Natural science foundation of Heilongjiang Province (QC2009C105)、Supported by the Fundamental Research Funds for the Central Universities (09034).

J. Luo (Ed.): Soft Computing in Information Communication Technology, AISC 158, pp. 183–188.
springerlink.com   © Springer-Verlag Berlin Heidelberg 2012

local convergence and poor results on the concave contour detection. 1997, Xu proposed the Gradient Vector Flow (GVF) which effectively solved the general problem of concave contour detection and the determination of the initial contour made the Snake model further improve. But the model has the large computation, many iterations, slow rapid and smaller capture range shortcomings. In order to overcome the defaults above, we proposed the ant colony optimization to improve the GVF Sanke model for image segmentation.

2   Principle of Ant Colony Algorithm

Ant Colony Algorithm, 1992 M.Dorigo and V.maniezzo who were affected by the route choice in the process of ant foraging proposed a bionic algorithm[5]. Through the observation, ants can find the optimal path when searching food sources by information exchanges and mutual cooperation behaviors between ant colonies, they can also quickly avoid obstacles and find the optimal path again although the former path is blocked. Each ant would release a pheromone which was volatile as time continues in the process of random walk. If there are more ants which have the ability to sense this pheromone, they will select the stronger pheromone path with the greater probability, so the number of the ants in this path would increase. This forms a positive feedback process.

Mathematical description of the ant colony algorithm:

There is a given image, each pixel is considered as an ant in the image and each ant is a two-dimensional vector which is characterized by intensity and gradient of image.

$$X = \{X \mid X_i = (x_{i1}, x_{i2}), i = 1, 2, \ldots, N, N = m \times n\} \tag{1}$$

First, we initialize the algorithm and set the pheromone of each path 0, that is $\tau_{ij}(0) = 0$, then we set the cluster radius r and the statistical error ε. The formula of calculating the pheromone concentration of each path is:

$$\tau_{ij}(s) = \begin{cases} 1 & d_{ij} \le r \\ 0 & d_{ij} > r \end{cases} \tag{2}$$

η_{ij} is guidance function which is reciprocal of d_{ij}, d_{ij} is the weighted Euclidean distance between X_i and X_j. The formula of the guidance function is:

$$\eta_{ij} = \frac{1}{d_{ij}} = \frac{r}{\sqrt{\sum_{k=1}^{m} p_k (x_{ik} - c_{jk})^2}} \tag{3}$$

There c_{jk} is the cluster center, r is cluster radius. The distance between pixel and c_{jk} is smaller, the value of the guidance function is larger and the probability of changing c_{jk} is larger; the distance between pixel and c_{jk} is larger, the value of the guidance function is smaller and the probability of changing c_{jk} is smaller.

$\tau_{ij}^{\alpha}(s)\eta_{ij}^{\beta}(s)$ is the attract between X_i and X_j. If $P_{ij} > P_0$, we merge X_i into the neighborhood of X_j.

$$C_j = \{x_k | d_{ij} \le r, k = 1,2,...J\} \tag{4}$$

C_j is the data collection, which were integrated into the neighborhood of X_j. The ideal cluster center which is equivalent to the food source is:

$$O_j = \frac{1}{J}\sum_{k=1}^{J} X_k \qquad X_k \in C_j \tag{5}$$

There co is the number of cluster centers.

3   Principle of Improved GVF Snake Model

The basic idea of Snake model is firstly set a closed curve ($v(s) = (x(s), y(s))$, $s \in [0,1]$) around the target of the image by human's recognition ability. Then make the closed curve move towards to target location under the combined action of internal force and external force. as well as with constantly updating the curve's energy through an energy function acted on each control points and make the points move toward to its value reduction area. Finally, when the energy function can be no longer reduced, it reached the target outline, at the same time the carve has minimum energy.

3.1   Traditional GVF Snake Algorithm

In 1997,xu Chenyang and Prince put forward the GVF Snake algorithm based on traditional Snake algorithm[6].Mainly for improvement of external force field, it successfully solved the detection of concave contour and determination of initial contour. The GVF Snake algorithm is to use the Gradient Vector Flow (GVF) as new external force, drawn Gradient Vector Flow active contour model (GVF Snake).

The GVF vector external force field defined by FGVF should be satisfied the following minimum of energy function:

$$F_{GVF} = w(x, y) = \{u(x, y), v(x, y)\} \tag{6}$$

$$E = \iint \mu(u_x^2 + u_y^2 + v_x^2 + v_y^2) + |\nabla f|^2 |w - \nabla f|^2 dxdy \tag{7}$$

The f(x,y) is the edge image received from the gray image I(x,y), ux,uy,vx,vy are the first-order Partial derivatives of u,v to x,y respectively. ∇f is the gradient of f(x,y), μ is the adjustment parameters. If the value of $|\nabla f|$ is larger, the energy E is mainly controlled by the $|\nabla f|^2 |w - \nabla f|^2$. When the energy E is acquired the minimum, w= ∇f ; If the value of $|\nabla f|$ is smaller, the energy E is mainly controlled by the $\mu(u_x^2 + u_y^2 + v_x^2 + v_y^2)$ .when the minimum energy

E acquired, the changes of gradient flow W is relatively flat along all directions, thus the action range of gradient vector can be spreading to the relatively flat region in the image.

Euler equations make the above-mentioned energy function to minimize:

$$\mu\nabla^2 u - (u - f_x)(f_x^2 + f_y^2) = 0$$
$$\mu\nabla^2 v - (v - f_y)(f_x^2 + f_y^2) = 0 \qquad (8)$$

Among them, ∇^2 is the gradient operator, fx, fy are the partial derivatives of x,y to edge image f(x,y). After discretization of the above equation, we obtain:

$$u_{t+1} = u + \mu\nabla^2 u_t - (u - f_x)(f_x^2 + f_y^2)$$
$$v_{t+1} = v + \mu\nabla^2 v_t - (v - f_y)(f_x^2 + f_y^2) \qquad (9)$$

Among them, The numbers of iterations is t.

3.2   Improved GVF Snake Algorithm

Paulownia has the soft material and the smaller texture chromatics, using the GVF Snake model directly to extract the texture is not effective. To solve this problem, this article targets initial texture of the pipa panel images using ant colony algorithm and the advantages of robustness, then breads up the panel images by improved GVF Snake model, the full target object is got ultimately. It is closer to the target to determine the initially texture contours using Ant colony algorithm, consequently reduce iteration times to improve computing speed, and to expand the capture range.

The steps of improved GVF Snake algorithm are as follows:

Step 1: Initialized the parameters of the ant colony algorithm: $\alpha=2$, $\beta=3$, $r=2$。 The numbers of iterations is n, n=0;

Step 2: Based on the formula of ant colony algorithm to calculate the maximum concentration of information among the nodes;

Step 3: n=n+1 ;

Step 4: If n is not in the intended final value, then go to Step 2; otherwise, go to Step 5;

Step 5: Calculate the numbers of cluster centers co and cluster centers, as the initial control points of optimal GVF snake model;

Step 6: Instead of external energy (external force) ,we use the modular of gradient vector field calculated for each control point w_i. Calculate the gradient vector field of edge image f (x, y) , $\mu=0.2$ is the adjustment parameters. The numbers of iterations is t, t = 0;

Step 7: Calculate the internal energy (the elastic energy and bending energy) by 8-neighborhood greedy algorithm:

$$E_{in}v(s) = \delta\left|\frac{\partial}{\partial s}\vec{v}\right|^2 + \zeta\left|\frac{\partial^2}{\partial^2 s}\vec{v}\right|^2 \xi$$

Obtained by experiment:δ=0.1,ζ=0.3

Step 8: t=t+1;

Step 9:if t does not reach the predetermined value, then go to Step 7; otherwise, go to Step 10;

Step 10: synthesize all energy to make the changes in curve, and get the ultimate convergence results of goal contour;

4   Experimental and Simulation Results

In order to validate the effectiveness of the improved GVF Snake algorithm for the pipa panel image segmentation, we can use MATLAB software to do the simulation experiment. Compare the proposed algorithm with the traditional GVF Snake model algorithm as well as canny operator and LoG operator, from the following simulation results, we can see that it obtain the smoother textures and edges of panels, better continuity, no false edges, and less time-consuming when we use the proposed algorithm to detect the textures and edges.

Table 1. The comparison of the traditional GVF Snake model and the improved GVF Snake model.

Size of the mage	GVF Snake		Improved GVF Snake	
	Number of the iterations (t)	Time-Consuming (s)	Number of the iterations(t)	Time-consuming (s)
420×315	130	12.36	35	3.52

Note: time-consuming means that the time of processing the images on the same computer.

Fig. 1. Original image of the pipa panel

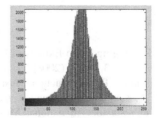

Fig. 2. Image histogram of the pipa panel

Fig. 3. Image detected by the canny operator   **Fig. 4.** Image detected by the LoG operator

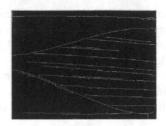

Fig. 5. Image detected by the traditional GVF   **Fig. 6.** Image detected by the improved
Snake algorithm   algorithm

5   Summary

The proposed algorithm improved the defects of the traditional GVF Snake algorithm that the big numbers of external force field iterations and the powerless to approximate complex texture contour. Combine with the gray, gradient and other information of the pipa panel image, this paper put forward an improved GVF algorithm to detect the texture and edge. The above experimental results show that the improved GVF Snake algorithm is more efficient than other algorithms, and it can eliminate the impact of noise better and get accurate and continuous images of texture and edge, and it has a strong practical.

References

1. Sheng, X.: Singing bump-Explore of Ethnic Musical Instruments Pipa Production Material. Chinese Timber 3, 19–21 (2008)
2. Gao, Z., Lin, S.: Pipa Production. Musical Instrument 6, 12–14 (1981)
3. Liu, Z.: Present research situation and the development tendency of acoustic quality of lumber in resonance in national Musical Instruments. Forestry Science 46(8), 151–156 (2010)
4. Kass, M., Withkin, A., Terzopoulos, D.: Snakes Active Contour Models. International Journal of Computer Vision 1(4), 321–331 (1987)
5. Dorigo, M., Maniezzo, V., Colorni, A.: Ant system:optimization by a colony of cooperating agents. IEEE Trans. on Systems Man and Cybernetics-Part B 26(1), 29–41 (1996)
6. Xu, C.-Y., Prince, J.L.: Snakes, shapes and gradient vector flow. IEEE Transactions on Image Processing 7(3), 359–369 (1998)

Design of Digital Servos Control System Based on ATmega128

Xiaoguang Zhu[*], Qingyao Han, and Zhangqi Wang

School of Mechanical Engineering
North China Electric Power University
071003 Baoding, China
{zhuxiaog,ncepuzxg}@sina.com, qingyao_han@163.com

Abstract. Digital Servos with good performance and high accurate control via digital signal is the future development direction of robot. The form, the function and the principle of the digital servos are expatiated in the paper. System design section is composed of hardware design and software design. Each part of the hardware design complies with the modular design principles, every block is given in detail. And the software design mainly introduces data packet format of AX-12 and how to achieve half-duplex asynchronous serial communication between controller and AX-12. The application shows this digital servos controller can control AX-12, and it has been applied to obstacle avoidance for wheeled mobile robot.

Keywords: Digital Servos, Half Duplex Asynchronous Serial Communication, AX-12, ATmega128.

1   Forewords

Servo is a position servo drive. It takes a certain degree of control signals and outputs certain angle. So it applies to the control systems which need keep changing the angle and maintain it, such as the arms and legs of humanoid robots, the direction control of model plane and model car. Servo is divided into: analog servo and digital servo. Analog servo need to be sent PWM signal continuously to keep the specified location or rotate on a given speed. Digital servo only needs to be sent a signal to do that. In addition, the biggest difference between the analog servo and digital servo is the way of input signal.[1] Compared with traditional Servo 50 pulses / sec PWM signal demodulation method, digital Servo used the signal preprocessing method and put the frequency up to 300 pulses / sec. By the high frequency, digital servo acts more precise, more responsive, acceleration and deceleration more smoothly.

2   Digital Servo Construction and Working Principle

This paper use Dynamixel Series Digital Servo Digital Servo AX-12. It is a Intelligent digital servo. Servo is composed of microprocessor, a precise DC motor, a reduction gear, power and position sensor. The internal schematic diagram is shown in Figure 1.

---

[*] Zhu Xiaoguang, (1978-), Male, PhD Student, engaged in the research of robotics.

J. Luo (Ed.): Soft Computing in Information Communication Technology, AISC 158, pp. 189–194.
Springerlink.com   © Springer-Verlag Berlin Heidelberg 2012

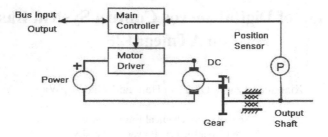

Fig. 1. Internal schematic diagram of digital servo

Its main features are: each Servo has a unique ID and be controlled by Daisy bus; Data transfer by half-duplex serial communication, the fastest transmission speed up to 1Mbps; Not only has a position feedback system, but also the speed feedback, temperature feedback and torque feedback. When used as a joint motor, can rotate 0 ~ 300 °; When used as a normal motor, can choose 360 ° continuous rotation mode. It has fast response and rotates 60 ° in just 0.269s. AX-12 can provide up to 16kg • cm of torque in the working voltage of 10V. Based on these advantages, AX-12 is a truly dedicated robot servo.

3   Control System Hardware Circuit Design

AX-12 servo is not controlled with PWM like normal motor. Its control signal is digital signal, use TTL-Daisy bus connection between main controller and servo and half-duplex asynchronous serial communication protocol[2]. Host controller controls the servo by sending and receiving data packets. There are two packets: one is the instruction packet, which is a control instruction from main controller to the servo; the other is the state package, which is back to the main controller from the servo. If the main controller sends the servo instruction with ID number N, only the corresponding servo will act or return the status. Control principle is shown in Figure 2.

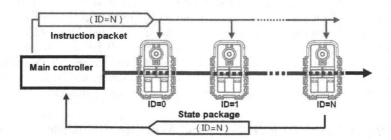

Fig. 2. Schematic of control system control

3.1 MCU and Its Peripheral Circuits

The servo control system uses ATMEL's AVR series 8-bit microcomputer as the master chip, The microcontroller is a low-power CMOS 8-bit microcontroller based on enhanced AVR RISC structure. It has 128KB Flash, 53 programmable I / O ports used for various servo controls. The external circuit is shown in Figure 3, in which the jumper JP1 can be used to select ADC voltage reference.

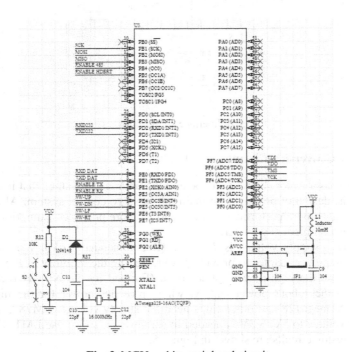

Fig. 3. MCU and its peripheral circuit

3.2 Interface Circuits

Interface circuits include RS232 serial communication interface circuits, JTAG emulation interface, and ISP download interface. JTAG and ISP interfaces are used for downloading and debugging program to ATmega128[3], RS232 serial port is used for communication between PC and controller. The interface circuits are shown in Figure 4.

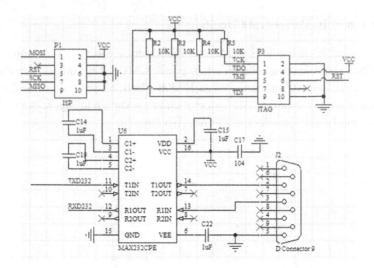

Fig. 4. Interface circuit diagram

3.3   Power Conversion Circuits

Digital servo has a major shortcoming behind the obvious advantage that it need more power. So the design of power is the main job in servo control system. Although the digital servo gets more performance at the cost of more power consumption, now the battery is not a problem any longer.

AX-12 servo supply voltage is 7 ~ 10V (Recommended 9.6V). We can use the corresponding power adapter to provide power. Taking the mobility of the robot into account, we use the high-power battery pack (Formed by the Ni-MH rechargeable battery). It is the robot's power supply, which provides power for the high power motor. For ATmega128 and other chips need 5V supply voltage, LM7805 is used to convert the voltage to 5V. 3.3V is needed in somewhere, so we use LM1117 to do it. Power conversion circuits are shown in Figure 5:

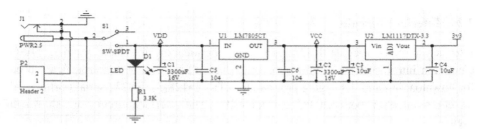

Fig. 5. Power conversion circuit

3.4   Servo Control Circuits

Servo adopts half-duplex asynchronous serial communication protocol.[4] Therefore to achieve communication between the controller and servo, controller must transform the serial signal into Half-duplex. That is to say, controller only receives or sends data alternately, not at the same time. Servo control circuits are shown in Figure 6.

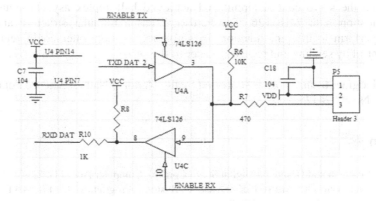

Fig. 6. Servo control circuit

AX-12 servo interface has three wires. Controller provides power to servo with the first and second wires. The third wire is the data line which be used to accept commands and send status. The direction of half-duplex asynchronous serial communication is decided by the ports PE2 and PE3 of ATmega128. When PE2 (ENABLE TX) is high, that is the 1st pin of 74HC126 is high, the 3rd pin is enabled, that is the TXD DAT is the output signal to the servo; When PE3 (ENABLE RX) is high, that is the 10th pin of 74HC126 is high, the 8th pin is enabled, that is the servo return the input signal to the RXD DAT. Therefore, we achieve the half-duplex asynchronous serial communication with 74HC126 four buses buffer, achieve data communication between the controller and the servo with one data line.

4   Control System Software Design

The communication form between controller and servo is asynchronous serial communication which is organized in 9-bits and 8-bist for data 1-bit for stop sign without parity. Controller sends commands to the servos in the form of packets and then the servo returns the corresponding status packet to controller to feed back the servo status after receiving instruction packet.

Servo adopts half-duplex asynchronous communication protocol. That is to say it can not receive and sand data at the same time. In the servo control circuits, we use bus buffer 74HC126 to achieve half-duplex asynchronous communication. The direction of the data transmission is decided by the pins PE2 and PE3. When PE2 is high, PE3 is low, data is sent. Otherwise the data is received. The system uses C language to write software by modular, including communication module and control module etc.

5   Conclusions and Summary

Digital servo has higher accuracy, fewer non-reaction zone, more accurate positioning, faster control response and stronger acceleration. This makes the digital servo the first choice for the robot game. In this paper, the design of the digital servo control system is based on ATmega128. We designed the power circuit; interface circuit and the servo control circuit and achieved half-duplex asynchronous serial communication with 74HC126. The hardware has a modular structure and each module perform a different function. In software, the controller could realize the servo control by sending and receiving order to the servo.

Acknowledgment. This work is supported by the National Natural Science Foundation of China. No.60974125.

References

1. Li, B.: Research and Design on Digital Servo Control. Computer Age (11), 22–24 (2008)
2. 8-bit Microcontroller with 16K Bytes In-System Programmable Flash-ATmega128 Datasheet. Atmel Corporation (2009)
3. Ren, Z.: Research on Servo Drive Circuit Based on AVR Microcontroller. Techniques of Automation and Applications (6), 89–91 (2008)
4. Li, S., Lei, J.: Digital Servo Control System Design and Implementation Based on DSP. Computer Measurement & Control 17(3), 484–486 (2009)
5. Dynamixel AX-12 User's Manual. Robotis Inc. (2006)

Feature Fusion Using Three-Dimensional Canonical Correlation Analysis for Image Recognition

Xiaogang Gong[1], Jiliu Zhou[2], Huilin Wu[1], Gang Lei[2], and Xiaohua Li[2]

[1] School of Electronics and Information Engineering, Sichuan University, Chengdu, China
[2] College of Computer Science, Sichuan University, Chengdu, China
{scugxg,cdleigang}@gmail.com, {zhoujl,lxhw}@scu.edu.cn,
whl0912@hotmail.com

Abstract. A new feature fusion method, namely three-dimensional canonical component analysis (TCCA), is proposed in this paper. It is an extension of traditional canonical correlation analysis (CCA) and two-dimensional canonical correlation analysis (2DCCA). The method can directly find the relations between two sets of three-dimensional data without reshaping the data into matrices or vectors, and dramatically reduces the computational complexity. To evaluate the algorithm, we are using Gabor wavelet to generate the three-dimensional data, and fusing them at the feature level by TCCA. Some experiments on ORL database and compared with other methods, the results show that the TCCA not only the computing complexity is lower, the recognition performance is better, but also suitable for data fusion.

Keywords: canonical correlation analysis, feature fusion, three-dimensional.

1 Introduction

Data fusion technology is one of the emerging technologies of data processing. It is a multilevel, multifaceted process dealing with the automatic detection, association, correlation, estimation and combination of data from single and multiple sources [1]. Fusion processes are often categorized as low (pixel fusion), intermediate (feature fusion) or high level fusion (decision fusion) depending on the processing stage at which fusion takes place. Feature fusion plays an important role in the process data fusion. The advantage of feature fusion is obvious. As a matter of fact, the different feature vector extract from the same pattern always reflects the different feature of patterns. By combining these different features, it not only obtains more mutual information of one pattern but also eliminates redundant information.

Recently, there has been increased interest in the use of Canonical Component Analysis (CCA) for feature fusion in various pattern recognition applications [2-5].

CCA is a way of measuring the linear relationship between two multidimensional variables. In this analysis, one finds a linear combination of the first set of variables and a linear combination of the second set of variables such that they both have unit variance and the Pearson correlation coefficient between them is maximum. Thus the obtained pair of linear combinations are called the first canonical variables and the correlation is called the first canonical correlation. This process is repeated to obtain

J. Luo (Ed.): Soft Computing in Information Communication Technology, AISC 158, pp. 195–201.
Springerlink.com

the second, third, ... canonical variables and correlations with the additional restriction that the pair of linear combinations currently being computed are uncorrelated with all the previously obtained pairs. Use of few canonical variables to perform data analysis is in fact a general way of dimension reduction. An important property of CCA is that they are invariant with respect to affine transformations of the variables.

But the CCA method faces some problems: small sample size(SSS) problem, where the dimensionality is higher than the number of vectors in the training set, singularity problem of the covariance matrix in the case of the high-dimensional space, and structural problem that samples vector operation loses spatial information. To address these problems, the two-dimensional Canonical Component Analysis (2DCCA) [6, 7], has been proposed.

However in the real world, most of the source data, such as video sequence, objects structure is the three -dimensional. So, it is essential to study the three dimensional version pattern analysis algorithms, which can be used on three dimensional objects directly. Inspired of [6], we propose a method namely three-dimensional Canonical Component Analysis (TCCA) and apply it to feature fusion for image recognition.

The rest paper is organized as follows: Section 2 introduces the traditional CCA briefly. Section 3 describes the proposed algorithm TCCA in detail. Section 4 gives the experiments on ORL database to evaluate algorithm, and Section 5 is the conclusions.

2   Canonical Component Analysis

Given two random variables $\{x_t \in R^m, t=1,2,...,N\}$ and $\{y_t \in R^m, t=1,2,...,N\}$, $\overline{x}$ and $\overline{y}$ denotes the mean value respectively, centered source data with $\tilde{x}_t = x_t - \overline{x}$ and $\tilde{y}_t = y_t - \overline{y}$ can obtain zero mean dataset $\{\tilde{x}_t \in R^m, t=1,2,...,N\}$ and $\{\tilde{y}_t \in R^m, t=1,2,...,N\}$. In computer vision, $\tilde{x}$ and $\tilde{y}$ can be seen as two views of one observation: Such as the images collected from different sensors, lighting conditions or different transformations (spacial or spectral). So features extracted from two views are correlated. Feature fusion using CCA is that to find a pair of projections u_1 and v_1, to make the canonical variables $\tilde{x}_1^* = u_1^T \tilde{x}$ and $\tilde{y}_1^* = v_1^T \tilde{y}$ have maximum correlation coefficient ρ_1. See (1).

$$\rho_1 = \frac{\text{cov}(u_1^T \tilde{x}, v_1^T \tilde{y})}{\sqrt{\text{var}(u_1^T \tilde{x}) \, \text{var}(v_1^T \tilde{y})}} \tag{1}$$

Then finding the second pair of canonical variables $\tilde{x}_2^*$ and $\tilde{y}_2^*$, which are uncorrelated with their first canonical variables $\tilde{x}_1^*$ and $\tilde{y}_1^*$, We only need to analyze a few pairs of canonical variables.

3   Three-Dimensional Canonical Component Analysis

Traditional CCA usually converts the matrix to vector first, such processing changes the space structure of source data and also causes dimension disaster. To overcome those problems, two-dimensional canonical correlation analysis (2DCCA) [6] was proposed which processing matrix directly. As much source data is three-dimensional, we expand CCA and 2DCCA to a three-dimensional version as follows:

Let $\mathcal{A} \in \mathbf{R}^{I_x \times I_y \times I_z}$ denotes a three-dimensional source data, I_x , I_y , I_z are the dimensions of x , y , z directions. $\mathcal{A}_{i,j,k}$ denotes an element of $\mathcal{A}$,where $1 \le i \le I_x$, $1 \le j \le I_y$, $1 \le k \le I_z$. The dot product of two three-dimensional variables is defined as $\langle \mathcal{A}, \mathcal{B} \rangle = \sum_{i,j,k} \mathcal{A}_{i,j,k} \mathcal{B}_{i,j,k}$, The norm of $\mathcal{A}$ is defined by $\| \mathcal{A} \| = \sqrt{\langle \mathcal{A}, \mathcal{A} \rangle}$. As the same as matrix, three-dimensional data in each direction can be flattened into their respective vector space. The x direction flattened matrix is denoted as $\mathcal{A}_{(x)} \in \mathbf{R}^{I_x (I_y \times I_z)}$ I_x and $(I_y \times I_z)$ are the rows and columns respectively of $\mathcal{A}_{(x)}$. The product of $\mathcal{A}_{(x)}$ and matrix U is defined as $U^T \mathcal{A}_{(x)}$ or $\mathcal{A}_{\times x} U$, and they are equal. Fig. 2 shows the data is flattened according x direction.

Consider two three-dimensional source data $\left\{ \mathcal{A}_t \in \mathbf{R}^{I_{x1} \times I_{y1} \times I_{z1}} \right\}_{t=1}^{N}$ and $\left\{ \mathcal{B}_t \in \mathbf{R}^{I_{x2} \times I_{y2} \times I_{z2}} \right\}_{t=1}^{N}$ which are real world of random variables $\mathcal{A}$ and $\mathcal{B}$. The mean of $\mathcal{A}_t$ and $\mathcal{B}_t$ is defined as $M_A = \frac{1}{N} \sum_{t=1}^{N} \mathcal{A}_t$, $M_B = \frac{1}{N} \sum_{t=1}^{N} \mathcal{B}_t$, then centered source data is denoted by $\tilde{\mathcal{A}}_t = \mathcal{A}_t - M_A$, $\tilde{\mathcal{B}}_t = \mathcal{B}_t - M_B$, as the same as traditional CCA, the purpose of TCCA is to find projections u_x, u_y, u_z and v_x, v_y, v_z , that makes correlation between $\mathcal{A}_{\times x} u_{x \times y} u_{y \times z} u_z$ and $\mathcal{B}_{\times x} v_{x \times y} v_{y \times z} v_z$ maximum. we construct objective function as follows:

$$J_1 = \underset{u_x, u_y, u_z, v_x, v_y, v_z}{\arg\max} \mathrm{cov}\left(\mathcal{A}_{\times x} u_{x \times y} u_{y \times z} u_z, \mathcal{B}_{\times x} v_{x \times y} v_{y \times z} v_z \right)$$
$$s.t. \mathrm{var}\left(\mathcal{A}_{\times x} u_{x \times y} u_{y \times z} u_z \right) = 1$$
$$\mathrm{var}\left(\mathcal{B}_{\times x} v_{x \times y} v_{y \times z} v_z \right) = 1 \tag{2}$$

The objective function (2) is non-linear constraints with the high-dimensional non-linear optimization problem. It is difficult to directly find the closed form solution. In this paper, we use alternative numerical iterative method to find the solution. We only discuss the solution of transforms $\boldsymbol{u}_x$ and $\boldsymbol{v}_x$, which belongs to x direction. And the transforms of y, z directions are the same.

Assuming that $\boldsymbol{u}_y, \boldsymbol{v}_y$ and $\boldsymbol{u}_z, \boldsymbol{v}_z$ are fixed. Defining matrices:

$$\Sigma_{AB}^x = \frac{1}{N} \sum_{t=1}^{N} \left(\tilde{\mathcal{A}}_{t \times y} \boldsymbol{u}_{y \times z} \boldsymbol{u}_z \right)_{(x)} \left(\tilde{\mathcal{B}}_{t \times y} \boldsymbol{v}_{y \times z} \boldsymbol{v}_z \right)_{(x)}^T$$

$$\Sigma_{AA}^x = \frac{1}{N} \sum_{t=1}^{N} \left(\tilde{\mathcal{A}}_{t \times y} \boldsymbol{u}_{y \times z} \boldsymbol{u}_z \right)_{(x)} \left(\tilde{\mathcal{A}}_{t \times y} \boldsymbol{v}_{y \times z} \boldsymbol{v}_z \right)_{(x)}^T \qquad (3)$$

$$\Sigma_{BB}^x = \frac{1}{N} \sum_{t=1}^{N} \left(\tilde{\mathcal{B}}_{t \times y} \boldsymbol{u}_{y \times z} \boldsymbol{u}_z \right)_{(x)} \left(\tilde{\mathcal{B}}_{t \times y} \boldsymbol{v}_{y \times z} \boldsymbol{v}_z \right)_{(x)}^T$$

Then $\operatorname{cov}\left(\mathcal{A}_{x \times x} \boldsymbol{u}_{x \times y} \boldsymbol{u}_{y \times z} \boldsymbol{u}_z, \mathcal{B}_{x \times x} \boldsymbol{v}_{x \times y} \boldsymbol{v}_{y \times z} \boldsymbol{v}_z \right)$ can be replaced by $\boldsymbol{u}_x^T \Sigma_{AB}^x \boldsymbol{v}_x$, and (2) can be formulated as the following optimization problem:

$$\begin{array}{c} \underset{\boldsymbol{u}_x, \boldsymbol{v}_x}{\arg\max} \, \boldsymbol{u}_x^T \Sigma_{AB}^x \boldsymbol{v}_x \\ s.t. \;\; \boldsymbol{u}_x^T \Sigma_{AA}^x = 1 \\ \boldsymbol{v}_x^T \Sigma_{BB}^x = 1 \end{array} \qquad (4)$$

This problem can be reduced to a linear equations by writing the Lagrangian of the optimization problem (let λ denotes the Lagrange's multiplier) and setting its gradient to zero. This results in:

$$\begin{aligned} J &= \boldsymbol{u}_x^T \Sigma_{AB}^x \boldsymbol{v}_x + \lambda_{u_x} \left(1 - \boldsymbol{u}_x^T \Sigma_{AA}^x \boldsymbol{u}_x \right) \\ &+ \lambda_{v_x} \left(1 - \boldsymbol{v}_x^T \Sigma_{BB}^x \boldsymbol{v}_x \right) \end{aligned} \qquad (5)$$

Solving $\dfrac{\partial J}{\partial \boldsymbol{u}_x} = 0, \dfrac{\partial J}{\partial \boldsymbol{v}_x} = 0$, then

$$\Sigma_{AB}^x \boldsymbol{v}_x - 2\lambda_{u_x} \Sigma_{AA}^x \boldsymbol{u}_x = 0 \qquad (6)$$

$$\Sigma_{BA}^x \boldsymbol{u}_x - 2\lambda_{v_x} \Sigma_{BB}^x \boldsymbol{v}_x = 0 \qquad (7)$$

Multiplying $\boldsymbol{u}_x^T$ and $\boldsymbol{v}_x^T$ to both sides of (6) and (7) respectively, we can obtain $\boldsymbol{u}_x^T \Sigma_{AB}^x \boldsymbol{v}_x = 2\lambda_{u_x}$ and $\boldsymbol{v}_x^T \Sigma_{BA}^x \boldsymbol{u}_x = 2\lambda_{v_x}$ by using the constraints of (4). Then we have $\lambda = 2\lambda_{u_x} = 2\lambda_{v_x}$. So u_x, v_x can be solved by (8), which is a generalized eigenvalue problem:

$$\begin{bmatrix} 0 & \Sigma_{AB}^x \\ \Sigma_{BA}^x & 0 \end{bmatrix} \begin{bmatrix} u_x \\ v_x \end{bmatrix} = \lambda \begin{bmatrix} \Sigma_{AA}^x & 0 \\ 0 & \Sigma_{BB}^x \end{bmatrix} \begin{bmatrix} u_x \\ v_x \end{bmatrix} \tag{8}$$

With the same method, the transforms u_y, v_y and u_z, v_z can be solved when u_x, v_x, u_z, v_z and u_x, v_x, u_y, v_y are given respectively. We compute the d pairs of transforms each has the maximum correlation (ρ) and consist of projection matrices: $U_x = [u_{x_1}, u_{x_2}, ..., u_{x_d}]$ and $V_x = [v_{x_1}, v_{x_2}, ..., v_{x_d}]$. Similarly, we can get the projection matrices U_y, V_y and U_z, V_z. In our proposed TCCA, the generalized eigenvalue problem for much smaller size matrices, compared to the traditional CCA, which can reduce the computation cost dramatically.

4   Experiments and Results

The three-dimensional data is generated using Gabor filter. The properties of Gabor wavelet are in accordance with the characteristics of human vision for the kernels of Gabor wavelet have the similar structure with the two-dimensional field profiles of the mammalian cortical simple cells. Because of this biological relevance, Gabor wavelet can capture the subtle details of an image, and render salient visual properties such as spatial localization, orientation selectivity and spatial frequency characteristic. All these features are useful for image understanding and recognition. We encode face images using Gabor wavelet from different scale and orientation. Fig. 1 shows a pair of three-dimensional data from Gabor wavelet feature extraction (4 scales, 5 directions and 4 directions, 5 scales for example).

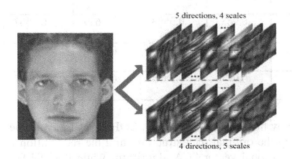

Fig. 1. Illustration of Gabor wavelet feature extraction

Some groups of experiments are designed to testify the effectiveness of the proposed approach on ORL database.

The ORL database includes 400 images with 119*92 dimensions from 40 individuals, each providing 10 different images. The series of 10 images presents variations in facial expression, in facial position (slightly rotated faces) and in some other details like glass/no glass. The Fig. 2 shows a subset of the ORL database.

Fig. 2. Some images belonging to ORL database

To reduce the computation complexity, each image is scaled down to the size of 28*23 pixels. We randomly select 5 to 9 samples from each class for training and remaining samples are used for testing. This process is repeated for 3 times and 15 different training and test sets are created to evaluate TCCA's robust. For ease of representation, the experiment is named as Gm/Pn which means that the m random images are selected for training and the n remaining images for testing. We extract 5 scales, 6 directions and 5 directions, 6 scales of total 30 Gabor filters to consist of a pair of 28*23*30 three-dimensional data. The nearest-neighbor classifier is used for recognition. For comparison, gray feature and fusion Gabor feature are chosen. Three methods: PCA, 2DPCA, 2DLDA also are tested. After 3 times of such experiments the average recognition rate is shown in Table 1.

Table 1. Recognition rate on different methods and different features

Recognition rate	G5/P5	G6/P4	G7/P3	G8/P2	G9/P1
PCA (gray)	0.9150	0.9313	0.9444	0.9708	0.9667
2DPCA (gray)	0.9417	0.9437	0.9635	0.9875	0.9750
2DLDA (Gray)	0.9483	0.9583	0.9639	0.9667	0.9833
PCA (Gabor)	0.9617	0.9708	0.9722	0.9875	0.9833
TCCA (Gabor)	0.9733	0.9875	0.9833	0.9917	0.9999

The results show that using gray feature, 2DPCA as to avoid the image matrix to quantify, can keep the original data structure, and the recognition rate has increased than the one-dimensional vector PCA algorithm, while the 2DLDA as a supervised learning method, which contains the class information has higher recognition rate than the non-supervised learning method 2DPCA. Gabor features have a stronger description ability of image information than the use of gray feature has a higher recognition rate, although this method as a unsupervised learning method, but because of the role of feature fusion, compared with only a single grayscale or Gabor features, the recognition rate of TCCA is highest.

In ORL database, image size is 28*23, after extracting 5 scales, 6 directions Gabor feature, the data is vector into 28*23*30=19320-dimensional feature in PCA, it encounters a 19320*19320 covariance matrix eigenvalue decomposition, that is high computational complexity and time-consuming. The matrix eigenvalue decomposition using TCCA transforms the data to three small matrices: 28*28, 23*23, 30*30, this process effectively reduces the computational complexity, and as the dimension of the covariance matrix is not high, reduces the possibility of singular matrix.

5   Conclusions

In this paper, we proposed a three-dimensional Canonical Component Analysis (TCCA) method and apply it to feature fusion and image recognition. Using of correlation features between two groups of features as effective discriminant information is not only suitable for data fusion, but also eliminates the redundant information within the features.

Besides, the method can effectively preserve the spatial structure of the source data and effectively decrease the possibility of singular problem of high dimensional matrix's eigenvalue decomposition. Be compared with other methods, experiments show that our proposed TCCA has better recognition rate as well as decreases the computation cost.

References

1. Worden, K., Dulieu-Barton, J.: An overview of intelligent fault detection in systems and structures. Structural Health Monitoring 3, 85 (2004)
2. Sun, Q.-S., Zeng, S.-G., Liu, Y., Heng, P.-A., Xia, D.-S.: A new method of feature fusion and its application in image recognition. Pattern Recognition 38, 2437–2448 (2005)
3. Sargin, M.E., Yemez, Y., Erzin, E., Tekalp, A.M.: Audiovisual Synchronization and Fusion Using Canonical Correlation Analysis. IEEE Transactions on Multimedia 9, 1396–1403 (2007)
4. Pengfei, Y., Dan, X., Hao, Z.: Feature level fusion using palmprint and finger geometry based on Canonical Correlation Analysis. In: 3rd International Conference on Advanced Computer Theory and Engineering, pp. 260–264. ASME Press, Chengdu (2010)
5. Jing, X., Li, S., Lan, C., Zhang, D., Yang, J., Liu, Q.: Color image canonical correlation analysis for face feature extraction and recognition. Signal Processing (2011) (in press, corrected proof)
6. Sun Ho, L., Seungjin, C.: Two-Dimensional Canonical Correlation Analysis. IEEE Signal Processing Letters 14, 735–738 (2007)
7. Sun, N., Ji, Z.-H., Zou, C.-R., Zhao, L.: Two-dimensional canonical correlation analysis and its application in small sample size face recognition. Neural Computing & Applications 19, 377–382 (2010)

Research on the Application of Program Design Pattern in Intelligent Optics Channel Analyzer

Yu Su[1], Huiyuan Zhao[1], Binghua Su[1], Xuedan Pei[1], and Guojiang Hu[2]

[1] Zhuhai Campus Beijing Institute of Technology, 519088 Zhuhai, China
[2] Advanced Fiber Resources (Zhuhai) Ltd. 519085 Zhuhai, China

Abstract. In order to provide a necessary basis for the intelligent management and control of network, the design of intelligent optic channel analyzer is developed based on Labview platform and the producer/consumer pattern, which focuses on the problem about surveying correlation parameters of optic signal. The formation and principle of system are introduced at first. Then through analyzing program design patterns, the producer/consumer pattern is adopted and the functions such as data acquisition, data analysis and process, display and so on are implemented by taking advantage of queue function. Finally, its superiority is summarized. The program has characters of good readability, expendability and maintainability.

Keywords: producer/consumer, program design pattern, analyzer, LabVIEW.

1 Introduciton

LabVIEW became quickly a very powerful programming language, having some characteristics which made it unique: simplicity in creating very effective User Interfaces and the G programming mode. Based on this platform, the measurement and control system supported by hardware and interface device can accomplish a series of functions, such as system control, data acquisition, data analysis, display and so on[1][2]. In order to provide a necessary basis for the intelligent management and control of network, the optical signal rate, wavelength, power and SNR need to be measured and monitored online. So the new intelligent optical channel analyzer is designed. Its workflow is firstly to extract a small amount of optical signal power (1%----5%) from the monitored network to enter optical channel analyzer, and secondly to carry out spectral decomposition and intensity measurement, at last to complete data acquisition, real-time processing and report's analysis. Fig .1 shows system principle.

2 Design Patterns

Design patterns represent solutions and techniques that have proved themselves useful time and time again. They typically have evolved through the efforts of many developers, and have been fine-tune for simplicity, maintainability and readability. Using design patters make it easier for others to read and modify code. The common design patterns include UI Event Loop, State Machine, Master/Slave, Produce/Consumer, Queued Message Handle etc[3]-[5]. The Produce/Consumer pattern is really just a subclass of the Master/Slave pattern where the main communicate between the loops is via queue. Fig 2 is main structures of Produce/Consumer Pattern.

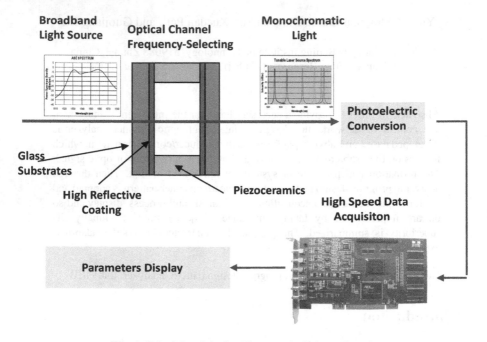

Fig. 1. Principle of the intelligent optical channel analyzer

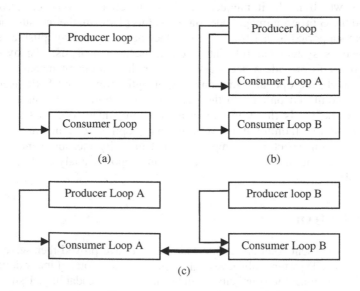

Fig. 2. Structures of Producer/Consumer Pattern

3   Producer/Consumer Loop Design

The whole system mainly consists of scanning interference module, data acquisition module, data analysis module. Its operation mode is successive data acquisition and real-time analysis. In particular, the analyzer collects two way signals (one is scanning voltage signal, the other is optical pulse signal.) via photoelectric conversion and analyses correlation parameters, such as peak value of optical pulse signal, corresponding amplitude of scanning voltage and position of optical pulse signal. Due to designing high-fineness of scanning interference module, the follow-up modules can be simplified and shorten computing time. As a result, the detecting speed is greatly improved.

We need to write an application that accepts data while processing them in the order they were received. So there are two processes. The first process performs data acquisition and the second process takes that data and analyses it for displaying. Because queuing up (producing) the collected data is faster than the actual processing (consuming), the Producer/Consumer design pattern is best suited for this application, which is used to decouple processes that produce and consume data at different rates[6]-[8]. This pattern approach to this application would be to collect the data in the producer loop, and have the actual processing done in the consumer loop. Data queues are used to communicate data between loops in this pattern. These queues offer the advantage of data buffering between producer and consumer loops. This in effect will allow the consumer loop to process the data at its own pace, while allowing the

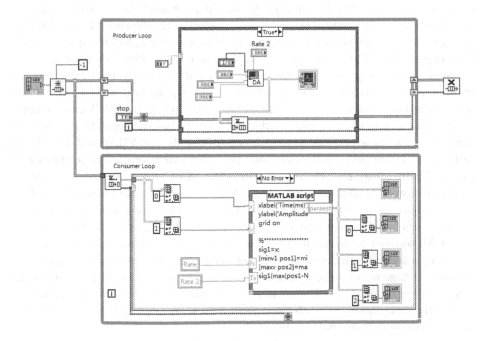

Fig. 3. Producer/Consumer Loop VI

producer loop to queue additional data at the same time. With a large enough communication queues (buffer), the data analysis process will have access to a large amount of the data that the data acquisition loop acquires. This ability to buffer data will minimize data loss. In this design (Fig.3), communication between producer and consumer loops is done by using data queues. LabVIEW has built in queue functionality in the form of VIs in the function palette.

Queues are based on the first-in/first-out theory. In the Producer/Consumer design pattern, queues can be initialized outside both the producer and consumer loops. Queue operations functions used in this block program involve dequeue element, enqueue element, obtain queue and release queue.

4   Implementation of Analyzer in Labview

The following functions that are Parameters Setup, Data Acquisition, Data Analysis, Graphic Display, Data Storage and Remote Monitoring have been achieved in the procedures. The important subroutines are data acquisition and data analysis.

Since the release of NI-DAQmx, users of NI data acquisition (DAQ) hardware have been taking full advantage of its many features designed to both save development time and improve the performance of the data acquisition applications [9]. In analyzer, we adopt NI PCI-6132 multifunction data acquisition (DAQ) board that features a dedicated analog-to-digital converter (ADC) per channel for maximum device throughput and higher multichannel accuracy. Through the following VIs, NI-DAQmx create virtual channel, NI-DAQmx timing, NI-DAQmx trigger, NI-DAQmx start task, NI-DAQmx read, NI-DAQmx clear task, we create subroutine DA to handle this problem. Before operation, the inputs are required to specify physical channels, minimum value, maximum value, scan period and sampling rate. Because MATLAB software and the associated m-file scripts can be developed in LabVIEW environment, the problem of data analysis is resolved by making the best of preponderance both of them. Some options involve moving data between the LabVIEW and MATLAB software environments [10]. Both environments offer extensive file I/O capability and can work with binary and ASCII (text) files. The MATLAB script node is a simple method that works with m-file scripts in LabVIEW. The node coexists with LabVIEW graphical code as a "script node," a rectangular region that we can add to LabVIEW programs and use to enter or load m-files. When the node executes, LabVIEW calls the MATLAB software to execute the m-file script. Moreover Right-click the node border to add input and output terminals and set their data type. The input terminals consist of scanning voltage signal, optical pulse signal, scan period and sampling rate. The data type of output terminal is 2-D Array of Real. Creating an indicator for the output and using index array function, we obtain analysis data at last. Processing steps within the data analysis module are shown in Figure 4.

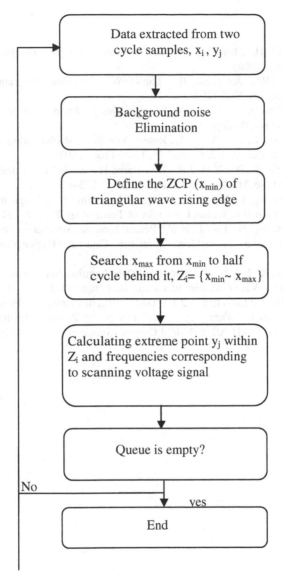

Fig. 4. Flow Diagram of Data Analysis Routine

5   Conclusion

The intelligent channel analyzer is developed based on LabVIEW. By adopting Producer/Consumer pattern, the critical problems of analyzer are resolved. As mentioned earlier its superiority is to give designer the ability to easily handle multiple processes at the same time while iterating at individual rates. In practical, the pattern can speed up software design and improve operation efficiency.

References

1. Yang, L., Li, H., Zhao, Y.: LabVIEW Advanced Program Design. Tsinghua University Press, Beijing (2003)
2. Johnson, G.W., Jennings, R.: LabVIEW Graphical Programming. McGraw-Hill Professional, New York (2001)
3. Chen, X., Zhang, Y.: LabVIEW Program Design From Entry to the Master. Tsinghua University Press, Beijing (2007)
4. Nie, Y., Feng, X., Liao, Y., Li, L.: Research on State Machine Model Based on LabVIEW. Computer Measurement & Control 15, 1166–1168 (2007)
5. Liu, D.D., Xu, R., Dai, X.D.: LabVIEW Based Hydraulic Master Brake Cylinder Assembly and Inspect Line. Machine Tool & Hydraulics 9, 125–126 (2005)
6. Chen, J., Huang, Y., Wang, Y.: Design of Real-time Data Acquisition System Based on LabVIEW. Journal of Wuhan University of Technology 29, 122–124 (2007)
7. Zhao, K., Xiong, H., Dai, J., et al.: Research on an Automatic Cigarette Sorting Control Algorithm Based on Virtual Queue Containers. Computer Engineering & Science 30, 67–69 (2008)
8. Ye, F., Zhou, X., Bai, X., et al.: Design of Data Acquisition Program Based on Queued State Machine in LabVIEW. Modern Electronic Technique 33, 204–207 (2010)
9. Long, H., Gu, Y.: LabVIEW 8.2.1 & DAQ. Tsinghua University Press, Beijing (2008)
10. Su, B., Wang, F., Wang, L.: Implementation of Zoom Spectrum Analysis Based on LabVIEW and MATLAB. Industrial Control Computer 21, 33–34 (2008)

Constructing Coarse-Grained System Dependence Graph Based on Ripple Effect

Lin Du, Haiyan Jiang, and Xiuqin Hu

School of Computer Science and Technology, University of Qilu Normal, Shandong 250014, China
dul1028@163.com, {haiyanjiang,huxq126}@126.com

Abstract. As described in this paper, we present a novel ripple effect based approach to construct coarse-grained system dependence graph by analyzing the defects which traditional system dependence graph have. Our approach could perfect object-oriented program semantics and reduce the computation complexity. Based on defining the coarse-grained, system dependence graph is simplified. Analysis of ripple effect is mapped to the dependence graph in order to embody semantic interaction among multiple objects. Finally, object-oriented program semantics are described in detail.

Keywords: Ripple effect, Coarse-grained, System dependence graph, Program semantics.

1 Introduction

Dependence analysis is the technology of program decomposing. It has been used in many areas of software engineering. Object-oriented program system dependence graph (OOSDG) can reflect the semantic characteristics of object-oriented programs, can deal with data flow and control flow between processes, can describe parameter transference and carry out inter-process analysis. However, there are the following questions in the actual construction of system dependence graph.

Firstly, the method to construct system dependence graph is complicated, what's more, lack of accuracy. Because of high complexity, on account of the different study, the actual construction ignores some of the semantic characteristics of object-oriented program. This would result in inaccurate. For example, in the OOSDG, only one parameter node is constructed for each corresponding class member variable. The class member variable has a separate copy in each class instance respectively. That is to say that a class member variable defined in one place is used in different places. This would result in the wrong data dependence among different class instances. Secondly, traditional construction method results in the loss of program semantic [1]. System dependence graph based on analysis of process, lack of semantic association, does not reflect the characteristics of object-oriented language completely [2]. The interactions among objects constitute the main framework of the object-oriented program. However this interaction doesn't base on the order of execution. So the construction of the traditional system dependence graph would not take into account all of relationship among objects. Therefore, the semantic of object-oriented is lost a

J. Luo (Ed.): Soft Computing in Information Communication Technology, AISC 158, pp. 209–214.

lot. Sometimes the use of traditional system dependence graph for program understanding is more difficult than a complete program.

In order to solve above defects that traditional system dependence graph have, the method based on ripple effect analysis is presented to construct coarse-grained system dependence graph. First of all, the meaning of coarse-grained is extended in order to make the size of grain come up to object-oriented program's semantic unit that is class, instance, member method and member variable. Ripple effects analysis plays the role of two aspects. First, the results of the ripple effect are mapped to the dependence graph in order to add semantic relationship among different objects. Second, the scope of analysis through ripple effect is narrowed in order to reduce the complexity of constructing graph.

The rest of the paper is organized as follows. In section 2, the meaning of coarse-grained is extended to simplify system dependence graph. The method of ripple effect analysis is proposed. In section 3, this paper designs the algorithms for analyzing ripple effects and constructing system dependence graph.

2   Constructing Coarse-Grained System Dependence Graph Based on Ripple Effect Analysis

2.1   Expanding the Meaning of Coarse-Grained and Simplifying System Dependence Graph

For understanding of object-oriented programming, analyzing the interaction relationship among multiple units is better than analyzing single statement. The meaning of coarse-grained is extended in order to make the size of grain come up to object-oriented program's semantic unit that is class, instance, member method and member variable [3].Coarse-grained expanded is defined as follows.

Definition 1. The graph G which meets the following characters is referred to as coarse-grained. (1) Graph G contains statement and predicate in the main(),class, instance, member method and member variable.(2) If a statement which is in the member method M belongs to G, then M also belongs to G. (3) If the instance, member method or member variable in the class A belongs to G, then class A also belongs to G.

On the basis of defining the coarse-grained, system dependence graph is simplified. Describing the process dependence doesn't need to enter the process inside but indicate process prelude node only. The data dependence which belongs to parameter nodes of different methods is indicated by data dependence among multiple methods. It is achieved by data dependence edge which point to the call directly.

2.2   Ripple Effects Analysis

Ripple effects analysis plays the role of two aspects. First, the results of the ripple effect are mapped to the dependence graph in order to add semantic relationship among different objects [4]. Second, the scope of analysis through ripple effect is narrowed in order to reduce the complexity of constructing graph.

2.2.1   Method to Analyze Ripple Effects

Analyzing ripple effect is to record the units involved by the ripple of one unit which is called the source of ripple. The following method is used. First step, the complete ripple graph which reflects corresponding object-oriented programming is constructed. Starting from the source of ripple, the direct and indirect ripple unit can be found through traversal all the ripple edges. However above method has problem as follows. The method to construct the complete ripple graph is complicated whose computation complexity is same as constructing system dependence graph. The result does not match our original intention we want to reduce the analysis scope through ripple effects analysis. Object-oriented program has the following properties. The interaction among the various units is either direct or indirect [5-6]. In particular, the indirect relationship can be expressed by the direct relationship among multiple units. This is called transitive. We can draw on the experience of the method to process transitive in the cluster. Ripple effect can be recorded through the use of matrix. Thereby the complete ripple effects can be calculated through matrix transitive operations.

2.2.2   Experimental Analysis

Example 1. Let us consider the following example.L1...L13 is the number of the corresponding statement.

```
L1: classA                          L6:  A *ca;
     { public:                      L7:  int V( )
L2:     virtual int F( ){...}        L8:  { A *a;        }// end L8
     }                              L9:  a=new A( ); }//end L5
L3: class B: class A                L10: a->F( );
     { public:                      L11: ca=a;
L4:      virtual int F( ) {...}      L12: delete a;
     }                              L13: ca->F( );
L5: class C
     { public:
```

The matrix is defined as follows. It is the first to make sure the units participated in ripple effect. If the number of all the units is n, then n * n matrix is constructed. In the matrix, the elements can be used 1 or 0 to represent ripple or not. Each row (or column) corresponds to the ripple unit. If the row number is the same as the column number, then its corresponding ripple unit is same. For each unit which is located in row i and column j, if ripple effect exits because corresponding ripple unit of row i acts on column corresponding ripple unit of column j, the unit's value is 1.On the contrary, the unit's value is 0. All the diagonal units' values are 1 because ripple effect influences units themselves. The matrix defined as above records ripple effect of the object-oriented program.

For above experimental codes, the unit B is the source of the ripple which launches the ripple. Both REO and REA are 10 * 10 matrix. The units corresponding to the row (or column) in accordance with row number (or column number) are arranged in order of size as follows. A,B,C,AF, BF,CV, C.ca,C.V.a,C.ca.F,C.V.a.F. The following matrixes are derived from above matrix definitions and algorithms.

0000000000	1111001100	0000000000
0100000000	0100100000	0100110011
0000000000	0010011000	0000000000
0000000000	0001000011	0000000000
0000000000	0000100011	0000000000
0000000000	0000010000	0000000000
0000000000	0000001010	0000000000
0000000000	0000010101	0000000000
0000000000	0000010010	0000000000
0000000000	0000010001	0000000000
(a)	(b)	(c)

Fig. 1. (a) The initial REO; (b) REA; (c) the final REO

For the final REO, the value of each unit doesn't change after the calculation is done over. If the unit value is 1, the unit is involved by ripple whose ripple source is class B. These units are the following(B,B.F,C.V,C.ca.F,C.V.a.F). It means that these units change with class B together.

3   Semantic Description

Object-oriented program semantics is described in detail in the system independence graph based on ripple effect as follows [7-8].

3.1   Description of Class, Instance, Member Method, the Relationship among Member Variables

In order to express membership, instance node, member method prelude node and member variable node are connected to the accessory class prelude node. Method and process have the same status. For method and process we do not achieve internal processing but provide prelude node. The meaning of method prelude node is expanded through hiding data transference among multiple parameter nodes. The expression of data dependence among parameter nodes of different methods relies on data dependence among methods. Then three kinds of nodes should be increased as follows. Member variable node should be increased because member method refers to member variable. Instance node should be increased because member method refers to class instance node. When a method is called by the class instance, instance node expressing message receiver object is increased in the method node. The aim is to reflect the change of object's state.

3.2   Description of Class Inheritance

In order to express inheritance, different class prelude nodes which have inheritance are connected. When one class interacts with another class, it is convenient to couple each other through class prelude node and class member edge. In order to reflect the inheritance hierarchy clearly and reduce backtracking, we take the following approach. If a virtual method in the child class which inherits from the parent class is modified, the method is described only in the child class. Meantime, associated edge should be increased between class prelude node of the parent class and method prelude node of the child class. This makes the expression of inheritance mechanism and virtual method doesn't require increasing associated edge between method of the parent class and method of the child class. Only the associated edge is increased between class prelude node and method prelude node.

3.3   Description of Polymorphism and Dynamic Binding

Polymorphism can be expressed completely by the virtual method prelude node and polymorphism call edge. Through multiple call edge, call node can be connected to each method node which is called by object possibly. The dynamic selection can be expressed by multiple polymorphism nodes which have the same protocol. This method can express all the possibilities.

4   Conclusions

This paper analyzes the defects that traditional system dependence graph have. The defects include high computation complexity, deficiency of accuracy and loss of program semantic. The method based on ripple effect is proposed to construct coarse-grained system dependence graph. Coarse-grained is extended and defined in order to make the size of grain come up to object-oriented program's semantic unit that is class, instance, member method and member variable. Ripple effects analysis plays the role of two aspects. First, the results of the ripple effect are mapped to the dependence graph in order to add semantic relationship among different objects. Second, the scope of analysis through ripple effect is narrowed in order to reduce the complexity of constructing graph. Finally, the algorithms for analyzing ripple effects and constructing system dependence graph are designed. Furthermore object-oriented program semantics are described in detail.

Acknowledgments. This work is partially supported by the Shandong Province High Technology Research and Development Program of China (Grant No. 2011GGB01017); Research Foundation of Qilu Normal University; Project of Shandong Province Higher Educational Science and Technology Program; Soft Science Research Program of Shandong Province (Grant No. 2011RKB01062).

References

1. Ap Xu, B.W., Zhou, Y.M.: Comments on a cohesion measure for object-oriented classes. Software–Practice and Experience 31(14), 1381–1388 (2001)
2. Chen, Z.Q., Zhou, Y.M., Xu, B.W., Zhao, J.J., Yang, H.J.: A novel approach to measuring class cohesion based on dependence analysis. In: IEEE International Conference on Software Maintenance, pp. 377–383 (2002)
3. Chen, Z.Q.: Slicing object-oriented Java programs. ACM SIGPLAN Notices 36(4), 33–40 (2001)

4. Xu, B.W., Chen, Z.Q., Zhou, X.Y.: Slicing object-oriented Ada95 programs based on dependence analysis. Journal of Software 12(12), 208–213 (2001)
5. Chae, H.S., Kwon, Y.R.: A cohesion measure for classes in object-oriented systems. In: Proceedings of the 5th International Software Metrics Symposium, pp. 158–166. IEEE Computer Society Press (1998)
6. Briand, L.C., Morasca, S., Basili, V.R.: Defining and validating measures for object-based high-level design. IEEE Transactions on Software Engineering 25(5), 722–743 (1999)
7. Korel, B., Tahat, L., Bader, A.: Slicing of state based models. In: Proceedings of the IEEE International Conference on Software Maintenance, pp. 34–43 (2003)
8. Harrold, M.J., Jones, J.A.: Regression Test Selection for Java Software. In: OOPSLA 2001, pp. 313–326 (2001)

A Survey on Applications of Program Slicing

Lin Du and Pingsheng Cai

School of Computer Science and Technology,
University of Qilu Normal, Shandong 250014, China
{du11028,caipingsheng2003}@163.com

Abstract. With the rapid increasing of software size and complexity, program slicing is widely applied in software engineering. Program slicing is a technique for simplifying programs by focusing on selected aspects of semantics. In this paper, four kinds of applications of program slicing are reviewed: program debugging, program testing, software measurement and software maintenance. Several theoretical and empirical contributions in the current decade related to applications of slicing are discussed. Based on the analysis of what has been achieved in recent years, we believe that program slicing will be paid more and more attentions in the future.

Keywords: Program slicing, Program debugging, Program testing, Software measurement, Software maintenance.

1 Introduction

Recent years have witnessed the increasing of software size and complexity. Thus several software engineering tasks require reducing the size of programs or decomposing a larger program into smaller components. Program slicing is a technique for simplifying programs by focusing on selected aspects of semantics. The process of slicing deletes those parts of the program which can be determined to have no effect upon the semantics of interest. As a viable method to restrict the focus of a task to specific sub-components of a program, program slicing has extensive applications in software engineering. In this paper, four kinds of applications are discussed: program debugging, program testing, software measurement and software maintenance.

The rest of the paper is organized as follows to survey applications of slicing. Section 2 introduces some related works about program debugging. Section 3 presents some pioneering works in program testing. In section 4, we survey the works in software measurement. Some methods for software maintenance are discussed in Section 5. In Section 6, we conclude the whole paper.

2 Program Debugging

Program debugging is a difficult task when one is confronted with a large program, and few clues regarding the location of a bug. Program slicing is useful for debugging, because it potentially allows one to ignore many statements in the process

J. Luo (Ed.): Soft Computing in Information Communication Technology, AISC 158, pp. 215–220.
springerlink.com

of localizing a bug. If a program computes an erroneous value for a variable x, only the statements in the slice x have contributed to the computation of that value. In this case, it is likely that the error occurs in the one of the statements in the slice. The application of debugging also motivated the introduction of dynamic slicing. The related works have been done are listed as follows.

During debugging processes, breakpoints are frequently used to inspect and understand runtime behaviors of programs. Although most development environments offer convenient breakpoint facilities, the use of these environments usually requires considerable human efforts in order to generate useful breakpoints. Before setting breakpoints or typing breakpoint conditions, developers usually have to make some judgments and hypotheses on the basis of their observations and experience. To reduce this kind of efforts, Zhang et al.[1] uses three well-known dynamic fault localization techniques in tandem to identify suspicious program statements and states, through which both conditional and unconditional breakpoints are generated.

Jiang et al.[2] present a new approach for locating faults that cause runtime exceptions in Java programs due to error assignment of a value that finally leads to the exception. The approach first uses program slicing to reduce the search scope, then performs a backward data flow analysis, starting from the point where the exception occurred, and then uses stack trace information to guide the analysis to determine the source statement that is responsible for the runtime exception.

Horwitz et al.[3] present callstack-sensitive slicing, which reduces slice sizes by leveraging the series of calls active when a program fails. It describes a set of tools that identifies points of failure for programs that produce bad output and apply point-of-failure tools to a suite of buggy programs and evaluate callstack-sensitive slicing and slice intersection as applied to debugging. Callstack-sensitive slicing is effective: On average, a callstack-sensitive slice is about 0.31 time the size of the corresponding full slice, down to just 0.06 time in the best case.

In paper [4], the authors propose a dynamic path slicing technique for object-oriented programs. Given an execution trace of an object-oriented program and an object created during the execution, a path slice per object with respect to the object, or PSPO, is a part ofthe trace such that (1) the sequence of public methods invoked on the object in the trace is same as the sequence of public methods invoked on the object in the slice, and (2) given a method invocation in the slice, the state of all objects accessed by the method is same in both the trace and slice.

3   Program Testing

Program testing is an important part of software engineering as it consumes at least half of the labor expended to produce a working program. Program slicing, which is based on the internal of the code, can be applied to structural testing technique. The related works about testing are as follows.

Duesterwald et al.[5] propose a rigorous testing criterion, based on program slicing: each def-use pair must be exercised in a successful test-case; moreover it must be output-influencing, have an influence on at least one output value. A def-use pair is output-influencing if it occurs in an output slice. It is up to the user, or an automatic test-case generator to construct enough test-cases such that all def-use pairs are tested.

Regression testing consists of re-testing only the parts affected by a modification of a previously tested program, while maintaining the "coverage" of the original test suite. Gupta et al. [6] describes an approach to regression testing where slicing techniques are used. Backward and forward static slices serve to determine the program parts affected by the change, and only test cases that execute "affected" def-use pairs need to be executed again. Conceptually, slices are computed by backward and forward traversals of the CFG of a program, starting at the point of modification.

In the paper [7], a kind of property extraction method is presented. Property model and dynamic slicing are combined to generate test sequence. As an example, the system structure of Minix3 is introduced. Exec, one of key system callings of Minix3, is modeling, slicing and its test sequences are generated. Minix3 provides open interfaces and modular. The results of slicing can be used to improve the process of software reuse.

Rupak et al. [8] present an algorithm that combines test input generation by execution with dynamic computation and maintenance of information flow between inputs. It iteratively constructs a partition of the inputs, starting with the finest (all inputs separate) and merging blocks if a dependency is detected between variables in distinct input blocks during test generation. Instead of exploring all paths of the program, it separately explores paths for each block (while fixing variables in other blocks to random values.

Seoul [9] applies hierarchical slicing technique to regression test selection in order to improve the precision of regression test selection and address the problem of level. The approach computes hierarchy slice on the modified parts of program, then selects test cases from different levels in terms of test case coverage. This approach can select test cases from high level to low level of program.

4 Software Measurement

The motivation for assessing the cohesiveness of a program or a part of it rests upon observations and claims that highly cohesive programs are easier to maintain, modify and reuse. A slice captures a thread through a program, which is concerned with the computation of some variable. If we took several slices from a function, each for a different variable, and we found that these slices had a lot of codes in common, then we would be justified in thinking that the variables were related in some way. The related works have been done are as follows.

San et al. [10] present a novel program execution based approach to measure module cohesion of legacy software. The authors define cohesion metrics base down definition-use pairs in the dynamic slices of the outputs. The approach significantly improves the accuracy of cohesion measurement.

In the paper [11], base-line values for slice-based metrics are provided. These values act as targets for reengineering efforts with modules having values outside the expected range being the most in need of attention. The authors show that slice-based metrics quantify the deterioration of a program as it ages. This serves to validate the metrics: the metrics quantify the degradation that exists during development; turning this around, the metrics can be used to measure the progress of a reengineering effort.

"head-to-head" qualitative and quantitative comparisons of the metrics identify which metrics provide similar views of a program and which provide unique views of a program.

In the paper [12], the authors explore the use of several metrics based on such slice profiles to give a quantitative estimate of the level of cohesion in a module. Example modules are used to analyze the behavior of the metric values as the modules are modified such that the cohesion is changed. The sensitivity of the metrics to the various types of changes is discussed.

Bieman et al. [13] examine the functional cohesion of procedures using a data slice abstraction. The authors analysis identifies the data tokens that lie on more than one slice as the "glue" that binds separate components together. Cohesion is measured in terms of the relative number of glue tokens, tokens that lie on more than one data slice, and super-glue tokens, tokens that lie on all data slices in a procedure, and the adhesiveness of the tokens. The intuition and measurement scale factors are demonstrated through a set of abstract transformations.

5   Software Maintenance

Software maintenance is often followed by reengineering effort, whereby a system is manipulated to improve it. One of the problems in software maintenance consists of determining whether a change at some place in a program will affect the behavior of other parts of the program. Program slicing can decompose a program into a set of components, each of which captures part of the original program's behavior. The related works have been done are listed as follows.

Gallagher et al. [14] use static slicing for the decomposition of a program into a set of components, each of which captures part of the original program's behavior. The authors present a set of guidelines for the maintainer of a component that, if obeyed, precludes changes in the behavior of other components.

Emily et al.[15] present and evaluate a technique that exploits both program structure and lexical information to help programmers more effectively explore programs. The approach uses structural information to focus automated program exploration and lexical information to prune irrelevant structure edges from consideration. For the important program exploration step of expanding from a seed, the experimental results demonstrate that an integrated lexical-and structural-based approach is significantly more effective than a state-of-the-art structural program exploration technique.

In the paper [16], the foundation of program slicing is introduced and a backward program slicing is implemented effectively using a cross reference and its lookup functions. The cross reference is a data structure that stores the relationships among a program's statements by parsing program sources. The lookup functions are operations on this cross reference data structure to retrieve desired information.

Liu et al. [17] introduce a variant of control flow graph, called validation flow graph as a model to analyze input validation implemented in a program. The authors have also discovered some empirical properties that characterizing the implementation of input validation. Based on the model and the properties discovered, they propose a method that recovers the input validation model from source and use program slicing techniques to aid the understanding and maintenance of input validation.

6   Conclusions

With the rapid development of software industry, there has been an amount of increasing of software size and complexity. Many areas in software engineering need to simplify programs by focusing on selected aspects of semantics. Program slicing is a simplification process defined with respect to a slicing criterion. The idea behind all approaches to program slicing is to produce the simplest program possible that maintains the meaning of the original program with regard to this slicing criterion. Therefore, program slicing has extensive applications in software engineering. In this survey, we have summarized the existing applications of program slicing and discussed theoretical and empirical methods for slicing. Recent works about program slicing are concerned on improving the precision of slicing methods and computing object-oriented program slicing. From above, we strongly believe that, in the near future, this research field will be paid more and more attentions by the researchers and will promote the fundamental theories research in the related fields.

Acknowledgments. This work is partially supported by the Shandong Province High Technology Research and Development Program of China (Grant No. 2011GGB01017); Research Foundation of Qilu Normal University; Project of Shandong Province Higher Educational Science and Technology Program; Soft Science Research Program of Shandong Province(Grant No. 2011RKB01062).

References

1. Zhang, C., Yan, D.: An Automated Breakpoint Generator for Debugging. In: Proceedings of the 32nd ACM/IEEE International Conference on Software Engineering (2010)
2. Jiang, S., Zhang, C.: A Debugging Approach for Java Runtime Exceptions Based on Program Slicing and Stack Traces. In: Proceedings of the 10th Quality Software International Conference (2010)
3. Horwitz, S., Liblit, B., Polishchuk, M.: Better Debugging via Output Tracing and Callstack-Sensitive Slicing. Software Engineering 36(1), 7–19 (2010)
4. Juvekar, S., Burnim, J., Sen, K.: Path Slicing per Object for Better Testing, Debugging, and Usage Discovery. Technical Report No. UCB/EECS-2009-132 (2009)
5. Duesterwald, E., Gupta, R., Soffa, M.L.: Rigorous data flow testing through output influences. In: Proceedings of the Second Irvine Software Symposium, ISS 1992, California, pp. 131–145 (1992)
6. Gupta, R., Harrold, M.J., Soffa, M.L.: An approach to regression testing using slicing. In: Proceedings of the Conference on Software Maintenance, pp. 299–308 (1992)
7. Seoul, Korea: Test Sequence Generation from Combining Property Modeling and Program Slicing. In: Proceedings of the 34th Annual Computer Software and Applications Conference (2010)

8. Majumdar, R., Xu, R.-G.: Reducing Test Inputs Using Information Partitions. In: Bouajjani, A., Maler, O. (eds.) CAV 2009. LNCS, vol. 5643, pp. 555–569. Springer, Heidelberg (2009)
9. Seoul: An Approach to Regression Test Selection Based on Hierarchical Slicing Technique. In: Proceedings of 2010 IEEE 34th Annual Computer Software and Applications Conference (2010)
10. Program Execution-Based Module Cohesion Measurement. In: Proceedings of the 16th IEEE International Conference on Automated Software Engineering, San Diego, California (2001)
11. Meyers, T.M., Binkley, D.: Slice-based cohesion metrics and software intervention. In: Proceedings of the 11th Working Conference on Reverse Engineering (2004)
12. Ott, L.M., Thuss, J.J.: Slice based metrics for estimating cohesion. In: Proceedings of the First International Software Metrics Symposium (1993)
13. Bieman, J.M., Ott, L.M.: Measuring functional cohesion. Software Engineering 20(8), 644–657 (1994)
14. Gallagher, K.B., Lyle, J.R.: Using program slicing in software maintenance. IEEE Transactions on Software Engineering 17(8), 751–761 (1991)
15. Hill, E., Pollock, L.: Exploring the neighborhood with dora to expedite software maintenance. In: Proceedings of the Twenty-second IEEE/ACM International Conference on Automated Software Engineering (2007)
16. Hoang Viet, M.S.: Software maintenance: A program slicer using cross referencer. ProQuest Dissertations & Theses 48(4), 72–76 (2009)
17. Hui, L., Hee, B.: An approach for the maintenance of input validation. Information and Software Technology 50(5), 449–461 (2008)

Recognizing Gait Using Haar Wavelet and Support Vector Machine

Libin Du[1] and Wenxin Shao[2]

[1] Institute of Oceanographic Instrumentation, Shandong Academy of Sciences
Qingdao, P.R. China
dulibinhit@yahoo.com.cn
[2] Information Development Center, Xi'an Dongfeng Instrument Factory
Xi'an, P.R. China
tomato1128@126.com

Abstract. This paper presented a new gait identification method based on Haar wavelet and Support Vector Machine. It solves the problem how to select key points and employ features to classify gaits. Firstly, images from video sequences are converted into binary silhouette. Haar wavelet transform is employed to obtain key points for distinct features, and the key points are analyzed. A subimage is utilized to represent gait features in each image, and employ Principal Component Analysis to reduce its dimensionalities. Finally, Support Vector Machine is employ to train and test, and it is helpful in analyzing features. Consequently, we can not only simplify the process, but also improve the recognition accuracy.

Keywords: feature extraction, gait recognition, Haar wavelet, Support Vector Machine.

1 Introduction

Biometric identification techniques allow identification of a person according to some physiological or behavioral traits that are uniquely associated with him/her. Common biometrics are fingerprints, hand geometry, iris, face, speech and handwriting. Biometrics are regarded as the state-of-the-art technology that provide secure solutions to the task of person identification and verification. Limitation of most current biometric identification systems is that these systems require the cooperation of individuals that are to be identified. Gait recognition [1], as an emerging biometric technology, aims to identify individuals based on their walking style. In comparison to other biometrics, an apparent advantage of gait recognition is that it can be applied unobtrusively. For example, using this technique, the attention or cooperation of the observed subject is not required.

Several interesting approaches for gait recognition have been reported to date. The identification using principal component analysis (PCA) was performed by Liu et al [2] using patterns derived from the horizontal and vertical projections of silhouettes. The projections constitute periodic signals that were subsequently analyzed using the mathematical theory of symmetry groups. By studying the correlation between symmetry groups and gait viewing direction, they developed a practical technique for

J. Luo (Ed.): Soft Computing in Information Communication Technology, AISC 158, pp. 221–227.
springerlink.com   © Springer-Verlag Berlin Heidelberg 2012

classifying imperfect patterns. Kale et al [3] proposed width of silhouette as a suitable gait feature. The width of silhouette is the horizontal distance between the leftmost and rightmost foreground pixels in each row of the silhouette. However, algorithms that use this feature are vulnerable to spurious pixels that often render the identification of the leftmost and rightmost pixels inaccurate.

The methods [2-4] try to extract the important information from the silhouettes in a gait sequence by measuring quantities related to the projections of the silhouettes in certain directions. Zhang et al present several new approaches [5-7], where they employ $\Re$ transform and key points to recognition activities and obtain better experiment results. This paper aims to extract the most important feature by transforming the Haar wavelet and identify the subjects by Support Vector Machine (SVM) in gait recognition. Consequently, this method yields distinct vectors of feature for each gait sequence, it strengthens major feature.

2   Haar Wavelet Transform and Support Vector Machine

2.1   Haar Wavelet Transform

Alfred Haar [8] proposed Haar wavelet function, which is a function set consisting of piecewise-constant function. Its defined field is [0,1), and in certain field every piecewise-constant function is 1 while in other field it is 0. Wavelet function is usually described as $\psi_i^j(x)$. Compared with frame function, it is called Haar wavelet function and defined as

$$\psi(x) = \begin{cases} 1 & 0 \leq x < \dfrac{1}{2} \\ -1 & \dfrac{1}{2} \leq x < 1 \\ 0 & \text{other} \end{cases}$$

The measuring function of Haar wavelet is defined as

$$\psi_i^j(x) = \psi(2^j x - i), \quad x = 0, 1, \cdots, 2^j - 1 \tag{1}$$

The vector space represented in W^j consisting of wavelet function is described by

$$W^j = sp\{\psi_i^j(x)\}, \quad i = 0, 1, \cdots, 2^j - 1 \tag{2}$$

where sp represents line production. The measuring factor is j, which varies with the size of graph function. The shift parameter is i, which can bring the function to shift along x-axis.

It hypothesizes that $\{c_{n,m}^L\}$ (n,m=0,1,...,N-1) is the signal of 2-D and discrete image, in which $c_{n,m}^L$ is between 0 and 255. The distance between pixels is N^{-1}, in which N is 2L and L is (j+1). It is supposed that $\tilde{h}$, $\tilde{g}$, h and g is a filter set of biorthogonal Haar wavelet transform, whose algorithm of Mallat in convolution form is defined as

$$c_{k,m}^{j} = \sum_{l,n} \tilde{h}_{l-2k}\tilde{h}_{n-2m}c_{i,n}^{j+1}$$

$$d_{k,m}^{j,1} = \sum_{l,n} \tilde{h}_{l-2k}\tilde{g}_{n-2m}c_{i,n}^{j+1}$$

$$d_{k,m}^{j,2} = \sum_{l,n} \tilde{g}_{l-2k}\tilde{h}_{n-2m}c_{i,n}^{j+1} \qquad (3)$$

$$d_{k,m}^{j,3} = \sum_{l,n} \tilde{g}_{l-2k}\tilde{g}_{n-2m}c_{i,n}^{j+1}$$

It consists of 4 parts after first rank transform of Haar wavelet, as shown as follows.

$$\begin{bmatrix} c_{k,m}^{j} & d_{k,m}^{j,1} \\ d_{k,m}^{j,2} & d_{k,m}^{j,3} \end{bmatrix}$$

where each subimage is one quarter of mother image in size, whose structure is as follows.

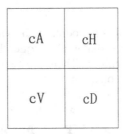

Fig. 1. The structure of Haar wavelet transform

After the filter of Haar wavelet transform is replaced and simplified, the equations are as follow.

$$c_{k,m}^{j} = \frac{1}{4}(c_{2m,2k}^{j+1} + c_{2m,2k+1}^{j+1} + c_{2m+1,2k}^{j+1} + c_{2m+1,2k+1}^{j+1})$$

$$d_{k,m}^{j,1} = \frac{1}{4}(c_{2m,2k}^{j+1} + c_{2m,2k+1}^{j+1} - c_{2m+1,2k}^{j+1} - c_{2m+1,2k+1}^{j+1})$$

$$d_{k,m}^{j,2} = \frac{1}{4}(c_{2m,2k}^{j+1} - c_{2m,2k+1}^{j+1} + c_{2m+1,2k}^{j+1} - c_{2m+1,2k+1}^{j+1}) \qquad (4)$$

$$d_{k,m}^{j,3} = \frac{1}{4}(c_{2m,2k}^{j+1} - c_{2m,2k+1}^{j+1} - c_{2m+1,2k}^{j+1} + c_{2m+1,2k+1}^{j+1})$$

2.2   Support Vector Machine

Support Vector Machines [9] are well known for their strong theoretical foundations, generalization performance, and ability to handle high-dimensional data. In the binary classification setting, let $((x_1, y_1),(x_2, y_2),\cdots,(x_n, y_n))$ be the training data set where xi are the feature vectors representing the instances and $y_i \in \{-1, +1\}$ are the labels of those instances. Using the training set, SVM builds an optimum hyperplane—a linear discriminant in a higher dimensional feature space—that separates the two classes by the largest margin. The SVM solution is obtained by minimizing the following primal objective function:

$$\min_{w,b} J(w,b) = \frac{1}{2}\|w\|^2 + C\sum_{i=1}^{n}\xi_i \tag{5}$$

with $\forall i \begin{cases} y_i(w \cdot \Phi(x_i) + b) \geq 1 - \xi_i \\ \xi_i \geq 0 \end{cases}$

where w is the normal vector of the hyperplane, b is the offset, y_i are the labels, $\Phi(\cdot)$ is the mapping from input space to feature space, and ξ_i are the slack variables that permit the nonseparable case by allowing misclassification of training instances.

3   Our Approach

It can be seen that process of gait is periodic, as shown in Fig. 2, so we can employ a gait cycle to represent the whole walking. Consequently, it is a key issue that distinct gait features are selected in recognition. There are two stages for gait recognition, i.e. feature representation and gait recognition.

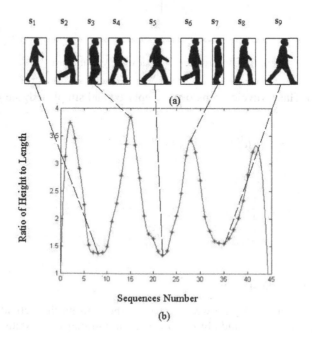

Fig. 2. Gait cycle

3.1   Feature Representation

We adopt action videos as gallery sequences. Gaussian Mixture Model (GMM) is used as background modeling and motive object detecting. Binary images of human silhouette are extracted by background subtraction. The connective contour within bounding rectangle is obtained by pre-process of binary images.

After obtaining the contour, we change its size to $l \times l$ for Haar wavelet transform. Then we implement Haar wavelet transform as well as [10] and obtain 4 subimages, as shown in Fig. 3. It concludes that the subimage in cH region, i.e. $\frac{l}{2} \times \frac{l}{2}$, highlights motional information, so they are thought as key points. Principal Component Analysis (PCA) is employed to select vectors, before they are formed another vector.

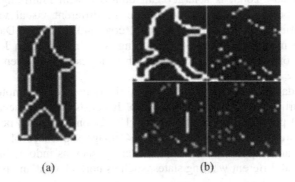

(a)   (b)

Fig. 3. Contour with Haar wavelet transform

3.2   Gait Recognition

We assume that C is one subimage and defined by

$$C_{\frac{l}{2} \times \frac{l}{2}} = \begin{pmatrix} c_{11} & c_{12} & \cdots & c_{1 \times \frac{l}{2}} \\ c_{21} & c_{22} & \cdots & c_{2 \times \frac{l}{2}} \\ \vdots & \vdots & \ddots & \vdots \\ c_{\frac{l}{2} \times 1} & c_{\frac{l}{2} \times 2} & \cdots & c_{\frac{l}{2} \times \frac{l}{2}} \end{pmatrix} = (C_1' \quad C_2' \quad \cdots \quad C_{\frac{l}{2}}') \tag{6}$$

We reduce the dimensionalities of C with PCA and obtain the principal components as

$$C'' = (C_1'' \quad C_2'' \quad \cdots \quad C_{\frac{l}{2}}'') \tag{7}$$

So we select m components and form a new vector, $D = (C_1'' \quad C_2'' \quad \cdots \quad C_m'')^T$, $1 < m < \dfrac{l}{2}$ to model every subject and recognize them in SVM.

4   Experimental Results and Discussion

4.1   Experiment Data

The CASIA database is initialized by the institute of automation in Chinese Academy of Sciences. There are three datasets in CASIA present, in which they are Dataset A, Dataset B and Dataset C. Dataset A, which is NLPR gait database and consists of 20 persons, was initialized in 2001. It consists of 3 walking directions, each of which has four image sequences whose sequence number is between 37 and 127. Dataset B initialized in January 2005, is a larger gait database with multi-angles of view. It consists of 124 persons, each of which is in 11 different visual views, and it is acquired in 3 different cases, such as normal, coat and briefcase. Dataset C, which was recorded by infrared camera in the evening, was initialized in July and August 2005. It consists of 153 persons, each of which acquires in 4 different cases, such as normal, fast, slow and bag.

The SOTON database is initialized by the University of Southampton in 2002 and supplemented further to be a normal database. It consists of two sections, the large dataset with 100 persons and the small with 12 persons. Each of persons walks in three different directions, in which there are 4 image sequences of 25 frames. The video sequences are acquired in different scenarios, such as indoors and outdoors, in different speed and different walking states, such as normal, coat and briefcase.

4.2   Experiment Discussion

The training and testing data of gait model are utilized from CASIA and SOTON database, respectively. We choose a cycle of gait images and obtain their vectors. The experiment results are 91% in Dataset B of CASIA, 90% and 94% in small and large dataset in SOTON, and its average achieves 91.6%. It concludes that we employ discriminative classifier rather than generative classifier, and avoid the data dependency in sequence, and obtain higher recognition accuracy than [11,12].

5   Conclusions

This method represents a distinct feature on the basis of Haar wavelet and Support Vector Machine. We extract feature and train gait SVM models from subjects, thus this method meets the application in video surveillance system. We will next study on the construction of gait database with brief representation and choice of image transform with lower computational cost.

References

1. Boulgouris, N.V., Hatzinakos, D., Plataniotis, K.N.: Gait recognition: a challenging signal processing technology for biometric identification. IEEE Signal Process. Mag. 22(6), 78–90 (2005)
2. Liu, Y., Collins, R., Tsin, Y.H.: Gait Sequence Analysis Using Frieze Patterns. In: Heyden, A., Sparr, G., Nielsen, M., Johansen, P. (eds.) ECCV 2002. LNCS, vol. 2351, pp. 657–671. Springer, Heidelberg (2002)
3. Kale, A., Cuntoor, N., Yegnanarayana, B., Rajagopalan, A.N., Chellappa, R.: Gait analysis for human identification. In: Proc. 4th Int. Conf. Audio- and Video-Based Person Authentication, Guilford, U.K., pp. 706–714 (2003)
4. Boulgouris, N.V., Plataniotis, K.N., Hatzinakos, D.: An angular transform of gait sequences for gait assisted recognition. In: Proc. IEEE Int. Conf. Image Processing, Singapore, pp. 857–860 (2004)
5. Zhang, H., Liu, Z., Zhao, H.: Automated Classification of Two Persons' Interactive Activities. Journal of Software 5(8), 810–817 (2010)
6. Zhang, H., Liu, Z., Zhao, H., Cheng, G.: Recognizing Human Activities by Key Frame in Video Sequences. Journal of Software 5(8), 818–825 (2010)
7. Zhang, H., Liu, Z., Zhao, H.: Human Activities for Classification via Feature Points. Information Technology Journal 10(5), 974–982 (2010)
8. Alfred, H.: Zur Theorie der orthogonalen Funktionensysteme. Mathematische Annalen 69(3), 331–371 (1910)
9. Cortes, C., Vapnik, V.: Support Vector Networks. Machine Learning 20, 273–297 (1995)
10. Liu, Z., Sarkar, S.: Improved Gait Recognition by Gait Dynamics Normalization. IEEE Trans. Pattern Anal. Mach. Intell. 28(6), 863–876 (2006)
11. Chen, S., Tian, Y., Huang, W., et al.: Automatic human gait recognition using temporal template. Journal of Xidian University 34(4), 605–610 (2007)
12. Ye, B., Wen, Y.: Gait recognition based on DWT and SVM. Journal of Image and Graphics 12(6), 1055–1063 (2007)

How to Do Contemporary College Communist Youth League's Work of "Recommending the Excellent as the Party's Development Object" Better

Ding Mingjun and Ye Chunqing

School of Information and Communication Engineering, Tianjin Polytechnic University,
Tianjin, China
dingmjh@126.com

Abstract. "Recommending the excellent as the party's development object" is an important responsibility that the party endows the communist youth league, which is the need to strengthen party construction and league building in a new age, to optimize the structure of league members, and to enhance the overall quality of league members team. In a case study of School of Information and Communication Engineering, Tianjin Polytechnic University, this paper analyzes the weak link and obvious problems that exist in the college's basic level communist youth league's work of "recommending the excellent as the party's development object", and proposes how to enhance pertinence and effectiveness of "recommending the excellent as the party's development object".

Keywords: College, Communist youth league, League building, "Recommending the excellent as the party's development object", Party construction.

1 Introduction

Recommending the excellent as the party's development object is a glorious task of communist youth league organizations that the party endows. As early as 1982, communist youth league the eleventh congress had formally written this task into youth league constitution. The sixteenth congress of the youth league writes: To educate the basic knowledge for league members and to recommend the excellent as the party's development object. The party's organizations at all levels pay high attention to leading the communist youth league's work of "recommending the excellent as the party's development object". The new party constitution of seventeenth congress clears the basic requirements of "recommending the excellent as the party's development object": to educate and cultivate the party activists, do better to regularly develop party members, and make a point of developing party members from among those in the forefront of work and production and young people [1].

With the development of the situation, the "recommending the excellent as the party's development object" faces many new problems. In order to deeply implement the "The Suggestions on Further Strengthening and Improving Ideological and Political Education of College Students" proposed by CPC Central Committee and State Council and adopt to the working requirements of youth league construction under the

J. Luo (Ed.): Soft Computing in Information Communication Technology, AISC 158, pp. 229–235.
springerlink.com   © Springer-Verlag Berlin Heidelberg 2012

influence of party construction in new period, it is necessary to develop investigation and research, deeply understand the working status of "recommending the excellent as the party's development object", constantly to sum up experience, to discover new problems, and to propose new countermeasures [2].

School of Information and Communication Engineering, Tianjin Polytechnic University developed a questionnaire of "recommending the excellent as the party's development object" for 768 students of 07 and 08 grades in 2009-2010. According to this survey, it reflects the basic situation and the existing problems of college "recommending the excellent as the party's development object" from multi-sides. The students' whole understanding of "recommending the excellent as the party's development object" is correct, their attitude is right, and their evaluation of the working effect is pertinent. However, simultaneously the investigation results also reflect that there exist some problems of "recommending the excellent as the party's development object" in propagating the implementation of education and principle of democratic centralism, system construction, organizing and guiding, etc.

2   The Students Have the Enthusiasm in Participation, But Their Understanding Need to Be Further Improved

From the investigation results, the students' approval degree and contribute degree for "recommending the excellent as the party's development object" are higher. 83.2% of the students understand the work of "recommending the excellent as the party's development object", 95.3% of the students concerned the "recommending the excellent as the party's development object" of their class, while 94.7% of the students actively participated in the democratic evaluation meetings of "recommending the excellent as the party's development object". It shows that the deeply development of "recommending the excellent as the party's development object" has better mass basis. However, the students' comprehensive degree and significance understanding for the "recommending the excellent as the party's development object" should improve, some students for the participation of "recommending the excellent as the party's development object" only stay in the passive and "spontaneous" stage. 19.6% of students are subjective and random in democratic appraisal and voting, which shows the work of "recommending the excellent as the party's development object" lacks enough depth propaganda.

3   The Principle of Democratic Centralism Has Been Implemented, But the Democratic Level Need to Be Deepened

The investigation shows that students' attitude for the implementation of the principle of democratic centralism of "recommending the excellent as the party's development object" is positive. Majority of league members have the right to know, right of supervision and voting right in the work of "recommending the excellent as the party's development object", 73.8% of the students can more fully and objectively show their personal will in democratic evaluation meetings. However, 26.2% of the students can not fully have the right to vote. The unfavorable factors which prevent from further deepening democracy of the "recommending the excellent as the party's development object" still exist. First, the system construction is still relatively weak, especially lack of system criterion in micro operation level, resulting in the larger subjective random. Second, the organization ability level of basic level league cadres for "recommending the excellent as the party's development object" still can't satisfy the masses completely. 45.8% of the students thought that the organization function of class league cadres didn't play completely in "recommending the excellent as the party's development object". Due to the subjective factors, such as lack of working experience, simple working method, etc. it affects the effective realization of the democracy of "recommending the excellent as the party's development object" during executing the work.

4   League Cadres Have Strong Sense of Responsibility, But it Need to Give the Necessary Guidance

The responsibility and work level of basic level league cadres directly affect the working effect of "recommending the excellent as the party's development object". The base league cadres mainly include students' league cadres and teachers' part-time league cadres. The investigation shows that the students' league cadres generally have high working enthusiasm and strong sense of responsibility, but their comprehension level and executive level of related system need guide and improve. The teachers' part-time league cadres are generally department department department secretary of league general branch, they have strong sense of responsibility, and higher theoretical level and working level. However, due to the constraints of time and effort, to some extent it affects the specific guidance of "recommending the excellent as the party's development object", which only get 67.7% of students' higher evaluation. Under such circumstances, it is particularly important to enrich and stabilize the communist youth league cadres, strengthen the education and training of base league cadres. Simultaneously, how to fully move all the favorable factors and forces of counselors, head teachers, fore-class party members, etc. to common concern and guide "recommending the excellent as the party's development object", which is a working direction need to strive.

As the assistant and reserve force of Chinese communist party, the communist youth league members should play their own role in the process of the party members' development and cultivation, aiming to complete the transportation of qualified talents for the party, and cultivate the reserve force of the party [3]. The author believes that the role of the communist youth league around the "recommending the excellent as the party's development object" in the development of university party organization can be summarized in four aspects:

(1) Laying a good foundation, to play a role in education. The communist youth league plays an important role in the education of the party activists by using their unique channel from political thought, humanistic quality, psychological quality, etc.

(2) Practical test, to play a role in exploration. The communist youth league form a long-term exploration for the party activists through many activity carriers, such as social practice, extracurricular activity, amateur party school, party constitution learning group, etc.

(3) Evaluation and appraisal, to play a role in holding the pass. In the process of implementing the "recommending the excellent as the party's development object", the communist youth league has deep and detail understanding and recognizing of each party activities' thought, words and deeds, quality, performance, etc. it can hold the pass for the development objects through league members' Evaluation and appraisal.

(4) *Inheritance* duty, to play a role in recommending. According to the form of By "recommending the excellent as the party's development object", the communist youth league recommends the excellent young league member to party organization, in order to the further training and study of party organizations, and lay a solid foundation for the party's organization and development.

The above four aspects supplement each other and none of the four is dispensable, recommending is the goal, cultivation *is* the base of the recommendation work, observation is the test for the training work, holding the pass is the key of recommending, cultivating and observing, all of them form a dynamic process, and play an irreplaceable role in the development of party organization.

In addition, the author believes that it is necessary to further straighten out the whole flow of the college communist youth league organization "recommending the excellent as the party's development object", clear the working emphasis of "before *recommending* the excellent", "by recommending the excellent", "after recommending the excellent", strengthen organic linkage among links, thus, it can really improve the effectiveness of college communist youth league "recommending the excellent".

5   Emphasis on Prophase Education: Strengthening Thought-Leading, and Reflecting Educational Function

In college, after the young League member submitting the party membership application to the party organization, the league organizations should strengthen to train and guide them from the new height, make them become maturity in thinking, organization, style, ability. In the current international and domestic situation, it is particularly important to guide the youth league members in the comparison and distinction to distinguish wrong and right, distinguish primary-secondary, master ideological weapon, improve political sensitivity and recognition ability, consolidate ideal and belief, and further enhance the comprehensive quality. League organization must put the training and education of the young League member when developing "recommending the excellent as the party's development object" first, enhance the training and cultivating strength for the party activities, and fully play the educational function of "recommending the excellent as the party's development object". First, perfect the training system of students' party school learning and league school learning, continuously improve the political theory quality of training objects; second, establish ideological reporting system, understand the ideological trend of training objects, and develop educational activities; third, establish theoretical learning group, in order to achieve the goal of mutual learning, mutual assistance, mutual supervision, mutual improvement; fourth, establish talk system, take the initiative to understand and solve the confusion of training objects in thought and work actively; five, increase the training and education for training objects in various activities, through the activities to enhance the sense of social responsibility and sense of mission of the training objects, close the feelings between the party and people, improve their overall quality, thus continuously to deepen the training and education [4]. According to the education leading of "before recommending the excellent", it makes the young league member have firm ideal and belief, have the right motivation of joining CCP, stick to use Marxism, Leninism, Mao Zedong Thought, Deng Xiaoping Theory, and the "Three Represents" to arm their mind. They are all round qualities, and they have both ability and moral integrity, and can take on heavy responsibilities.

2 Highlight process guidance: enrich subject and method, to reflect scientific evaluation

Innovative organizational settings, and enrich the main body of "recommending the excellent" subject. With the emergence and promoting of credit system, because of the students' free controlling of the learning content and speed, it makes the original more stable class league branch appear the larger personnel flow, difficult to concentrate organize the education activities, the unfamiliar with each other among students, etc. In addition, the extracurricular activities on campus has become increasingly active, the young league members belong to various organizations of different types are very common. The environment of "educating the excellent" is unstable, so it is difficult to achieve the systematic cultivation and objective evaluation of the young league members. Relying on the league branch of various classes to develop the traditional model of specific "recommending the excellent" which is hard to adapt to increasingly diversified study and life model of college students, naturally the scientific quality suffers greater question. Therefore, if the college communist youth league's "recommending the excellent" really wants to implement student-oriented, it should be based on the study and life reality. When evaluating the recommending objects, it can be considered to combine their main positions of work and study with their main positions of "recommending the excellent". The basic league organization is the stay point of various work of communist youth league, and also is the big stage for the young league members to accumulate experience, exercise ability and show their talent. Therefore, we should select the basic level league branch closely related with the recommending objects' work and study as the main position of recommending work. Because only they have the condition to make long-term training and objective observation for the recommending objects, really to recommend the excellent youth to become the development object of the party organization. The college communist youth league organizations should put the innovation's effort point of "recommending the excellent" in innovating the basic league organization form, build some basic level league branch which is more suitable for recommending objects, form a beneficial supplement for the league organization form in unit of class, lay a better organizational guarantee for the scientific development of "recommending the excellent". Innovative evaluation index, and enhance the scientific quality of "recommending the excellent". In the past work of "recommending the excellent", the standard of "recommending the excellent" is rather vague. Because the guide of school's and department's league members to the basic level league branch is limited, and the regulation system is more principle. In addition, because there is no specific regulation for the standard of "the excellent", it is just according to the standards of party member in "Party Constitution", this operation always makes the league members consider "recommending the excellent" equals to "selecting the excellent". Thus the students who have good marks, who are popular or good are easy to be recommended. In addition to the lack of the great propagation of "recommending the excellent", the work is not deep, there exists randomness and blindness. Some league members consider that the "recommending the excellent" refers to "selecting the excellent leader", "commenting three good student" and some even consider it is "electing the party member", and some consider that it is just a form, and some branches even use the ten minutes between classes to vote, all of them lack objective and real evaluation of object of investigation [5]. Currently the college students' quality program which is developed by college communist youth league organizations can provide important combining platform and ideas for establishing the scientific league member evaluation system. The college students' quality development plan is an important platform to comprehensively record the students' participation of extracurricular activities at school, it comprehensively shows the growth and progress of each youth league members during college from six aspects, such as ideology and politics and moral quality, social practice and voluntary services, academic technology and innovation and pioneering, culture and art and physical and mental development, community activities and social work, technical training, etc. The league organization can combine with such a platform to build a scientific and reasonable league evolution system, and quantize the examination index of league members to make the standard of excellent league members more operational, change the one-sided emphasis on the result to emphasis on both the result and process, then to promote the scientific of examination index. It should create the evaluating method and enhance the fairness of "recommending the excellent". If there are the multiple main body of "recommending the excellent" and relatively scientific and reasonable evaluating system of "recommending the excellent", the method of democratic appraisal will become more important in the process of "recommending the excellent". The democratic appraisal method of open, just and fair, is not only the objective evaluation of a youth league member, but also is the realistic guide to other youth league members, so to select the scientific and reasonable, innovative and fair appraisal method will help to improve the effectiveness of "recommending the excellent". The traditional way of league branch democratic appraisal is that the league branch holds all league members meeting, introduce the practical conditions of appraisal objects, and submit it to the meeting to vote by ballot, and then public. This method should be the important form that majority league branches develop the work of "recommending the excellent" currently. This method has its scientific, but the author believes that, except the current form, each league members can explore some forms that more meet the actual characteristics of modern university students. For example, in order to avoid the embarrassment of face to face appraisal, the written comments can be increased; in order to more real to investigate the motivation of joining CCP of the party activities, the personal reply can be into increased based on the situation introduction of the branch; In order to minimize the possibility of human operation as possible, the each performance of the party activities can be showed in form of credit to quantize objectively; in order to improve the public knowledge degree, the public can be put in proper network, etc. In short, the quest for the democratic appraisal forms which are suitable for majority youth league members, the greatest purpose is to improve the effectiveness and impartiality of democratic appraisal through this form to select the good students by using the good form and good system.

6   The Anaphase Training: Continue to Track and the Investigate, to Realize the Connection of Party and League

The combination of party and league in "recommending the excellent" is not close enough, it doesn't form the better work situation that league organization actively promotes, and the party organization actively receives. Most colleges' league organizations think that it is the end after recommending the excellent to party organization. Once recommending the excellent, the party organization is responsibility for educational management; to a much greater extent there exist the phenomenon of emphasis on recommending and lighting training. In fact, the party after "recommending the excellent as the party's development object" still remains.

League membership and has league member identity. For the league members who don't develop, the base league organization still should develop evolution and education for them, and should propose higher requirement, and put them in a bigger space and higher platform for further test. It form a long-term exploration for the party activists through many activity carriers, such as social practice, extracurricular activity, amateur party school, party constitution learning group, etc. Timely report their performance to party organization from time to time, feedback their various data and files to party organization timely, and change the "recommending the excellent" into the important carrier that college party and league organization builds better interaction relationship, to really realize the seamless connection of party and league organization in party construction.

"*Recommending* the excellent as the party's development object" is the important work platform of college youth league construction under the influence of party construction, is the important carrier for communist youth league to strengthen self-construction and play educational function. In the new historical period, league's and party's organizations at all levels should strengthen linkage, actively integrate all educational resources, together discuss the work rules of "recommending the excellent as the party's development object", promote the benign development of "recommending the excellent as the party's development object".

References

1. Liang, Y., Liu, S.: Thinking on the "Recommending the Excellent as the Party's Development Object" Organized by College Communist Youth League in new period. Theory Horizon (01) (2008), TV University (03) (2008)
2. Ma, B., Xu, Z.: Thinking on the Several Important Links of College's Work of "Recommending the Excellent as the Party's Development Object". Journal of Liaoning Radi
3. The Suggestions of CPC Central Committee and State Council on Further Strengthening and Improving Ideological and Political Education of College Students. Printed by CPC Central Committee General Office (14) (February 8, 2004)
4. Wang, Z.: Study on the Work of College Students Party Construction in New Period. Journal of Liaoning Technical University: Social Sciences 8(2), 223 (2006)
5. Yan, M.: Thinking on the Improvement of the "Recommending the Excellent as the Party's Development Object" Organized by College Communist Youth League. Journal of Guangxi Youth Leaders College (01) (2006)

On the Work of Contemporary College Counselors

Ding Mingjun and Ye Chunqing

School of Information and Communication Engineering, Tianjin Polytechnic University,
Tianjin, China
dingmjh@126.com

Abstract. College counselors, as an important component of teachers in colleges, are the backbone of moral education and ideological and political education of college students. Meanwhile, they are instructors and pathfinders of college students for their healthy growth. Therefore, it is of great importance to be a qualified counselor.

Keywords: college counselor, mental health, student leader.

1 Introduction

As the main force of ideological and political work, counselors are the backbone of ideological work and political work for students, and they are guarantee of cultivating students' healthy personality, good personality, entrepreneurial spirit, and creative consciousness. In the new historical period, the ever-changing new situation is a new challenge for counselors. This paper makes a discussion on how counselors carry out their daily work in new situation based on the writer's own experience for several years.

2 Counselors Must Have a Firm and Correct Political Position

College counselors are the backbone of ideological and political education of college students, and they are instructors and pathfinders of college students for their healthy growth. Whether ideological and political education base for college students is strong or not and whether the campus environment stable or not largely depend on the political quality level of counselors. Therefore, college counselors should have high political quality, firm and correct political position, so as to ensure the ideological and political work.

First, the counselor should strengthen the study of political theory, and improve their quality and theoretical level. Although counselor team is a relatively young group among college teachers, they must make sure the overall situation in their work. Therefore, the political nature is the most essential characteristics of counselors, and it is also the fundamental guarantee of being a good counselor. For contemporary college students, the counselor can make practical research on students' situation deeply, and foster their values, world outlook, and outlook on life to the point. Therefore, it is an extremely important job. College counselors have great responsibilities of cultivating

J. Luo (Ed.): Soft Computing in Information Communication Technology, AISC 158, pp. 237–242.

and shaping talents for the state, so it is impossible for a counselor without high political quality to have a good performance on his position.

Secondly, during the process of strengthening the study of political theory and improving their political theory level, counselors should effectively improve their policy level which requires mastering relevant documents spirit of superior organizations or college Party committee. For relevant documents, counselors should not only have a comprehensive study and systematic control, but also convey the spirit to students with simple and clear expressions. Except for that, counselors need to make research on students' growth in the light of their professional features and specific conditions, and formulate simple rules widely accepted by students with pellucid language to help students grow and become useful persons, which is also one of the aims and objectives of mastering document spirit for counselors. Only mastering the spirit, can counselors get a really meticulous in practical work, and can they really understand the policy and do pragmatic work, which are the most important prerequisite and foundation of being a good counselor.

3   Counselors Must Have a Heart of "Love Students"

Love can warm every heart, and love can save every soul. Among the entire counselor's work, love is the constant theme. Counselors should not only be a messenger of love, but also the evangelist of love who sow the seeds of love in students' hearts and make them root. It can be said that love is the basis of counselor's work.

First of all, counselor should accept each student with a maternal heart. Since the 21st century,   more and more students who are "the generation after 90s" in city get into the university campus, and the majority of these students are only child at home who are treated as "little emperors" or "Little Princess" and basically living a comfortable and wealthy life. While, we should note that a large percentage of students are from rural areas, and they are not rich, even some of them have difficulty in their fees, so they have to borrow money from relatives and friends or apply for student loans. Facing with these kinds of students and all single individuals, as a counselor, this paper thinks counselors should at least play dual role, that is, a teacher and a friend of students, and treat them with a maternal heart to make them feel at home and feel the warmth from family.

Secondly, counselors should do everything well for students with philanthropy. "Sometimes students will tell you their secrets and trouble to you to air their feelings, and seek your instruction. When students do this, your education is successful, because when students tell you their difficulties or secret sorrows, you have virtually played a role of reliable intimate friends. Making good friends with students, and doing everything well for them with philanthropy, are the premise and foundation of completing the counselors work comprehensively and splendidly." This paper cannot agree the above points more, and insists that before they turn to you for help, they must have mental struggle. Therefore, after knowing their difficulties, the counselor should try his best to help them in a responsible manner for students, parents and society. Try to solve problem for students and consider their difficulties as your own, only this can counselors establish a good image in students' heart and become their reliable friend.

4   Counselors Must Pay Attention to Mental Health Education of Students

In recent years, some serious incidents of students in universities happened frequently, and they are increasing year by year, so mental health education of college students quickly becomes the focus of community. In this paper, author believes that, for making college mental health education have a healthy development, the college counselor should mobilize the students' potential, develop their strong will, and further strengthen the preventive and developmental mental health education and setbacks education, meanwhile, active value interference is necessary during the education.

First, help students establish consciousness of enhancing psychological quality, which is the basis of carrying out mental health education.

The prerequisite for overcoming the negative emotion is sober awareness of their emotional state, and guiding students to establish consciousness of adjusting emotions. In human mental activity, there are two levels of consciousness and unconsciousness. Consciousness refers to clear awareness of things in an awakening state. If they can rise to the level of consciousness when facing something, they will square up to it, analyze its causes and consequences, and make right judgments of its effect and damage. Then they can seek specific methods to solve problems according to their purpose and desire. Wisdom of human not only can make analysis and evaluation on objective things and environment, but also can point the edge of wisdom to ourselves, which enable us to recognize, evaluate and reflect on our physical and mental state. Through our thoughts and will, human can make interference on the state and change it toward the direction that is good for its survival and development. Emotional fluctuations are of frequent occurrences. If we can guide students to establish consciousness of adjusting emotions, when having negative emotions, they can find ways to adjust them actively, and become masters of their own emotions, rather than be enslaved by negative emotions.

Second, the emphasis on cultivating the good volitional quality of students is the key point of mental health education.

The aim of mental health education in colleges is to help students improve their ability to withstand setbacks, and foster good volitional quality. The good volitional quality: first, it refers to the independence of will. The independence shows that man is good at presenting the purpose of their action according to their own reasonable creation; it is a kind of volitional quality that man can take effective methods steadily, and is responsible for outcomes. Second, it refers to assertiveness and boldness of will. assertiveness and boldness is a kind of quality that man is good at making decisions quickly and effectively in complex situations, and put into action in time and bravely. What is contrary to this quality is infirm and haste. Third, it refers to fortitude of will. Fortitude is a kind of quality that man believes his decision is reasonable for a long time, overcome all difficulties persistently, and go toward the established goal tirelessly. On the one hand, an individual with fortitude shows firm belief; on the other hand, he has fierce determination, that is, the individual is good at keeping actions which comply with his goal in long term, and overcome the difficulties persistently and tirelessly. What is contrary to fortitude is vacillation and obstinacy. Fourth, it is self-government of will. It is a kind of quality that the individual can manage himself, overcome his desires and emotional disturbance, and force himself to implement well-founded decisions, or curb some actions firmly.

5 Counselors Must Respect Student Leaders and Display Their Subjective Initiative

Counselors should rely on students fully in the work. Especially pay attention to improving the power of student backbone, such as Party members, student cadres in general Youth League branch of faculty and student union and class, even dorm director. And fully mobilize and develop the enthusiasm and initiative of every student cadres.

Student leaders are not only organizers and leaders of activities, but also a tie of counselor and student. Among students, they have a certain prestige, and have a strong influence and appeal on other students. Therefore, for doing the work well, counselors should organize an excellent and active student leader team.

In most universities at present, class is still the basic unit of students' life and study, and it's a necessity for their growth. Only give a full play of student leaders' model role, can make the class lively and positive; only have a class with good atmosphere, can students grow up healthily. Therefore, the members of the class leadership should be observed and selected carefully, which cannot be taken lightly. Counselors need to give student leaders in class and League a special and regular training, provide them clear responsibility, and help them to carry out the work actively, so as to develop their capabilities of self-education, self-management, and self-service, assume stressful task consciously when they are working, and enable them to participate in the student management work actively. Facts show that, only counselors give a full play of student leaders' subjective initiative, and let student leaders and relative students deal with the student management work together, can both cultivate students' abilities and make smooth progress of the work.

6 College Counselors Should Be a Mentor of Students for Entrepreneurship and Choosing Career

Entrepreneurship, a major event in life, is not only charismatic, but also challenging. Especially in the new millennium and the new century with a background of economic globalization, information networking, technology socialization, and knowledge capitalization, entrepreneurship becomes an important way to achieve the maximum value in their limited life. In universities, with the further development of educational system reform, students choose jobs and create business on their own, which has gradually become an important form of employment. In recent years, since employment pressure is gradually increasing, guiding college students to choose jobs rationally, and encouraging them to start companies, build entities, and create businesses, have become an important measure for universities to ease employment pressure and widen the employment channels. However, the present students' entrepreneurship on campus is failure more but successful less. The main reasons lie in the poor entrepreneurial atmosphere for college students, low degree of social recognition, weak entrepreneurship endurance, venture funding difficulty, and inadequate mental preparation for entrepreneurship. However, the much more important reason is that college students haven't fully understood entrepreneurship, and there is no enough intelligence support on college students' entrepreneurship from faculty. Therefore, the counselor should be a good entrepreneurship mentor of students. Through the courses of "Entrepreneurship Economics", "Entrepreneurship Philosophy", "Entrepreneurship Environmental Studies", "Entrepreneurship Management", counselors can provide them entrepreneurship theory, and actively build an entrepreneurship platform to help them design their business and guide them to put into practice, which can lay a solid foundation for students to get into society successfully.

7   Counselors Should Be a Facilitator and Creators of Campus Culture

Campus culture, an important part of popular community culture, has become the pioneer that guide social mainstream culture, that is to say, getting hold of campus culture trend means that we can grasp the ideological pulse of students to some extent, and from which we also can accurately grasp the trend of community culture.   Various campus cultures are not only a tempered stage for students' growth, but also a critical platform for students' quality development education in colleges. Healthy cultural activities on campus are a key carrier for students to cultivate their temperament, condense their mind and sublimate thoughts. Therefore, developing the health and progressive campus culture has become an important aspect of the ideological and political work. Counselors should actively guide students to engage in group activities, students quality development activities, technology innovation, and social practice activities, etc. and enable them to sublimate thoughts, show wisdom and practice abilities, which can help students to be the main body of developing community culture and to be the faithful representative of the advanced culture.

8   College Counselors Should Be the Designer of the Future Career for College Students

Living in the world, everyone wants to be successful in a career. However, not all men can achieve it, so how to make it successful? This is a question that every student cares about. Helping college students conduct career design can enable them to fully recognize themselves and analyze environment objectively, set scientific goals. Of course, there are also successful persons who have no career design, but if you have, career development and achievements will be greater. For newly graduated students, they are like a jade that has just been mined, while the counselor plays a great role in carving the jade. Counselor should be concerned about the healthy growth of students by considering the actual situation of them. In addition, counselor should design a series of training programs which are good for students' practice and improvement according to their interests, expertise, hobbies, personality, knowledge, skills, IQ, EQ and their ability of organization activities, management and coordination to help them make clear mission, and establish correct values and outlook on the world and life, and help students complete career planning as early as possible which can enable them to accomplish all tasks of different targets in stages in four years, and, continuously realize their short-term goals, medium-term goals and long-term goals. Therefore, college counselors should continue to study diligently, continuously improve their qualities, and strive to become a career designer of college students.

These are the author's brief idea about the work of college counselors. The work of counselors is trivial and complicated, which requires counselors to keep love, patience, care and perseverance. It also needs counselors to make friends with the students with a sincere heart, think what students are thinking and meet what students need, respect and appreciate students from the heart and give them more encouragement and support, more concern and help. Meanwhile, the counselor should also strengthen their self-cultivation and improve their inner quality constantly for the development needs of times to be a qualified college counselor in new era.

References

1. Han, L.: On problems and solutions of college counselor team's construction. Journal of Inner Mongolia Normal University (Social Sciences Edition) 3 (2008)
2. Zhao, S., Fang, L.: Professional construction problems of college counselor team. Data of Culture and Education 3 (2009)
3. Ding, Y.: Measures to strengthen the construction of counselor team. Time Education 2 (2009)
4. Jade, B.: On the recognition of the role of political counselors in colleges. Journal of Shanxi Agricultural University 4(2) (2005)
5. Wang, Z.: On the management work of students in new period. Journal of Beijing Vocational College 9 (2004)
6. Chen, M.: Application of service philosophy in student management in colleges. Science and Technology Innovation Herald (2007)
7. Li, C.: Application of motivation theory in student nurses management. Journal of Nursing Administration 2 (2004)

On the Enrollment Education of Post-90s College Freshmen

Li Dong and Ding Mingjun

School of Information and Communication Engineering, Tianjin Polytechnic University,
Tianjin, China
tjgydxld@126.com

Abstract. Post-90s group has entered college campuses ,which is as a special group in colleges, therefore, their school education is clearly very important. This paper illustrates some existing problems about 90 freshmen on college life, learning, interpersonal relationships, and so on. At the same time, the enrollment education of freshmen is explored. Finally, as for the freshman counselors' work, a few new working methods are put forward to guide freshmen to adapt to the new environment of universities, and promote the healthy growth and talent of Post-90s.

Keywords: counselor enrollment education, Post-90s, work, Freshmen, adaption.

1 Introduction

In the Report of the Seventeenth National Congress Party, to strengthen and improve ideological and political work, do good work about young people's ideological and moral education is considered as an important requirement to promote the development and prosperity of socialist culture. According to the survey, about 56% students on campus life and learning are not suited. As a Post-90s college freshmen just entering campus, how to do well on their enrollment education, to change roles as quickly as possible, to quickly adapt to university life is not only particularly important, but also higher demands for the new counselors.

2 Widespread Problems and the Cause of Post- 90s College Freshmen

On learning, learning methods and educational forms of change which led to the lost of learning objectives. Freshmen in high school generally have a clear objective, which is went into college. This goal is also told by the teachers and parents in advance, therefore, they do not need to think independently, just study hard. A variety of pressure around students from the community, schools and families does not allow them to have slack in the study. But in the university, the targets of high school have been achieved, while being used to follow parents, teachers, university students, the Post-90 has became very confused and helpless. The goal of many students like a kite without line, erratic, only following feeling, and what to do in next step has became a

J. Luo (Ed.): Soft Computing in Information Communication Technology, AISC 158, pp. 243–250.
springerlink.com © Springer-Verlag Berlin Heidelberg 2012

troubled issue of Freshmen. Target problems are not resolved, which could be muddleheaded, so that they may waste good times, or seemingly let themselves busy to cover the emptiness and helplessness.

At the same time, many new students neglect the exercise of comprehensive quality due to different college and high school education system, which most of the early high school are subject to make the evaluation of students by school achievement. Many students are only one child at this stage, especially the Post-90s freshmen. They having difficulties at home basically solved by their parents, while they lack the ability to analyze and solve problems independently. Despite in the opening period, teachers and counselors of freshmen have repeatedly stressed the need to familiarize themselves with "Students Read" and "Student Handbook", but many students have adopted an indifferent attitude towards the school's rules and regulations.

About Life, poor self-care capacity, poor interpersonal relationships, and generally low psychological qualities. As is mentioned above, most Post-90s freshmen are the only children, and the daily life previously arranged by the parents, the family spoil so many children, so that they are fed and dressed by others. After entering the campus, they come to a strange environment and begin to live collective life. Environment has been changed a lot, which need them to plan their own lives: how make the arrangements for basic necessities, how to spend money, how to arrange time, how to solve the problems. Individual students do not wait in a queue to eat at canteen, and even the basic necessities of life are difficult to take care of themselves. Lack of ability make them face enormous challenges in life, which requires a process of adaptation and change.

Thinking, self-centered, poor team spirit, and lack of self-discipline. Mostly Post-90s students are the only child, previous life before the University generally has a strong sense of superiority, they have key protection at home, and in school learn the leader , creating such an environment that they cannot help but self-centered, strongly self-esteem, feisty, strong sense of honor, but lack of team work spirit.

Recommending the excellent as the party's development object is a glorious task of communist youth league organizations that the party endows. As early as 1982, communist youth league the eleventh congress had formally written this task into youth league constitution. The sixteenth congress of the youth league writes: To educate the basic knowledge for league members and to recommend the excellent as the party's development object. The party's organizations at all levels pay high attention to leading the communist youth league's work of "recommending the excellent as the party's development object". The new party constitution of seventeenth congress clears the basic requirements of "recommending the excellent as the party's development object": to educate and cultivate the party activists, do better to regularly develop party members, and make a point of developing party members from among those in the forefront of work and production and young people [1].

With the development of the situation, the "recommending the excellent as the party's development object" faces many new problems. In order to deeply implement the "The Suggestions on Further Strengthening and Improving Ideological and Political Education of College Students" proposed by CPC Central Committee and State Council and adopt to the working requirements of youth league construction under the influence of party construction in new period, it is necessary to develop investigation and research, deeply understand the working status of "recommending the excellent as the party's development object", constantly to sum up experience, to discover new problems, and to propose new countermeasures [2].

School of Information and Communication Engineering, Tianjin Polytechnic University developed a questionnaire of "recommending the excellent as the party's development object" for 768 students of 07 and 08 grades in 2009-2010. According to this survey, it reflects the basic situation and the existing problems of college "recommending the excellent as the party's development object" from multi-sides. The students' whole understanding of "recommending the excellent as the party's development object" is correct, their attitude is right, and their evaluation of the working effect is pertinent. However, simultaneously the investigation results also reflect that there exist some problems of "recommending the excellent as the party's development object" in propagating the implementation of education and principle of democratic centralism, system construction, organizing and guiding, etc.

3 The Students Have the Enthusiasm in Participation, But Their Understanding Need to Be Further Improved

From the investigation results, the students' approval degree and contribute degree for "recommending the excellent as the party's development object" are higher. 83.2% of the students understand the work of "recommending the excellent as the party's development object", 95.3% of the students concerned the "recommending the excellent as the party's development object" of their class, while 94.7% of the students actively participated in the democratic evaluation meetings of "recommending the excellent as the party's development object". It shows that the deeply development of "recommending the excellent as the party's development object" has better mass basis. However, the students' comprehensive degree and significance understanding for the "recommending the excellent as the party's development object" should improve, some students for the participation of "recommending the excellent as the party's development object" only stay in the passive and "spontaneous" stage. 19.6% of students are subjective and random in democratic appraisal and voting, which shows the work of "recommending the excellent as the party's development object" lacks enough depth propaganda.

4 The Principle of Democratic Centralism Has Been Implemented, But the Democratic Level Need to Be Deepened

The investigation shows that students' attitude for the implementation of the principle of democratic centralism of "recommending the excellent as the party's development object" is positive. Majority of league members have the right to know, right of supervision and voting right in the work of "recommending the excellent as the party's development object", 73.8% of the students can more fully and objectively show their personal will in democratic evaluation meetings. However, 26.2% of the students can not fully have the right to vote. The unfavorable factors which prevent from further deepening democracy of the "recommending the excellent as the party's development object" still exist. First, the system construction is still relatively weak, especially lack of system criterion in micro operation level, resulting in the larger subjective random. Second, the organization ability level of basic level league cadres for "recommending the excellent as the party's development object" still can't satisfy the masses completely. 45.8% of the students thought that the organization function of class league cadres didn't play completely in "recommending the excellent as the party's development object". Due to the subjective factors, such as lack of working experience, simple working method, etc. it affects the effective realization of the democracy of "recommending the excellent as the party's development object" during executing the work.

5   League Cadres Have Strong Sense of Responsibility, But It Need to Give the Necessary Guidance

The responsibility and work level of basic level league cadres directly affect the working effect of "recommending the excellent as the party's development object". The base league cadres mainly include students' league cadres and teachers' part-time league cadres. The investigation shows that the students' league cadres generally have high working enthusiasm and strong sense of responsibility, but their comprehension level and executive level of related system need guide and improve. The teachers' part-time league cadres are generally department department department secretary of league general branch, they have strong sense of responsibility, and higher theoretical level and working level. However, due to the constraints of time and effort, to some extent it affects the specific guidance of "recommending the excellent as the party's development object", which only get 67.7% of students' higher evaluation. Under such circumstances, it is particularly important to enrich and stabilize the communist youth league cadres, strengthen the education and training of base league cadres. Simultaneously, how to fully move all the favorable factors and forces of counselors, head teachers, fore-class party members, etc. to common concern and guide "recommending the excellent as the party's development object", which is a working direction need to strive.

As the assistant and reserve force of Chinese communist party, the communist youth league members should play their own role in the process of the party members' development and cultivation, aiming to complete the transportation of qualified talents for the party, and cultivate the reserve force of the party [3]. The author believes that the role of the communist youth league around the "recommending the excellent as the party's development object" in the development of university party organization can be summarized in four aspects:

(1) Laying a good foundation, to play a role in education. The communist youth league plays an important role in the education of the party activists by using their unique channel from political thought, humanistic quality, psychological quality, etc.

(2) Practical test, to play a role in exploration. The communist youth league form a long-term exploration for the party activists through many activity carriers, such as social practice, extracurricular activity, amateur party school, party constitution learning group, etc.

(3) Evaluation and appraisal, to play a role in holding the pass. In the process of implementing the "recommending the excellent as the party's development object", the communist youth league has deep and detail understanding and recognizing of each party activities' thought, words and deeds, quality, performance, etc. it can hold the pass for the development objects through league members' Evaluation and appraisal.

(4) Inheritance duty, to play a role in recommending. According to the form of By "recommending the excellent as the party's development object", the communist youth league recommends the excellent young league member to party organization, in order to the further training and study of party organizations, and lay a solid foundation for the party's organization and development.

The above four aspects supplement each other and none of the four is dispensable, recommending is the goal, cultivation is the base of the recommendation work, observation is the test for the training work, holding the pass is the key of recommending, cultivating and observing, all of them form a dynamic process, and play an irreplaceable role in the development of party organization.

6   Emphasis on Prophase Education: Strengthening Thought-Leading, and Reflecting Educational Function

In college, after the young League member submitting the party membership application to the party organization, the league organizations should strengthen to train and guide them from the new height, make them become maturity in thinking, organization, style, ability. In the current international and domestic situation, it is particularly important to guide the youth league members in the comparison and distinction to distinguish wrong and right, distinguish primary-secondary, master ideological weapon, improve political sensitivity and recognition ability, consolidate ideal and belief, and further enhance the comprehensive quality. League organization must put the training and education of the young League member when developing "recommending the excellent as the party's development object" first, enhance the training and cultivating strength for the party activities, and fully play the educational function of "recommending the excellent as the party's development object". First, perfect the training system of students' party school learning and league school learning, continuously improve the political theory quality of training objects; second, establish ideological reporting system, understand the ideological trend of training objects, and develop educational activities; third, establish theoretical learning group, in order to achieve the goal of mutual learning, mutual assistance, mutual supervision, mutual improvement; fourth, establish talk system, take the initiative to understand and solve the confusion of training objects in thought and work actively; five, increase the training and education for training objects in various activities, through the activities to enhance the sense of social responsibility and sense of mission of the training objects, close the feelings between the party and people, improve their overall quality, thus continuously to deepen the training and education [4]. According to the education leading of "before recommending the excellent", it makes the young league member have firm ideal and belief, have the right motivation of joining CCP, stick to use Marxism, Leninism, Mao Zedong Thought, Deng Xiaoping Theory, and the "Three

Represents" to arm their mind. They are all round qualities, and they have both ability and moral integrity, and can take on heavy responsibilities.

2 Highlight process guidance: enrich subject and method, to reflect scientific evaluation

Innovative organizational settings, and enrich the main body of "recommending the excellent" subject. With the emergence and promoting of credit system, because of the students' free controlling of the learning content and speed, it makes the original more stable class league branch appear the larger personnel flow, difficult to concentrate organize the education activities, the unfamiliar with each other among students, etc. In addition, the extracurricular activities on campus has become increasingly active, the young league members belong to various organizations of different types are very common. The environment of "educating the excellent" is unstable, so it is difficult to achieve the systematic cultivation and objective evaluation of the young league members. Relying on the league branch of various classes to develop the traditional model of specific "recommending the excellent" which is hard to adapt to increasingly diversified study and life model of college students, naturally the scientific quality suffers greater question. Therefore, if the college communist youth league's "recommending the excellent" really wants to implement student-oriented, it should be based on the study and life reality. When evaluating the recommending objects, it can be considered to combine their main positions of work and study with their main positions of "recommending the excellent". The basic league organization is the stay point of various work of communist youth league, and also is the big stage for the young league members to accumulate experience, exercise ability and show their talent. Therefore, we should select the basic level league branch closely related with the recommending objects' work and study as the main position of recommending work. Because only they have the condition to make long-term training and objective observation for the recommending objects, really to recommend the excellent youth to become the development object of the party organization. The college communist youth league organizations should put the innovation's effort point of "recommending the excellent" in innovating the basic league organization form, build some basic level league branch which is more suitable for recommending objects, form a beneficial supplement for the league organization form in unit of class, lay a better organizational guarantee for the scientific development of "recommending the excellent". Innovative evaluation index, and enhance the scientific quality of "recommending the excellent". In the past work of "recommending the excellent", the standard of "recommending the excellent" is rather vague. Because the guide of school's and department's league members to the basic level league branch is limited, and the regulation system is more principle. In addition, because there is no specific regulation for the standard of "the excellent", it is just according to the standards of party member in "Party Constitution", this operation always makes the league members consider "recommending the excellent" equals to "selecting the excellent". Thus the students who have good marks, who are popular or good are easy to be recommended. In addition to the lack of the great propagation of "recommending the excellent", the work is not deep, there exists randomness and blindness. Some league members consider that the "recommending the excellent" refers to "selecting the excellent leader", "commenting three good student" and some even consider it is "electing the party member", and some consider that it is just a form, and some branches even use the ten minutes between classes to vote, all of them lack objective and real evaluation of object of investigation [5]. If there are the multiple main body of "recommending the excellent" and relatively scientific and reasonable evaluating system of "recommending the excellent", the method of democratic appraisal will become more important in the process of "recommending the excellent". The democratic appraisal method of open, just and fair, is not only the objective evaluation of a youth league member, but also is the realistic guide to other youth league members, so to select the scientific and reasonable, innovative and fair appraisal method will help to improve the effectiveness of "recommending the excellent". The traditional way of league branch democratic appraisal is that the league branch holds all league members meeting, introduce the practical conditions of appraisal objects, and submit it to the meeting to vote by ballot, and then public. This method should be the important form that majority league branches develop the work of "recommending the excellent" currently. This method has its scientific, but the author believes that, except the current form, each league members can explore some forms that more meet the actual characteristics of modern university students. For example, in order to avoid the embarrassment of face to face appraisal, the written comments can be increased; in order to more real to investigate the motivation of joining CCP of the party activities, the personal reply can be into increased based on the situation introduction of the branch; In order to minimize the possibility of human operation as possible, the each performance of the party activities can be showed in form of credit to quantize objectively; in order to improve the public knowledge degree, the public can be put in proper network, etc. In short, the quest for the democratic appraisal forms which are suitable for majority youth league members, the greatest purpose is to improve the effectiveness and impartiality of democratic appraisal through this form to select the good students by using the good form and good system.

7   The Anaphase Training: Continue to Track and the Investigate, to Realize the Connection of Party and League

The combination of party and league in "recommending the excellent" is not close enough, it doesn't form the better work situation that league organization actively promotes, and the party organization actively receives. Most colleges' league organizations think that it is the end after recommending the excellent to party organization. Once recommending the excellent, the party organization is responsibility for educational management; to a much greater extent there exist the phenomenon of emphasis on recommending and lighting training. In fact, the party after "recommending the excellent as the party's development object" still remains league membership and has league member identity. For the league members who don't develop, the base league organization still should develop evolution and education for them, and should propose higher requirement, and put them in a bigger space and higher platform for further test. It form a long-term exploration for the party activists through many activity carriers, such as social practice, extracurricular activity, amateur party school, party constitution learning group, etc. Timely report their performance to party organization from time to time, feedback their various data and files to party organization timely, and change the "recommending the excellent" into the important carrier that college party and league organization builds better interaction relationship, to really realize the seamless connection of party and league organization in party construction.

"Recommending the excellent as the party's development object" is the important work platform of college youth league construction under the influence of party construction, is the important carrier for communist youth league to strengthen self-construction and play educational function. In the new historical period, league's and party's organizations at all levels should strengthen linkage, actively integrate all educational resources, together discuss the work rules of "recommending the excellent as the party's development object", promote the benign development of "recommending the excellent as the party's development object".

References

1. Liang, Y., Liu, S.: Thinking on the "Recommending the Excellent as the Party's Development Object" Organized by College Communist Youth League in new period. Theory Horizon (01) (2008); o and TV University (03) (2008)
2. Ma, B., Xu, Z.: Thinking on the Several Important Links of College's Work of "Recommending the Excellent as the Party's Development Object". Journal of Liaoning Radi
3. The Suggestions of CPC Central Committee and State Council on Further Strengthening and Improving Ideological and Political Education of College Students. Printed by CPC Central Committee General Office (14) (February 8, 2004)
4. Wang, Z.: Study on the Work of College Students Party Construction in New Period. Journal of Liaoning Technical University: Social Sciences 8(2), 223 (2006)
5. Yan, M.: Thinking on the Improvement of the "Recommending the Excellent as the Party's Development Object" Organized by College Communist Youth League. Journal of Guangxi Youth Leaders College (01) (2006)

On the Class Collective Construction of Fine Study Style in College

Li Dong[1] and Jin Shangjie[2]

[1] School of Information and Communication Engineering, Tianjin Polytechnic University,
Tianjin, China
[2] Military Transportation Institute of the General logistics Department, Tianjin, China
tjgydxld@126.com, 376005054@qq.com

Abstract. Class is the basic environment for students' learning and living and the cradle for students' development, the study and life of class collectivity directly affect students' life quality in school, students' personality development and level, and students' following life quality and development. In this paper, it combines with the activity of "creation of the class collectivity of fine study style" in our school, the existing problems of colleges in the construction of study style are analyzed, and some feasible methods and recommendations are put forward.

Keywords: Style to create, Moral education, dormitory, Typical demonstration.

1 Introduction

The report of the Seventeenth Party Congress pointed out: "Education is the cornerstone for national rejuvenation; educational equity is the important foundation for social justice. We should fully implement the Party's education policy, adhere to cultivation-orientation and moral education first, implement quality education, aiming to improve the level of *educational* modernization, cultivate socialist builders and successors with moral, intellectual, physical and aesthetical development, and run the education satisfied people." The construction of study style is a very good important part in education, the fine construction of study style relates to the improvement of students' of a very important part of building a good of related to the students' comprehensive quality. Thus, it is an eternal topic to create the class collectivity of fine study style in college.

Study style is the attitude and behavior reflected by school's all levels of organization and staff in the political, ideological, organizational, work, and life, etc. Study style is the soul of a university, the fine style is a school's precious wealth and major educational resources, and is a tremendous spiritual force. General Secretary Hu Jintao had proposed when inspected in Beijing University: "We should cultivate fine study style; teachers should take the lead in creating the fine academic atmosphere" [1]. There are some new situations in the current construction of study style, showing some new *characteristic*, which requires us to timely pay attention and adopt new and effective ways to promote the construction of fine style study. However, class is a basic unit of student group, the study style of each class greatly affects each member in the class, and thus it is necessary to create class collectivity of fine study style in college.

J. Luo (Ed.): Soft Computing in Information Communication Technology, AISC 158, pp. 251–255.

2   The Existing Problems and Reason in the Construction of Study Style in College

After the hard study of many years, the students who just entered the university campus finally achieved their college dream, *starting* a new life and standing at a new starting point in college. There are many students who ultimately enter into college in the in the urging of their teachers and parents to study. However, in college there is no urging of teachers and parents, and their study environment is relatively relax, so lots of students think they are free, and then relaxes the requirements of themselves and cultivates lots of bad study style. There are some existing problems of the bad study style in college, mainly in the following aspects:

Vague study purpose. Many students believe that study is to "finish the job", and even consider study as "a kind of pain". The lack of scientific and *correct* study purpose make it easy to become the quick benefits of study.

Lack of motivation. According to statistics, in research topic "What is your learning motivation", there are 34.6% of "specialization interest", 3.03% of "pressure from family, school and society", 33.1% of "personal future", and 2.0% of "scholarship and honor".

Slackness of learn discipline and poor study habits. There are many students who don't fully grasp the characteristics of the education in the university, and they don't timely adjust *their* study habits and study methods, so they often are in a passive state. They don't cultivate positive, conscious and active study habits, and learning methods are inappropriate.

First, they don't listen to the teacher carefully in class. Second, they do not consciously learn after lesson. Third, they do not pay *attention* to the cultivation of practical ability. There are few students who complete the experiment item dependently and correctly, and they ignore and take the chance of practical experiment for granted. Fourth, there is little awareness for students to improve their overall quality in their spare time, so they don't learn the university resources to enhance and improve themselves, preparing for the following work and life.

Negative examination discipline. As lots of students is lack of integrity of honesty character, lack of personality cultivation. As they usually do not study hard and their knowledge isn't solid and comprehensive, they are opportunistic and cheat in examination.

3   Reason Analysis

The impact of the change of social environment on the study style. Currently, there are many bad tendencies, such as quick benefits, materialism, hedonism, etc. which serious impact on the school culture atmosphere and students' learning psychology, leading to the negative phenomena of style study, such as impetuous, utility, undisciplined, lack of motivation, etc.

Their knowledge for the status and role of construction of study style in school's development is still not enough. Currently, many schools usually focus on hardware, but not software. The construction of study style grasps relatively loose and the construction of study style is considered as the thing of a certain section, which isn't included in focus of the school building, which is obviously inappropriate.

The construction of study style is lack of effective measures. In recent years, many schools have clearly recognized the importance of study style, but there is lack of systematic measures in how to grasp the study style, many of which have become a mere formality. For example, every college takes the construction of study style as their aim of running school and characteristics, when students just enter college, they say "grasp study style" loudly, but there are only several meetings and some lectures till the graduation of students. Thus the study style can't be grasped better obviously.[2]

Student's own factors. Most of the college students are only child, they can accept new things and new information because of their active thinking, but their self-centered sense is strong. They are diligent in thinking, desire to succeed and have high self-evaluation, but lack of hard-working spirit. It is found that it is a collective behavior for students in the same dormitory or some student group to play computer games and play cards together, but not study.

4   The Methods and Recommendations for Strengthening Class Collective Construction of Fine Study Style

Correct study style by solid moral education. Students should correct their study style from their study motivation and study objectives, to correct study attitude, and to grasp their study enthusiasm, effort degree, volitional quality, their inquiring ability, etc. however, the set of learning objectives, the inspiration of motivation, the correct of attitude and the cultivation of volitional quality all need be helped and improved through moral education. The moral education in our school is from the following aspects: first, it is a prominent problem of the lack and confusion of learning objectives for the current students. The lecture was held by our school's vice secretary, who helped students to establish lofty ideals, to plan their career and university life, to solve the problems of "why we learn?", "for whom we study?" etc. Second, it helps students to improve their learning cognitive, reinforce their learning emotion, from study seeking to study knowing, and then study happy; change "let met study" to "I want to study". Only by solving students' ideological problems, make them study hard and study seriously, and ultimately form a good study style. Third, it helps college students to learn how to be a person, and learn to respect the fruits of others; requiring students get their real results through their own efforts. Fourth, it can build the inside and outside linkage mechanism between school education and family education. In family education, the school should give students' parents one letter to tell them the characteristics of the college life at the beginning of the college, and tell the parents what students should pay attention in school about the study, life, etc. In addition, for specific learning difficult students, there should be a learning warning for them, then to form organic interaction between schools and families.

Guide study style through good institution. Study style is not only just the style of students in learning, but also it should be a comprehensive atmosphere which integrates atmosphere of school education, teachers' educating and fostering students, service atmosphere of staff for students, and students' exploring atmosphere of study. Therefore, the study style relates to all aspects of school work, party groups, students and staff all have responsibility. It should be unified command, assigned responsibilities, made coordinate operations, and make concerted efforts and layered advance. First, there should be set a leading group for the construction of study style in school, the office of the construction of study style should also be set which is led and coordinated by student affairs office, and the other sections and each secondary college as member units implement the related work. Secondly, there should be set a leading group for the construction of study style in college, which implements specific work steps, combing the actual existing problems and summarizing the gained experience to make pointed rules and programs for the implementation of study style. The leaders in charge of the relevant matters should be set in college, tutors (head teacher) should deep into the classrooms and student dormitories to know students' study situation and supervise rules of study style. It should play "self-education, self-management, self-service" in each class, and set up a inspector group which takes the relevant teachers as leaders, and students as the main body, to check students' class attendance, study situation of evening classes.[3]

Encourage study style through better measures of rewards and punishments. While constructing the study style, we should also establish rules of rewards and punishments, using the combination of incentives and encouragement to promote students' learning enthusiasm, and play an active guiding and promoting role in formation of fine study style.

For student leaders. The student leaders who have warned above probation warning in one semester should be removed their existing duties. Students should put study at a very high position, so the students leaders who absent classes without any permission, the tutors or head teachers must give them some guiding education after confirmation; the student leaders who don't correct after education, their positions in all student organizations must be cancelled. For the student leaders who are addicted to playing computer games and playing other games in bedroom which disturb other students, if they don't correct after education, at least give them a warning and remove all their positions. Student leaders should play active leading role in the construction of study style, they should make friends with poor study students actively. For the student leaders who play an active role in the construction of study style and helping students with learning difficulties should be praised and would priority be recommended all kinds of rewards and employment training opportunities.[4-6]

For ordinary students. The students who absent class without any permission, they must be criticized the first time, and their selection qualifications of outstanding members, outstanding students and merit student would be cancelled the second time, and selection qualification of their scholarship would also be cancelled. The students who have don't pass the examinations more than one subject (including one subject), their qualification as an Excellent League to join the party would be removed. For the student who are addicted to playing computer games and playing other games in bedroom which disturb other students, if they don't correct after education, they must be criticized, and their selection qualification of scholarship would be cancelled the second time. The students who make significant progress (their grade ranking increases by 20%) in each semester should be praised. For the students who play an active role in the construction of study style should be praised and would priority be recommended all kinds of rewards and employment training opportunities.

Modeling study style through typical example. The power of examples is endless, some of model students can impress and infect others through their own words and deeds, and motivate other students to identify and model, and then to promote the construction of study style. It can be specified into three aspects: first, focus on the leading role of outstanding students, and actively implement "party leads class project", "the learning mode of one by one", etc. according to the promoting project to enhance the relationship between students, and increase the mutual spirit between students, and then to achieve mutual progress and realize win-win. Second, it should carry out the activity of "report group of deeds of college students". It should fully play the "star" effect of advanced students, change persuasion education to model education, and change classroom education to imperceptible education, using the advanced deeds to inspire students, and using the excellent typical examples to guide students, and then to promote the formation of study style. Third, it should select the students who have expertise in professional development and interest training to form "student instructors", and organize the topic report of students' self-education. The lecture and topic report are held by college students can stimulate students' learning enthusiasm, and it can also set a good example for the students who attend the report and make the study atmosphere strong.

Study style is like the invisible backbone of college, which always supports all aspects of our teaching and life, so it is necessary to create class collectivity of fine study style. Under the new situation, the construction of study style has a long way to go.

References

1. Hu, J.: The talk in the Forum on behalf of Teachers and students at Peking University. Guangming Daily (May 4, 2008) (A1)
2. Yu, L.: Several problems studied by higher education and science. Research in Educational Development (January 1982)
3. Sun, C., Li, W., Li, H.: On survey of the reasons why students go to college - A Case Study of Students in seven Colleges. China Youth Study (11), 63–65 (2007)
4. Li, Y.: Rethinking on the construction of students' study style. Student Work (1), 119–120 (2005)
5. Cai, Y.: Status and Countermeasures of the construction of College's study style in new period. College Management (January 2005)
6. Kerr, C.: The Use of the University. Howard University Press, Massachusetts (1995)

Some Discussion about the Team-Building of University Student Cadres in a New Situation

Li Dong[1] and Jin Shangjie[2]

[1] School of Information and Communication Engineering, Tianjin Polytechnic University,
Tianjin, China
[2] Military Transportation Institute of the General logistics Department, Tianjin, China
tjgydxld@126.com, 376005054@qq.com

Abstract. Students' cadres is a bridge between university and students, they play a vital role in the work of university students, so how to exert a better ability of them is of great importance to the development of university. This paper analyzes a lot of problems existed in the team-building of university student cadres and presents my own view on how to improve the executive ability of student cadres and how to enhance the team-building of student cadres.

Keywords: student cadres, executive ability, effect.

1 Introduction

Student cadres mainly include party branch secretary, cadres of Youth League Committee, student committee, class as well as student organizations, they all play a very important role in the work of university students and are a significant and powerful part in successful quality education. The executive ability of student cadres can affect the overall quality of students to a certain extent. So improve the executive ability of student cadres and enhance the team-building of student cadres are of great significance to promote the university's ideological and political work.

2 The Role of Student Cadres in University Education Work

Student cadres have double identity in the management of university education. On one hand, they are ordinary students, but on the other hand, they are a supervisor of class and at the same time a bridge between teachers and students. Student cadres are the backbone of the Higher Education Management system, their main role reflected in the following three aspects: First is as a role in organizing and leading. Student leaders at all levels are leaders and organizers of different activities. Under the leadership of party and government, with the Mao Zedong Thought, Deng Xiaoping Theory and "Three Represents" as guidance, take study as a fundamental task and put improvement of the overall quality as a premise, student cadres make work plan and carry out activities with more ideological, edutainment and infotainment according to the operational requirements of the university and features of university students. They also lead students to implement "first to excellence", and comprehensively improve

J. Luo (Ed.): Soft Computing in Information Communication Technology, AISC 158, pp. 257–261.

their ideological and political qualities, professional and cultural qualities, physical and psychological qualities. Second is as the role of a bridge. Student cadres deliver some relevent university's policy-making, requirements, and information to students through a certain way, while in turn feedback the problems, opinions recommendations to university. They also report to the university leaders and functional departments with the students' thinking, learning, living conditions and views, requirements, recommendations in a timely manner, which supplies a correct decision for the leadership of departments and some effective information for improving the management of university education. The third is an exemplary role. Student cadres are recommend by students and teachers, or by self-appointment based on the principles of "voluntary, democratic, open, fair and just". In learning, they study hard and strive to set an example for good; in life, they care about the students, help each other and set an example of disciplined civilization; in the class building, they set a good example in motivate classmates to build a collective good class; in the students' work, they make active efforts to improve overall quality. By doing so, the student cadres can get endorsement and support from classmates, thus their role as an example are of great impact to students.

3   Reasons for Insufficient Execution of Student Cadres

Generally speaking, today's student cadres have a good executive ability, they can achieve a balance between study and work, they work hard and pragmatic, and serve for students with whole-heart . They try their best to help teachers and leadership with ordinary work and make a certain contribution to university. However, the phenomenon that the quality of student cadres rank in different levels, lacking of executive ability continues to appear. That is to say they can not well done or simply can not complete their task. The reasons are as follows:

First, objective is not clear: A clear objective is the key of executive ability. Many students think that they have already finished their life goal after complete the university entrance examination. They don't know that as a college student they still should have their own goals of learning and working, the same problems also existed in some student cadres. Therefore, students will lack working motivity when they really join in work, and as time goes by they will have a resistance or even a hatred to work. Which one is first, studying or working? That is a hard and confused questions to the majority of students. Some of them only focus on studying and ignore work, and those student cadres who just focus on work may affect their prestige to students because of their academic record. So in a word, have a clear objective is the key to the successful implementation of tasks.

Second, seriously benthamism thinking. Psychological research shows that human behavior is caused and determined by motivation. The motivation has a direct impact on behavior result of student cadres. Some students try best to be a student cadre just because they can feel superior to ordinary students and can become closer to teachers. Furthermore, they can enjoy some preferential policies when appraise and choose excellent, advanced, scholarship or join the party, they even enjoy the power to manage other students. Once these people achieve their goals they'll become arrogance and do nothing except enjoying, some even bully others and badly affect others.

Third, lacking of will, some students give up halfway in university.   Student cadres undertake a great many of important work, therefore their job is complex and heavy. All those require for a strong will, hard working, sacrifice and pay out of student cadres. But the present student associations usually have a common phenomenon that several dozens of fresh members will be taken in every year, few of them can hold on for half a year. That simply because some students feel like full of energy to do a big job at the beginning, but they cannot undergo work pressure and   setback after a long time, they are afraid of difficulties and give up their jobs halfway.

In fact, everyone will encounter some pressures and difficulties in work, but no matter what kind of issue it is, the key is to find a way to solve all the difficulties. And in this problem-solving process, there is no excuse for anyone, so just go ahead, break through all obstacles, and ultimately achieve goals.

Student leaders have a dual role, they are both public servants and a member and master of classes. On the one hand they help teachers with ordinary work and manage other students, on the other hand they must set a "self-management" model for students. However, some students cadres lack of self-discipline, responsibility and a serious attitude. They even can not finish some ordinary jobs while some other students can. Such kind of student cadres will definitely lose credibility and encounter a lot of difficulties in working.

4   Several Proposals on How to Enhance the Execution of Student Cadres

At first, we should enhance training, improve the quality of student cadres and guide student cadres set goals, because goals bring on motivity. However, student cadres neither have a pay nor powerful right, they must have a clear position on their own, that is to say the goal of working is aimed at simultaneously serves for students and increase their own capabilities. Youth League organizations and political workers should give student cadres a job guidance and training in order to let them have a clear identification of their jobs and at the same time guide them set a recent and long-term objective step by step by combining themselves with reality. They should know what to do now and in future, it is better for them to make a practicable career plan to achieve self-development and fully implement their social value.

Second, a good and firm executive ability is to strive to practice with highly-efficient. People are born with inertia, but as long as you keep insist in you can be succeed. Each success is only achieved through hard work. Student cadres are executors with practical responsibility. They should work hard to achieve goal as soon as possible   in the spirit of   practicality without asking too much why. Dripping wears away a stone. Nothing is impossible. As long as you own a clear objective and a passion serving for students, focusing on details, you can achieve any goals no matter how complex and trivial the problems are.

Third, timely monitoring, enhance self-management. Student cadres who have the power can be easily got "corruption", especially those who lack of self-control. Usually their work experience is insufficient, ability needed to be improved, so teachers should give them right guidance, appropriate criticism, and establish a system to constraint and management them. In order to let student cadres practice what they preach and strengthen self-discipline, we should organize an movement calling for "a cadre, a flag" activities to establish the typical student cadres model. By doing so can establish a mutual supervision between student leaders and students and meanwhile among student cadres. Teachers should called on them to help students around to establish credibility. At the same time it is good to unite students and consolidate the mass foundation, then they'll get supports and co-operations from students to promote the implementation of student tasks.

The establishment and improvement of appraisal system, in time, can be a combination of regular and random checks to ensure the fair assessment; In content, the assessment to leading organization of student activities and implementation process is the key; In methods, we adopt a combination of unannounced visits and checks. We teachers should give student cadres timely feedback, problems analysis,timely improvement and a appropriate praise to encourage them.

In addition to the trainings of working methods, psychological counseling, group training, enhancement of team awareness and persistent endurance are also necessary. In short, the execution upgrade of student cadres is not an easy problem, we can't solve it in one day. That requires a co-operation between teachers and student cadres by constantly correcting   mistakes and exploring methods.

5   Some Thoughts about How to Strengthen Team Construction of Student Cadres

We should focus on the selection part of student cadres to provide an organization assurance for team building and management of student cadres. Student cadres are both elites and backbone of students, their role are irreplaceable. Student cadres of today may be leaders or backbone of tomorrow. Therefore, when select student leaders, we must adhere to high standards and strict requirements and deal with this work in the purpose of training successors. First of all, fully achieve the work of the election of members in first league branch and class committee. A good cadres team can help university administrators do better in education and forming of study style. After the freshmen enter into university, counselors are required to recongnize and know students in the progress of military training. In addition they should know and observe some excellent students by a serious of activities in military training(such as internal health, song competition in military training, propaganda and report). As for those who submit an application to join the party, teachers may pay more attention to them. After military training, we can't set up the first class committee and League branch only because of some personal opinion or in case of troubles, we must adopt methods such as individual application, recommendation by teachers, organizational talk, selection of candidates and democratic elections of class. Secondly, the later selection of student cadres should follow the principles of democratic centralism and play up strengths and avoid weaknesses by the purpose of training and educating to select student cadres and make good use of their subjective initiative.

Second, we must be focused on training to provide effective guidance for works of student cadres. We can take the primary party school or the youth league school as training base, adopt the following several ways to train student cadres: First, invite some leaders, experts to make reports and lectures to preach the party's principles, policies, describes the domestic situation and subject developments as well as the recent work of the college and basic work methods of students to help student broaden their perspective, enhance their theoretical training and sense of responsibility and mission; Second, organize a debriefing and experience-sharing meeting of student cadres to make them learn from each other and improve together. Third, organize short training courses, organizations about student cadres learning political and business as well as a variety of social activities to help student cadres have the necessary theoretical knowledge and multiple skills to improve their overall quality. Fourth, we must improve and enhance the system of team building about student cadres. In order to improve the quality of student cadres, in addition to self reflection, hard study and diligent practice, we should also take measures to strengthen training and assessment to improve their quality. To establish the appraisal system of student cadres and improve the quality of student cadres, only by practice we can prove examination and analysis about "morality, ability, diligence, and achievements" of student cadres, and only by practice we can objectively evaluate the work of student cadres to effectively promote self-reflection, self-education and the quality improvement of all student cadres.

6   Conclusion

In summary, team building of university student cadres in new situation is a systematic project, with the help of party organizations and student committees at all levels, we should have a overall planning, and actively explore new ways and means. Team building of student cadres is undoubtedly a work of long-term significance, it is of great importance to university personnel training. Only if we fully play the role of enthusiasm and exemplary, we can better promote all students have a healthy growth.

References

1. Zhao, Z.: Successful Quality Oriented Education-New Breakthrough of University Ideas and Models. China Education (April 9, 2004)
2. Shi, Q.: Management and Training of College Student Cadres. Development 12, 123–124 (2006)
3. Yang, W.: Theory of University Quality Oriented Education. Nanhai Publishing Company, Haikou (2002)
4. Gao, C., Hui, O.: Training Manual of Teachers' Quality. Jiuzhou Book Publishing, Beijing (1998)
5. Xu, Z.: Outstanding Student Cadres in University-Research on Capacity Needs of Future Youth Leaders in Talent Management. Consultation and Guide on Technology (16) (2007)

Research on the Monopoly Management in Clothing Brand Communications and Innovative Ideas Construction

Peng Yong[1] and Teng Kunyue[2]

[1] Institute of Art and Fashion, Tianjin Polytechnic University, Tianjin, China
[2] Textile Institute, Tianjin Polytechnic University, Tianjin, China
httpheart@sina.com, tracy_weid@yahoo.com.cn

Abstract. Through study on clothing brand, this paper proposes monopoly management theory in brand communications and construction of innovative ideas. Using some successful cases, it also analyzes the route of transmission of clothing brand and marketing strategies. Besides, by analyzing the transition from wholesale to retail of monopoly management in clothing market and the transition of the internal management in clothing brand stores, it finally studies the innovative ideas of clothing brand communications, which has important research values.

Keywords: clothing brand, monopoly management, innovative ideas, construction.

1 Introduction

The development of clothing stores has a maximum limitation from the monopolized products. Because it is monopolistic, therefore it will face with different types of consumer demands and the choice of four seasons. Few domestic brands can launch products which are suitable and popular in four seasons. Some brand products have advantages in blouses and skirts, some in down coat or woolen sweater, and some in suits. For example, Erdos has strength in woolen sweater, Seven Wolves in casual clothes, Bosideng mainly in down jacket, JOE ONE in trousers, and etc. So a single brand store has limitations to some extent and can only meet a part of consumer demand, which has an obvious off and busy seasons. Many monopoly stores often use the profits in busy seasons to subsidize off season, as a result the pressure is relatively large. Sometimes, in order to make up product deficiencies, a lot of monopoly stores will introduce some products of other brands to supplement sales, but this move will inevitably have an impact on the integrity of monopoly system, which makes monopoly lose its meaning.

Some small brands are bound to find it hard to make a breakthrough relying on natural sales because of their insufficient influence, shortage in fame and needless to say reputation. Clothing dealer's overall operation level is not high and clothing is not with high barriers to entry, which results in uneven distribution of employees, and many enterprises are not strictly in selecting dealers. A great number of dealers simply do not possess the ability to run monopoly brand, still stuck in the management concept

J. Luo (Ed.): Soft Computing in Information Communication Technology, AISC 158, pp. 263–267.

of grocery stores, lack the basic knowledge on brand, price and image, and cannot keep up with the business development ideas of producers, which is also the reason that many stores have poor management.[1]

2   Clothing Brand Communication

The essence of marketing is to satisfy the target consumers' demand. Nowadays, with the great number of brands, the most important task facing by clothing business is to provide differentiated products and services, and establish a differentiated brand image in accordance with target consumers' demand. Here, the differentiation should be specially emphasized. As mentioned above, in the same mall, we are always carrying out design and service, price competition, so how can we highlight ourselves from so many brands in the mall? Of course there are lots of means can be available, but the focal point lies in brand building. First, we should often ask ourselves, do we really understand those target consumers? How does their mind change? What are their impressions on our brand? How to strengthen their loyalty to our brand? How to adjust the 4PS (product, price, place, promotion)? Certainly there are many other factors which may impact clothing brand image such political factor, economical factor, cultural environment; management, human resources, capital and so on. While the clothing brand building can start from the fundamental consumer demand. A product is the carrier of a brand to transfer appeal, so a product without story or a product without soul cannot own the fame and reputation, and it at best can represent only a symbol or a name.

Brand story is a kind of transmitting idea which is extracted from the brand development process and represents its advantages. It's clear, easy to remember, and full of imagination. Speaking of this, some people will think of the advertisement of many companies, however, consumers can only remember the brand name at the moment of seeing those advertisements, but cannot deeply understand the real concept of the brand. Brand story is another form of advertisement, and it is a successful emotional conveyance between brand and consumers. What consumers want to obtain is not just a piece of clothing, the emotional experience and related association are also expected. Besides, this kind of association can help consumers create connection and acceptance to brand. Therefore, a successful brand is made up of numerous moving stories, no story no brand. Now let's see some successful examples:

Giordano brand is founded in the early 90s, which brings us colorful clothing and uses a green frog as LOGO. The appearance of Giordano is like bright sunshine or a cool breeze, so refreshing. It has advocated a kind of dynamic and natural way of life and its brand concept is "Giordano, a world without strangers", from which it can be seen that Giordano's main products are all basic styles. Giordano is a colorful world where there is no gender gap and age difference, so anyone can choose his favorite product if he likes Giordano, which is just one of the final method that Giordano can move consumers to use its compatible image, and also the key to obtain the consumer acceptance.[2]

LV is a luxury brand with a history of 100 years, which has served for royalty and is famous for its leather products. From a common cobbler to serving for Louis XIII, LV has experienced a legendary background, however, only these stories are not enough to increase the core value of brand. There spread many popular stories in the industry or among consumer about LV's stringent requirements: the fastener of LV leather products will be tested for thousands of times before shipping; after processing, the leather products will also be conducted the destructive test such as infrared ray, ultraviolet ray, corrosion resistance and falling from high place. What should be specially mentioned is the procurement of raw materials, for LV strictly selects enclosed leather produced in UK and France, so that even the BMW claims that the seat in their car has applied LV leather.......In a certain way, it is these stories that help LV establish unique high quality image in consumer's heart.

It's not different to see that the growth of a brand is made up of countless number of stories, which include the legendary birth story, story of being strict on brand's quality as well as touching story about service for consumers. After experiencing 10 years, 20 years' development, any brand has its excellent story and each step is a story. However, eventually what really moves consumers and makes them bear the brand in mind is just the emotional resonance built by these stories rather than the product itself. Of course, brand story is not artificially made up or copied out, but it originates from business's concern about brand such as holding various events, promotional activities, social welfare activities and some human-oriented measures of product quality. Create brand story is an accumulation of service and attitude that a successful entrepreneur objectively treats brand, product and dealer. Therefore, if disseminate the brand story accumulated by these measures through advertisement and many other mediums, when the dissemination reaches saturation, the brand will touch the heart of consumers in a friendly and easily closing way. If reaching the highest level, the brand will become a kind of religion which may impact the consuming concept of consumers and eventually become a belief.

3   Clothing Marketing

Clothing monopoly is the most popular clothing sales model in recent years, and some people in clothing marketing circle even call it as "the final marketing model in clothing industry". But is the sales model of clothing brand monopoly stores really optimistic as it looks? A spacious and luxuriously decorated BUSEN store in a place put up a poster of ready to transfer and big clearance; a big ShanShan store in a city has little customers, and the female boss looks dejected; a Romon store in Guangzhou CTS business tower is ready to transfer and closed; K-Boxing stores sell other no-name products; bossini flagship store in Beijing Road, Guangzhou changes hands, etc. These brands are leading brands in china clothing industry which are very influential, but their stores have also suffered from declining market; of course, this may be just a minority phenomenon in some local market, however it at least tells us that the monopoly store model does not always look perfect; particularly for the majority of middle and small brands, what is the condition like of their marketing?[3]

Currently, in many clothing street in our country, nearly 80% monopoly stores are found at least not in good profit status, and there is a common phenomenon in each city: the mainstream brands of monopoly stores usually are making profit, while the middle and small brand stores mostly keep going painstakingly; the business of stores in busy streets appear to be prosperous, yet stores in less busy streets have very few visitors.

Another phenomenon also exists: a lot of clothing monopoly stores are not really "monopoly" with operations in some other brands, some women's boutique even sell man's wear.

All of these are signs of potential crisis. If stores which are considered as the sale terminal do not operate well, and cannot generate enough sales pull, how could they stimulate the development of upper enterprises? How to promote the healthy development of the clothing industry? For many brands, the close of one monopoly store means the declining of the brand in this region, and the resulting domino effect would spread around the city, and to boil down, the problem is that the marketing model of domestic monopoly stores is worrying.[4]

4   Innovative Concepts in Clothing Brand Communications

With the progressing of global integration, clothing market will become increasingly standardized so the competition will be more and fiercer. China's clothing business has experiencing a collective decline, which is due to its lack of momentum for the innovation of brand. After completing the original wealth accumulation and brand promotion, many businesses become ignorant for how to improve core competence of their own brand, grasp the market and judge the future developing direction, so they are always struggling to find a new breakthrough. In fact, the innovative breakthrough of a clothing brand cannot be blind, and a breakthrough point must be found. The key point is to consider the innovation of the clothing business. A man without distant care must have sorrow, in the same way, the development and innovation of clothing business needs to start from two types of transition.

Monopoly management transition from wholesale to retail in clothing market. As there exist issues such as homogenization, distinctiveness, management capacity and market demand, etc of clothing product, the profit of many clothing franchised stores has been declined, which leads to the transition from wholesale brand to retail brand. This process can be summarized as that using 10 years to complete the wealth accumulation, 2 years to conduct blind transition, 1year to declare bankruptcy. Is the transition really so hard to achieve?

On one hand, the transition should be conducted from the aspects of concept and execution, as for business, transition not only means a leap but also a barrier, if business can cross this barrier, it will head for another development trend. YICHUN, Koti are the successful cases for transition, and their success lies in the well prepared concept, technology, and idea.

On the other hand, be clear that wholesale and retail means 2 marketing models that are entirely different, mainly shown in the following 3 aspects.

(1) Different marketing ideas. Wholesale emphasizes less style, strong imitation, and flexible reflection for market, so grasping imitation, physical distribution and customer means success. However, retail is just at the reverse, the problem of which lies in that the design of the product should own personality and clear brand positioning. In addition to guarantee sales, the personality of product should be ensured to cultivate a kind of identity with consumers and finally make the brand be recognized by the market to the maximum extent.

(2) Different cash flow. The cash flow of wholesale is faster, while the retail cash flow is relatively slow.

(3) Different inventory management. The inventory of wholesale is usually an inventory structure of a single product, while the counterpart of retail, because of different sales channels, will generate diversified inventories, which gives enormous pressure on enterprises.

The transition of the internal management in clothing store. Such a transition like Metersbonwe, YiChun, Li Ning and other mature enterprises with scale, after the successful completion of the first transition, they gradually builds its own brand, and the next round of transition is from the internal management of enterprise, including management innovation, marketing innovation, and demanding profit from management and marketing. Through the brand modification, to meet new market demands.

5   Conclusions

You in future competition, our enterprises should learn how to enhance the responsiveness to the market and increase profit through management. Speed should be considered as the means to improve company's ability to meet market needs and cut inventory, brands like ZARA, GAP, and UNIQLA's quick responsiveness to market are good learning examples for us. Grasping popularity, brand positioning, and the leading edge information play a key role in clothing brand innovation, and the constant innovation is the key to enterprises' success. However, innovation is easy said that done. A successful innovation is real innovation, otherwise it is trauma. Without innovation, enterprise will be like frogs in boiled in hot pan, if they do not jump out, they will die slowly in comfort. Innovation can let us find out the growing point of future, and it reminds us that "I come from 2010", which requires us to treat today's market with future insight. Only in this way, can we keep fresh in continuing innovation and achieve greater success.

References

1. Zhang, J.: Theory and method of international competitiveness evaluation. Economic Science Press (2002)
2. Jin, B., et al.: On competitive economics. Guangdong Economic Press (2003)
3. Huang, Y.: Launch of green clothing by Chinese textile and garment green clothing for WTO. International Trade (6), 18–121 (2001)
4. Li, H.: Whether clothing business can cross the Green Trade. China Textile (5), 20–123 (2001)

Research of Value Segment Evaluation Index System Based on Campus Network

Xianmin Wei

Computer and Communication Engineering School of Weifang University, Weifang, China
wfxyweixm@126.com

Abstract. Campus building should focus on building networks, building applications, building teams, building mechanism, and must put the application in a prominent position. This paper from the start of all the value segment on quantitative assessment indicators and all general assessment indicators of the campus network, combining the actual situation, given the effective assessment and evaluation indicators from the parameters, the purpose is to promote the campus network hardware, software, information resources and other aspects of balanced development.

Keywords: campus network, value assessment, index system, evaluation method.

1 Introduction

In the digital campus network construction, the wave of campus construction is great. The actual effect of the latter campus network construction was often ignored by the people, the rapid construction and incredible waste of resources are in proportion to the development. How to make the campus network construction closely integrate with teaching, to change education thought and teaching into a fundamental change, to make teaching information promote the modern teaching, we need further efforts.

To assess the campus network construction, hardware and software technical indicators need to quantify, the input of funds to have continuous protection, the network database construction to plan, the average cost of interconnection to the appropriate machine, network traffic should be moderate and so the average machine. Now quantitative assessment indicators of each section of the campus network value evaluation system and assessment indicators system of all the value section need a brief overview.

2 Quantify Evaluation Index of Each Value Section in the Campus Network

Campus network should enable the institutions to establish the teaching, management, office, and information exchange to change from fundamental cases. In the support of software to achieve online work, to standardize management practices institutions, to improve management and efficiency. While the vast resources and information of institutions are online, using the campus network and Internet connection, to make the

J. Luo (Ed.): Soft Computing in Information Communication Technology, AISC 158, pp. 269–274.

institutions can use the network to achieve family, teachers, higher authorities, and other exchange of information between organizations, so that teachers and students have access more information resource for the service of teaching.

In order to make a scientific assessment of the campus network, we proposed the number of people in the region of campus network, the following is divided into a few stalls of 5,000, 5,000 ~ 8,000 8,000 ~ 10,000, 10 , 000 more than, from the hardware and software environment, remote connectivity, network applications, security management to conduct a comprehensive quantitative assessment. For 8,000 to 10,000 people as one stall, for example, to quantitate the assessment indicators of campus network system.

Hardware and Software Environment. Server, with CPU 3.0GHZ or above, 2.0GB memory or above, 198GB hard drive or above, should have the primary domain servers, file servers, database servers, application servers and video server, using a reliable data security mechanism.

Router, usually the router backplane throughput is key indicators of measuring performance, more than than 40Gbps are known as high-grade routers, backplane throughput between 25 and 40 Gbps are known as the mid-range routers, and lower than 25Gbps called low-grade router.

Switch, CPU of switch uses special purpose integrated circuit chip ASIC, RAM / DRAM main memory, NVRAM storage and backup files, Flash ROM image file storage system software, ROM diagnostics, guidance and operating system software storage.

Bandwidth, backbone with 1GBps, switching to the desktop 1000/100/10MBps adaptive.

Network nodes, 2,500 or more, or greater than 40% of the number of teachers and students.

Scope, administration, teaching spaces, library information center, and teacher and student hostels.

Network software, Windows, Linux, Unix, application software and education software.

Remote Connections. Connection model, broadband or xDSL line access, and having own home page and proxy server.

The Scope of Application Systems. Institutions management, opening the Senate, security, students, personnel, finance, party, logistics and other management systems.

Teaching and Learning, classroom teaching with good teaching information and resources, high quality courseware, there are material library and practice exam.

Research Management, a dedicated research information service system, information resources and with adequate maintenance of professional acquisition personnel.

Book application, libraries in network management, and with editorial, circulation and retrieval system, electronic reading room has 500 seats or more.

IC card applications, connections with the campus network.

Safety Management. Organizations, a hospital participating in the management of leading organizations, the park has full-time network management departments and hierarchical management.

Network center managers, a full-time manager 3 to 5 persons with high quality. 1 ~ 2 of them having Master's degree.

Network Management Center is equipped, the core area room is 200m² or above, a stable power supply systems, uninterruptible power supply, fire, lightning protection, anti-static and anti-electromagnetic interference, safety and prevention facilities.

Management system, a network management, security system, and to conscientiously perform.

3   General Assessment Indicators of Campus Network Value Segments

Now to make overview of the park section of general assessment indicators of campus network value segments, combined with the actual situation of the institutions, from the network database development, networking, the average cost of the machine, the average flow of machine networking, network bandwidth utilization, equipment, etc., and using charts.

Network Database Construction Assessment. To build campus networks normatively, must first plan and network construction later, first training and network construction later, first building database and network construction later, the only way to make campus network construction has made a high efficiency.

Campus Network Application database system should normally be the following components:

1) personnel file library, including teachers, staff and students of the file libraries.

2) financial Archive, including files from the repository, wages and income and expenditure of the file libraries.

3) file archives, including document library issued by the government, the institutions, a variety of award-winning Archive and plan and summary document libraries.

4) teaching library, including the disciplinary lesson plans database, test database, operating instructions, teaching reform, and teaching and research information database and other components.

5) education library, including basic students archives, student status management database, teachers, library records and student exchanges, educational information file libraries.

6) multimedia information base, including the sound libraries, images, material library, film and television material library and so on.

Only to plan, build and operate well some of the above information resource on the campus network. If this part of the information to be built, then the completed campus network after only a shell, without any actual running value of net capital investment in the campus can not receive the desired effect.

The Average Cost Evaluation of Online Machines. That is the total investment cost of the campus network/campus network the actual number of networked machines. Such as a total investment of 400 million of the Institute to build a network, the actual networking of its campus had a total of 2,000 machines, then the average cost of its network of 2,000 per machine. Such investments, and other supporting facilities not to reach the appropriate requirements. In order to make investments in campus network consistent with their actual effectiveness, the network should be at least the average cost of the machine 3,000, so as to achieve not only consistent with the actual institution, but not so frequent upgrades. If the network construction process, the excessive pursuit of high-profile, ignoring the development of user or configuration is too low, leading to frequent upgrades, will make the index too high.

Average Flow Assessments of Online Machine. Consisted of 2 parts, namely, the total export of the campus network traffic / number of actual network machines and the main campus network server traffic / number of actual network machines. We conducted a survey of hospital campus, the results shown in Table 1.

Table 1. The ratio of export flow

Total export flows (M)	Mainly server traffic (M)	Actual number of networked machines (unit)	Network machine average flow assessment	
			Total flow of the park / campus network actual machines (M / units)	*Within main server traffic / campus network actual machines (M / units)*
100	1000	2000	0.05	0.5

In the survey slow network speed of the machine in general was found, we believe that the total exports of campus network traffic / number of actual network machine (M / units) should be 1M; the main campus network server traffic / number of actual network machine (M / units) to 1.5M appropriate. In order to adapt to high-speed information society.

The average flow of networked machines is closely related with the information within a campus network applications, if the public information E-mail, FTP, Web services, the average flow is large, export traffic flow and public servers will be larger; if the management application rich, then the server will be more traffic.

The Assessment of Network Bandwidth Utilization. Mainly including the following aspects, the trunk network bandwidth, the center server farms for the export of campus network bandwidth and bandwidth utilization. Units of the institutions of our findings of the author as shown in Table 2.

Network bandwidth utilization can be used to assess the effectiveness of a campus network construction, we believe that this target of 60% to 75% for the best. If too low, that application is relatively small, the capacity of the equipment greatly exceeded the actual demand, investment waste; too high is that network capacity can not meet the actual demand, there is insufficient investment phenomenon.

Evaluation of Device Elimination Rate. Since some large computer business is constantly misleading speculation, the school in order to complete the task regardless of the actual situation of the school, the construction of large-scale campus networks, resulting in debt. The so-called network since the founding of the park is possible at that, because the computer every 3 to 6 months replacement time, CPU performance doubling every 18 months, half price, especially the prices of other accessories. Therefore, the campus network evaluation system should include equipment out rate, although the phase-out is inevitable, but the ratio should be as small as possible. Relationship between performance and price shown in Figure 1.

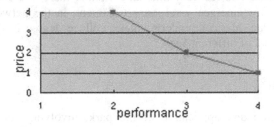

Fig. 1. Relationship between cost-effective

Assess the Proportion of Hardware and Software. Application of the campus network covers a wide range, software lack and low-cost bidding are the existing problems. According to experts, software products online should be at least of the total investment of 1 / 4 to 1 / 3. According to the survey, most of the campus network construction failed to meet the ratio. Reason is that the relevant administrative departments involved in the campus network without a unified software requirements, software vendors to develop products and colorful mixed; in the campus network is difficult tendering and implementation of unified organization, resulting in bidding on the campus network when the software to the general suspension of procurement. After the hardware bidding and construction, for other reasons of funds and awareness, teaching management and application configuration software is difficult to really put in place, the high price of genuine software, it does not want to put a lot of units.

Because China is pursuing government procurement, the construction of the campus network is no exception. Bidding in the form of evaluation by expert groups, which can guarantee the construction of campus networks in the open, fair and impartial. But in the tender process, the inevitable emergence of a vicious price competition, enterprises to take advantage of the situation. Corporate profit margins were too small, which is after-sales service, and system upgrade has left a serious risk.

Reasons that led that after the construction of campus network teachers and students was more disappointed, became a college campus departments passed all levels of inspection and acceptance by a senior artist. More importantly, the lack of application of the campus network, the network is basically idle, low utilization, a few years later, most of the investment is wasted.

The Proportion Assessment of Continuous Financial Input. Development and growth issues were the existence of the network itself, the normal operation of the campus network, the network expansion, the growth of applications, equipment and software upgrades and updates and need to follow-up investment funds. Network planning in the park, we must consider whether there is follow-up investment capacity. The proportion of continuous input of funds should also be assessed as an indicator of the campus network, the total annual investment funds should be put into 4% better.

Network Management Degree can be Assessed. Namely, the degree of network management. Whether some key data was in a timely manner, including the number of networked machines, number of Internet users, Internet application (type, traffic), etc.; network fault response and recovery time and whether there is a complete record of network operation; for emergency response treatment; in the network management staff, with a doctorate, master's, undergraduate, college degree in the proportion of persons and personnel of different ages.

4   Conclusion

Network construction and application of the park, involving capital, technology, management and many other aspects of this work is a sense of urgency and steady progress. This section of the park from all the value of net value of quantitative assessment indicators and all indicators of a general assessment of the campus network segment, combining the actual situation, giving the parameters for the effective assessment and evaluation indicators, the purpose is to promote the campus network hardware, software, information resources and other aspects of balanced development. I believe in the near future, the campus network evaluation system will become a hot topic in society.

References

1. Peng, C.: Computer Network. Hunan University Press (August 2006)
2. Zhong, W.: Standardized security information. China Education Daily (October 2002)
3. Wang, Y.: Moving a unified way: from education management information standards promulgated start. Network World 10 (2002)
4. Liu, Y.: Construction of optimal learning in school. Shanghai Normal University 3 (2001)

Study of Force Control in Ceramic Grinding Process

Pei-ze Li

College of science, Tianjin Polytechnic University, Tianjin Polytechnic University,
Tianjin, China
lipeize@263.net

Abstract. The international researches for the force control in ceramic grinding
are very rare. For this situation, the paper does the research on the grinding
force control which uses the workpiece material of Al2O3. In this paper, we
present the study results of influence of machining parameters on the grinding
force. To control the grinding force, the fuzzy control strategy is used in this
study. The effects of grinding force control on the stability is studied. It was
found that the machining process is more stable than that without it.

Keywords: Ceramic Grinding, Grinding force, Fuzzy control, Al2O3, Fuzzy
control rules.

1 Introduction

Ceramics has wide applications in the industry, particularly in the field of precision
manufacturing industry, because of its dimensional and thermal stabilities. However, it
is easy to generate cracks during machining and the stress concentration remains under
the machined surface due to the hardness and poor thermal conductivity of
ceramics.Grinding method is an effective way to process hard and brittle materials.
Usually, diamond grinding wheels are used to machine ceramics. During machining,
however, the grinding wheel is always jammed and passivated rapidly, weakening its
machining ability. Meanwhile, the grinding region creates a great deal of heat within a
short time and the transient temperature may reach 1000°C, resulting in surface burned
and micro-cracks generated on the surface or sub-surface. All those seriously affect the
use of performance parts and their work-life. The solution is to control the grinding
force to ensure grinding in ductile regime based on a control algorithm.

In this paper, the relationship between the grinding force and machining parameters
is studied. A neural networks control method is used to grind ceramics Al2O3. The
summary is given in the final section.

2 The Impact of the Grinding Parameters on Grinding Force During the Grinding Process

There are many factors affect the grinding force and the surface quality during the
grinding process such as the depth of grinding, feed speed, the size and materials of the
grinding wheel and so on. In this section, we will discuss these grinding parameters'

J. Luo (Ed.): Soft Computing in Information Communication Technology, AISC 158, pp. 275–283.
springerlink.com
© Springer-Verlag Berlin Heidelberg 2012

effects on the grinding force one by one, which provide a basis for a reasonable choice of processing parameters during the grinding process.

The impact of ultrasonic on grinding force.The grinding efficiency will increase and the grinding force will decrease after joined the ultrasonic vibration during the grinding process. Y.Wu in Shanghai Jiaotong University did research on it and did experiments to prove this conclusion. She found that the grinding force will decrease about 20%. We also did experiments to prove her conclusion. We grinded the Ceramic at a suitable feed speed and recorded of the grinding force. We did the experiment two times. In the first one, we did not join the ultrasonic vibration and in the second one, we join it which the amplitude is 2um. The experimental conditions provides in Table 1. After the experiment, we calculated the average of the grinding force.

Table 1. The experimental conditions

Granularity of grinding wheel	20 [um]
Grinding wheel speed.	3600 [rpm]
The depth of grinding	0.003 [mm]
Grinding fluid	Water
The amplitude of ultrasonic	0 [um],2 [um]
Feed speed	0.03 [mm/s]

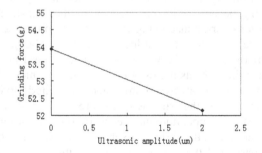

Fig. 1. The result of the experiment

The result of the experimentation brings forth in Fig.1. We find that the grinding force decrease after joining the ultrasonic vibration which proves that the grinding efficiency increases. But the ultrasonic amplitude is too small in this experiment, so we don't fully prove that conclusion.The impact of depth of grinding.The grinding force will increase as the depth of grinding increase, which proved by Y.Wu in Shanghai Jiaotong University . We also did several experiments to prove this conclusion. We grinded the Ceramic at different depth of grinding and recorded the grinding force. After that we calculated the average of the grinding force. The experimental conditions provide in the Table 2 and the result is in Fig.2.

Table 2. The experimental conditions

Granularity of grinding wheel	20 [um]
Grinding wheel speed.	3600 [rpm]
The depth of grinding	1 [um]/3 [um]/5 [um]
Grinding fluid	water
The amplitude of ultrasonic	0 [um]
Feed speed	0.03 [mm/s]

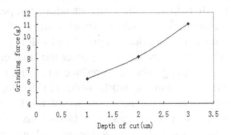

Fig. 2. The result of the experiment

The results of the experiment brings forth in Fig.2. We find that the grinding force increases as the depth of grinding increases.

The impact of the feed speed. The grinding force will increase as the feed speed increase, which proved by Y.Wu in Shanghai Jiaotong University and X.M.Rui in North China Electric Power University. We also did several experiments to prove this conclusion. We grinded the Ceramic at different feed speeds and recorded of the grinding force. After that we calculated the average of the grinding force. The experimental conditions provide in the Table 3 and the result is in Fig.3.

Table 3. The experimental conditions

Granularity of grinding wheel	20 [um]
Grinding wheel speed.	3600 [rpm]
The depth of grinding	3 [um]
Grinding fluid	water
The amplitude of ultrasonic	0 [um]
Feed speed	0.03 [mm/s],0.05 [mm/s], 0.08 [mm/s]

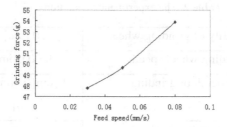

Fig. 3. The result of the experiment

The results of the experiment brings forth in Fig.3. We find that the grinding force increases as the feed speed increases.

The impact of the grinding fluid.During the grinding process, choosing a kind of suitable grinding fluid will help to reduce the grinding temperature, to reduce the grinding force and to extend the grinding wheel's work-life. C.Guo proved that different grinding fluids have a great effect on grinding by doing the experiments. They found the bigger of grinding fluid's penetrating power the smaller of the grinding force during the grinding. The grinding fluid can reduce the grinding force, prevent the generation of heat cracks and clean the work-piece. H.F.Ke had a different idea, he thought because of the ceramic's poor performance of the thermal conductivity, the grinding fluid's cooling effect can increase the brittleness of the ceramic and the surface of the ceramic will generate the cracks. He pointed out that if we wanted to get the good surface, we should not use the grinding fluid. So whether or not to use the grinding fluid, we should look the specific processing requirements. We did several experiments to determine the impact of the grinding fluid on the grinding force and the surface quality of the ceramic. We grinded the Ceramic two times. In the first time we join the grinding fluid and in the second we did not. The experimental conditions provide in the Table 4 and the results are in Table 5.

Table 4. The experimental conditions

Granularity of grinding wheel	20 [um]
Grinding wheel speed.	3600 [rpm]
The depth of grinding	3 [um]
Grinding fluid	Water/None
The amplitude of ultrasonic	0 [um]
Feed speed	0.03 [mm/s]

Table 5. The results of the experiment

Result	Join the water	None
Grinding force	36.3049 [g]	73.4346 [g]
Ra	0.125 [um]	0.234 [um]

By analyzing the result of the experimen, we find that the grinding force decreases and the surface quality improves after joining the grinding fluid. We can conclude that the grinding fluid can improve the grinding quality.

The impact of the grinding wheel.Because of direct involvement of the grinding process, The grain size of the grinding wheel and the bonding of the grinding wheel have great effects on grinding process. K. W. Lee grinded the   using the 100 # and 240 # grinding wheel. He found that as the grain size of the grinding wheel is bigger, the destiny of grinding wheel edge reduced and the grinding force decreased. Bi Zhang also achieved the same conclusions by using different size grinding wheels to grinding the ceramic. We also did the experiments to do research on the relationship between the grain size of the grinding wheel and the grinding force. We grinded the   in different grain size of the grinding wheels and compared the average grinding force. The experimental conditions provide in the Table 6 and the result is in Fig.4.

Table 6. The experimental conditions

Granularity of grinding wheel	20 [um]/10 [um]
Grinding wheel speed.	3600 [rpm]
The depth of grinding	3 [um]
Grinding fluid	Water
The amplitude of ultrasonic	0 [um]
Feed speed	0.03 [mm/s]

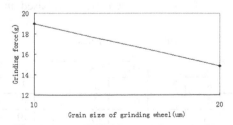

Fig. 4. The result of the experiment

The result of the experiments bring forth in Fig.4. We find that the grinding force decreases as the grain size of grinding wheel increases.

3   Application of Artificial Intelligence Control in Grinding of Al2O3

Artificial intelligence control system in grinding Al2O3. The grinding process is complex because many factors are involved. To realize the machining stability, it is necessary to apply a suitable control method. Artificial intelligence control simulates skilled operator's experience and knowledge which are suitable for controlling the non-linear complexity and uncertainty system that has multiple input and one output. The Artificial intelligence control system used in this study is shown in Fig. 5.

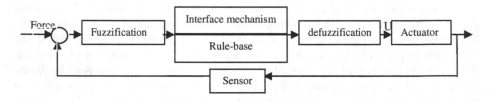

Fig. 5. Artificial intelligence control system

To control the grinding force, it is necessary to determine the adjusting parameter. In ceramics grinding, two parameters, depth of cut and workpiece feed speed, are easily adjusted by moving the working table on which the workpiece is fixed. During machining, the change of the workpiece feed speed has no influence on the machining accuracy, while the change of the depth of cut may result in the change of the surface profile of workpiece.

Based on extensive experimental results of precision grinding of ceramics Al2O3, it is found that the variation of grinding force around the setting value is from -0.09Kg to 0.09Kg. The input domain, E, is defined as {-0.09Kg, 0.09Kg} and its corresponding internal domain E* is {-11, 11}. The control value, U, is the variation of workpiece feed speed, whose domain is defined as {-0.009mm/s, 0.009mm/s}. The corresponding interal domain of control value, U*, is {-11, 11}. A larger or smaller variation of workpiece feed speed may lead to an unstable machining process. Based on the definitions of these domains, the Artificial intelligence control rules are listed in Table 7. Table 8 gives the variation of workpiece feed speed, U, based on U*.

Table 7. Control rules

E*	-11	-10	-9	-8	-7	-6	-5	-4	-3	-2	-1	0
U*	11	10	9	8	7	6	5	4	3	2	1	0
E*	1	2	3	4	5	6	7	8	9	10	11	
U*	-1	-2	-3	-4	-5	-6	-7	-8	-9	-10	-11	

Table 8. Variation of workpiece feed speed

U^*	ΔSpeed [mm/s]	U^*	ΔSpeed [mm/s]	U^*	ΔSpeed [mm/s]	U^*	ΔSpeed [mm/s]	U^*	ΔSpeed [mm/s]
-11	-0.009	-6	-0.004	-1	-0.0003	4	0.002	9	0.007
-10	-0.008	-5	-0.003	0	0	5	0.003	10	0.008
-9	-0.007	-4	-0.002	1	0.0003	6	0.004	11	0.009
-8	-0.006	-3	-0.001	2	0.0005	7	0.005		
-7	-0.005	-2	-0.0005	3	0.001	8	0.006		

When the grinding force, Factual, is measured, the difference, e, can be expressed as:

$$e = F_{actual} - F_{setting} \qquad (1)$$

where   is the actual grinding force;   is the setting grinding force.

The matching factor, Ke, which is used to determine the number in E*, is given

$$K_e = \frac{n}{X_e} \qquad (2)$$

where n—number in E* (here n=11), Xe=0.09, which is the biggest value in the input domain E.

Therefore, Ke=122.2.

Based on e in Eq. 1, the estimating number which is used to locate the control value of U* in Table 2, m, is calculated based on Eq. 3.

$$m = K_e e \qquad (3)$$

The rule used to determine the number in E* is defined as:

When m>0, IF   m is larger than k (the number in E*), Then m=k+1.

When m<0, IF   m is larger than k (the number in E*), Then m=k-1.

Based on m, the corresponding control value can be found in U* in Table 2. Therefore, the adjustment of workpiece feed speed can be located in Table 3 based on the control value U*.

Experimental verification. To verify the effect of the Artificial intelligence controller and the machining aided by ultrasonic vibration, experiments are performed. Experimental conditions are listed in Table 9. The recorded grinding force under different machining conditions is shown in Fig. 6. It can be seen that the machining process is much more stable with Artificial intelligence control in Fig. 6 (b), compared with the process without grinding force control in Fig. 6 (a).

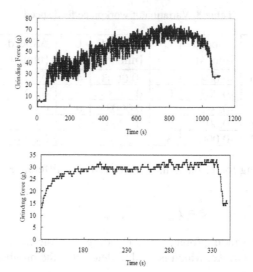

(a) Machining without control.   (b) Machining with Artificial intelligence control.

Fig. 6. The result of the experiment

Table 9. The experimental conditions

Setting of grinding force [g]	30
Workpiece material	Al_2O_3
Abrasive grain material	Diamond powder
Grain size of grinding wheel [μm]	20
Rotation speed of wheel [rpm]	3600
Depth of cut [μm]	3
Grinding fluid	Water
The amplitude of ultrasonic [μm]	0, 2
Feed speed [mm/s]	0.03

4   Conclusion

In this paper, the relationship between the grinding parameters and the grinding force in ceramics Al2O3 grinding is studied. The experimental results are analyzed. To stablize the machining process, a Artificial intelligence controller is proposed. Experimental results indicate that the application of Artificial intelligence control improves the stability of ceramics grinding process.

References

1. Hodge, E.J., Thomas, R.K.: Adaptive Pole-Zero Cancellation in Grinding Force Control. IEEE Transactions on Control Systems Technology 7(3), 363–370 (1999)
2. Yang, Z., Gao, Y., Zhang, D., Huang, T.: A Self-tuning Based Fuzzy-PID Approach for Grinding Process Control. Key Engineering Materials 238-239, 375–382 (2003)
3. Chen, S.P., Wang, M.S., Huang, Y., Mao, X., Liu, H.: Control of Grinding Force in Grinding Diamond. In: Proceedings of SPIE, vol. 5605, pp. 257–265 (2004)
4. Li, B.C., Ye, B.C., Jiang, Y.H., Wang, T.: A Control Scheme for Grinding Based on Combination of Fuzzy and Adaptive Control Techniques. International Journal of Information Technology 11(11), 70–78 (2005)
5. Liu, C.H., Chen, A., Chen, C.A., Wang, Y.T.: Grinding Force Control in an Automatic Surface Finishing System. Journal of Materials Processing Technology 170, 367–373 (2005)
6. Xu, C.Y., Shin, Y.C.: Creep-Feed Grinding Processes Using a Multi-Level Fuzzy Controller. Journal of Dynamic Systems, Measurement, and Control 129, 480–492 (2007)

Computer Network Technician Professional Group of the Construction and Discover

Xue-Ping Chen

Chongqing College of Electronic Engineering, Chongqing 401331, China
yuanju01@163.com

Abstract. The development of vocational education in China has a long history, then raises large quantities of intermediate technology workers in China's socialist economic construction plays an important role. Education is in and technicians under the new situation as China's economic construction, the talents with high skill and production. Education of professional technicians and relative technical education, it has made great development, many emerging and gradually develop specialty, which is one of the major of computer network, this paper technician education of computer network professional group of construction scheme, in order to explore the primary education development tide of engineer has pushed by wavelet function, and the molecules in the scheme for vocational computer network professional construction will also play demonstrative effect.

Keywords: Technician education, Professional groups, Skilled personnel, career-oriented, job requirements, skills identification.

1 Introduction

Technician education of our college has been 6 years, the technician education students through the 4-year study, the final examination after the theory and practice will be recognized by the Ministry of Labor National General National Vocational Qualification (Secondary) technician certificate, this will implementation of computer network technician education program of professional and clusters of preliminary exploration.

2 The Construction of Network Engineer Professional Goals

The technician education examination according to labor and employment demand evidence, cultivate higher network system construction and maintenance management and application development level of skilled talents.

Specific objectives are as follows:

(1) with the famous enterprise cooperation, continue to promote "set pattern" and "order", "work-integrated learning" training mode, according to the request of cultivating talents.

J. Luo (Ed.): Soft Computing in Information Communication Technology, AISC 158, pp. 285–290.
springerlink.com　　　　　　　　　　　　　　　© Springer-Verlag Berlin Heidelberg 2012

(2) with the co-operative oriented and "network" as the core capability, arrangement according to the actual position knowledge and skills, at the request of "learning network", "Application of network", "Organized network", "The network management ", "Constructing the network" five ability and learning in order to construct reasonable curriculum system and network modularity technician to establish professional skill level assessment standards and course (or course evaluation criteria of module).

(3) to expand the lateral ties, positive for society, especially small and medium-sized enterprise network construction, maintenance and technical development, product r&d providing services, based on the southwest becomes the small and medium-sized enterprise information technology service center.

(4) directly improve computer network of web development direction, network programming direction and network management level, professional direction course of curriculum construction, so as to improve the overall improve the whole network of professional level, professional construction.

3   Network Technician Professional Groups Main Construction Content

1.Strengthen and industry, enterprise, deepen professional talent training mode reform

(1) the deepening and Cisco, huawei 3com, Symantec, SUN microsystem leading enterprises such as cooperation, emphatically introducing Cisco, huawei 3com, Symantec, SUN microsystem, Microsoft's industry mainstream technologies, enhances the student to the relevant international standards of understanding. Increase the strength and talent cultivation, continue to promote set pattern, order, such training mode accords with the enterprise post of talents, students' ability to connect with the enterprise to seamless.

(2) through the student to the enterprise working practice, lets the student in the real business environment in post skills. Continue with the enterprises to develop students' term to work week, month working holiday enterprise training mode of the rotation, By way of students to participate in the school campus network project construction and maintenance of campus network, provide the real construction and maintenance jobs, and the practice complementary, ensure students before graduation is related to the enterprise is engaged in professional work network accumulated more than half a year.

(3) To deepen teaching reform and the way of evaluation. Through the teacher to learn, then theory-practice-retheory-repractice-summary teachers will be a real project, case and construction techniques in teaching contents, course content and the technical front against the production practice and further "teachers and students as the main body", In examining, comprehensive evaluation, and further implementation of formative assessment, the separation and network examination.

2.Construction of developing a "network building" as the core capabilities of high-quality online course

(1) the "network construction", based on the core competence of course, according to "learning network", "Application of network", "Organized network", "The network management", "Constructing the network "five aspects of construction of the curriculum, and make students according to "modular, combination, advanced type"

the knowledge and skills of the master schedule, deepens master knowledge, and through the network strengthening students' choice of curriculum modules, a consideration of other abilities.

(2) refer to Cisco, huawei 3Com, SUN microsystem JAVA industry leading enterprises of the intermediate examination authentication standard of IT, increasing requirement, build professional skill level evaluation criteria, and then decomposed into the curriculum, build each course (or course evaluation criteria of module). On this basis, try to establish test and online platform, operation evaluation theory and practical operation test combination, objectively evaluate students' ability.

(3) according to the network management of labor division (2), the network administrator (3 and 4) occupational qualification certification needs to offer technicians education 4-year curriculum. With computer network management, web development, web programming, database application, computer hardware maintenance and debugging, computer application, such as the main direction of the direction of six direction, is the main network engineer examination, and the professional group of network course direction. In the process of curriculum development, and strive to create complete professional courses, Internet, for all main course contests of students' autonomous learning, and also provides the high-quality service for the national teachers teaching resources. Figure 1 lists network education specialized technicians of the backbone of the curriculum.

Figure 1 in the main course setting, we increase occupational qualification certification, make students employment master various vocational skills. Each semester curriculum and the corresponding license certificate of skills such as shown in table 1.

3. Build a senior technician of high level double-quality teachers

(1) mainly introduce the technician, senior technician enterprise talented or senior lecturer.

(2) strengthen the contact with labor bureau, and strengthen and Cisco, huawei 3Com, SUN microsystem, Symantec international famous IT companies on the close cooperation, through training and qualification authentication and so on many kinds of ways to ensure professional teachers' knowledge is updated ceaselessly, make more close to the actual teaching content of market development, reflected the technology. Strive for all professional teachers are specific technician level, most of the teachers have senior technician level or senior lecturer or professor level.

4, resource sharing, and play a good role model and high professional resource sharing of computer network

(1) the main course teaching resources online, and finish the work of these courses,

(2) in cooperation with enterprises on the basis of consensus, Cisco, together 3Com huawei, Symantec's education resources online,

(3) complete relevant professional technical information, documents and standard of Internet.

Formation of network course, including professional standard and professional information WenXianKu sooty characteristic resource database, multimedia, KeJianKu, test and putted forward the computer network technology and professional information repository may add video animation library professional Figure 1.

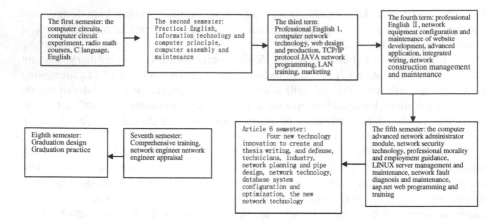

Fig. 1. Main curriculum construction

Table 1. Each semester curriculum and the skills of the corresponding license certificate

Certificate of textual name	ding course	time	note
A computer operator intermediate	Information technology	Semester 2	Necessary
Computer assembly and maintenance Intermediate	Computer assembly and maintenance	Semester 2	Necessary
Web Producer	Web production	Semester 3	Necessary
Internet Applications (ASP)	Advanced Application Development Website	Semester 4	Expand
Senior Network Administrator	Comprehensive Training Network Administrator	Semester 5	Necessary
Database Applications	Database system configuration and optimization	Semester 6	Expand
Network manager	Comprehensive Training Network	Semester 7	To test
English certificate		Each semester	Expand
Putonghua certificate		Each semester	Expand

4   Implementing Network Technicians Specialized Construction Scheme of the Expected Bennfits

First, the education from the perspective

Innovation with work-integrated learning is the core of professional talents training mode, will promote the healthy development of the professional training, further more practical skills mastering the network technology talents; a more solid foundation Professional talent training mode and professional construction experience, not only can promote the common development of the professional group, but also for other mechanic college and learn from it.

Second, from professional requirements

Figure 1 curriculum and table 1 skill appraisal certificate of corresponding form, we can see through the student in the school during the network technology and application of basic theory and technology of learning, and strict training, and practical skills to national vocational qualification (2) engineer.

1 and in the knowledge structure

(1) has the network technology and application to the basic theory of professional knowledge.

(2) has a strong professional basic knowledge of English.

(3) at least master a programming language.

(4) computer application and maintenance of the foundation of knowledge.

(5) master JAVA, asp.net program design and development of knowledge.

(6) master SQL Server database development of knowledge.

(7) master Dreamweaver website construction of knowledge.

(8) has the enterprise management and technical and economic analysis of the foundation of knowledge.

2, in the ability structure

(1) with network equipment installation and commissioning and testing and maintenance.

(2) with network management, network management software is used to optimize the ability.

(3) is the ability of website development.

(4) is a JAVA and my asp.net application design ability.

(5) has the ability of SQL Server database design.

(6) is the ability to work on technician.

Third, from the employment channels for watching (see)

Network engineer graduates in various enterprises and institutions, network equipment company, network application of network equipment company is engaged in the production, sales, testing and maintenance work. Can be engaged in the following career: the network administrator, network engineering and technical personnel, network equipment and debugging, maintenance personnel, Internet applications, computer maintenance personnel, computer programmers, commissioning, web site construction workers and staff, computer operator database applications.

5   Conclusion

We simply introduces the computer network education specialized technicians of construction scheme, we analyzed the network professional curriculum development, and with the professional qualification certificate (profession), affirmation have identified deficiencies, so as to obtain the criticism and the experts, also hope to develop education technicians in the good situation, education curriculum development engineer has suggested. Can also for higher vocational education development of computer network play demonstrative effect.

References

1. Chongqing Institute of Electronics Technician Network Professional Training Programme (December 2007)
2. Li, D.: Higher vocational computer network professional course construction of vocational and technical education of the (January 2005)

Modern Information Technology Education and Change of Learning Style Research

Xue-Ping Chen

Chongqing College of Electronic Engineering, Chongqing 401331, China
yuanju01@163.com

Abstract. Implementing the new curriculum most core and the most crucial link is how to promote students innovations on their ways of study, this kind of change is not one of the forms of the change, but a qualitative leap into modern information technology education curriculum reform, brought the ways of learning, the paper discusses the change of modern information technology and curriculum integration, analyzes the phase change of learning style strategy, and points out the core curriculum integration is a digital learning styles, and introduces digital way of learning type.

Keywords: Information technology, curriculum conformity, study way, study strategy, digitized study.

1 Introduction

The information technology the function which and the status lives in the modern humanity is getting more and more important. Information technology's high speed development, the network time's arrival, will give people's study, the life to bring a huge transformation. Is not only changing people's production method and the life style enormously, moreover is changing people's thinking mode and the study way enormously. The information technology supplies each kind of knowledge information by the software and the multimedia ways to the learner, its interactive and aspect and so on networking unique superiority had decided it in transforms in people's study way to play the strong character. In view of this, we proposed this topic's research, realizes the student by the time to study the way the radical transformation.

2 Information Technology and Curriculum Integration Stage

The first stage is enclosed, with knowledge centered curriculum integration.

The traditional teaching and most current teaching belong to this stage: all the teaching are in strict accordance with the syllabus, put students in teaching material or simple closed within the courseware, make its and rich resources, realistic complete isolation. According to the arrangement and class teaching material to the requirements of design all teaching activities, if the course content is less, arrange some discussion, many design some activities, if the course content are many, use "daunted, in the form of" facilities.at not overtime, many. Although by a certain counseling software, but the

J. Luo (Ed.): Soft Computing in Information Communication Technology, AISC 158, pp. 291–297.
springerlink.com

current counseling software were also in these ideas come out, so also compiled under no breakthrough. The whole teaching in "knowledge" as the center, under the guidance of teaching aim, teaching content, teaching form and teaching organizations and the traditional classroom teaching without any difference, the whole teaching process is still in teachers' teaching primarily, students are still is passive responders, knowledge is psionically objects. Information technology is introduced, in helping teacher ease teaching workload has made some progress, and for students' thinking ability and development, compared with traditional way, and there is no substantial progress.

The second phase is open mode, with resource center for curriculum integration.

The integration of information technology with curriculum first phase are basically closed, with individualized learning and teaching primarily. In the second phase, teaching concepts, teaching design guiding ideology, the role of teacher and student's character, etc, will be in great changes. Educators to pay more attention to the students have knowledge of significance construction, teaching design from knowledge as center for change in resources as the centre, regard learning as the center, the whole teaching of resource is open, the students in the study of an academic discipline within knowledge can be many other subject knowledge, students in the possessive based on the rich resources of complete all ability the raise, and the students become study main body, the teachers to become students learning guidance, aide, and organizers.

The third stage is the comprehensive curriculum integration.

The first two stages they never make the teaching content, teaching objectives, and teaching organization framework to conduct a comprehensive reform and informationization. When education theory and learning theory to obtain the full development and utilization, when information technology in the teaching application system to get more and more scientific discussing and refining, the inevitable will promote education happens once great innovation, promoting education content, teaching objectives, teaching organization structure reform so as to complete the whole teaching informationization, the information technology education of seamless integration to each link, achieve information technology and curriculum reform of higher goals.

3   Information Technology and Curriculum Integration Omni-Directional Bring Ways Learning Revolution

Learning style refers to students in task in the process of reflected the basic behavior and cognitive orientation. Learning style is not a kind of specific learning methods or learning method combined, but that they reflect a student with what kind of behavior and cognitive style to complete the study task of inertia. Therefore, the study does not refer to specific way of learning strategies and learning methods, but higher than strategies and methods, the effect and guide students to specific strategies and methods to make a choice about learning behavior of basic characteristics.

The new curriculum is required, the passive, accept and closed learning mode for initiative, found and cooperative study way, advocate independent and exploration, exert the students' subjectivity and creativity and practical ability, and makes students become the study master.

The integration of information technology with curriculum the mainest is bring ways of learning revolution. The rapid development of information technology, network information are required.this for humans learning styles produced profound transformation function. Learners from traditional accept in the learning into active learning, inquiry learning and the investigative study. Meanwhile digital learning will become the learners' future development.

4   Modern Information Technology Education and Change of Learning Style Research Contents and Strategies

Information technology and change of learning style the purpose of the research is in the modern information technology background, the full use of modern information technology, and actively guide student, lets the student in the independent environment free choice of interest to himself degree of learning contents, free will pick up learning methods, active participation, willing to explore, is industrious hands. Change put too much emphasis on learning by rote, machinery, accept the training of classroom teaching, to realize "teacher-student interaction, resources sharing, mutual, information sharing" student multi-dimensional traffic, and to study the interaction transmission method into ability and habits, so as to finally realize the basic transition of the mode of student learning.

(a) information technology for students to study mode change provide conditions

1, strengthen the construction of campus network, provide for the network teaching material foundation, specific requirements is continuously strengthen repository of construction, and the campus network and Internet connection so that students can obtain abundant information on the Internet.

2, the full establishment of modern information technology situation, cultivating students' interest, guiding students to master basic knowledge of computer, can skilled operate computer. At first, the students do not know about computers, but the student age is small, strong curiosity, we should actively cultivate students' interest in study. This point is vital to the foundation, therefore, should fully creating context, the students master the basic knowledge of computer, can skilled operate computer.

3, guiding students to master network technology, and guide students online learning, in online learning practice experience, master the characteristics and method of online learning. Through the application of modern information technology, and actively guide student participation, raises the student active learning awareness, finally realizes the students to learn the basic transition of the mode of.

(2) under the information technology environment change ways of learning strategies

1st, student learning concept transformation

Network environment changed students study way, it makes the student's study way by original passive become more active learning, and realized the true sense of interactive learning, cooperative learning, discovery-learning, innovation learning, in order to make students learn to study, improve their learning ability, Learn to create, improve found the problem and problem-solving skills; Learn to cooperation and communication, and improve the team cooperation ability.

2, student learning behavior changes: sharing - personality - willingness -- interaction

First of all students could pass network and teachers, students share all kinds of information resources, learning resources, and the whole world share resources and knowledge, the students' studying world became a can be extended unlimitedly world. Meanwhile, students can also utilize the network interactive, oneself the special problems of ask the teacher for advice, discussions, According to their interests are some ways more in-depth study and exploration, to ensure the development of personality. Network classroom compared with the traditional classroom, has the obvious "nonself" characteristics of ncos is, the suppressive "online learning" mode, increased the students to participate in the learning opportunities, but because the suppressive, teachers of classroom control ability opposite weaken, the student to obtain knowledge no longer rely on teachers' distribution and imparting, which requires the students to learn self plan, and self management, self evaluation and feedback, i.e., require students to higher levels of learning initiative. In addition, environment of network study make classrooms formed a one-on-one, a pair of many, and many of my "network" mode, interactive and cooperation in the learning becomes the main mode of learning.

3 and study atmosphere of liberalization - edu-tainment strategy.

In the teaching, the use of teaching computer game software to scientific, interest, educational sets a body, can stimulate students' interest in study and recreation, thus exercise students reaction speed, decision-making ability and manipulating ability. In addition, using the information technology media, conduct art appreciation, making games, student works display activities, also can stimulate the student's study enthusiasm, help students to master the knowledge, development ability, training innovation consciousness and improve the innovation ability.

4, the process of learning activities of incentive strategy - situation.

Situational incentive strategy is through the integration of information technology with curriculum, teaching situation, to establish in teaching should provide plenty of time and space, ensure students involved in learning activities of opportunities. And through the emotional exchange improing teacher-student such effective means, cause learning motivation, make students actively participate in the new knowledge learning, greatly stimulate students' enthusiasm of exploration and discovery. Let the student through image perception, independent comprehension, collaborative discussion, mutual evaluation, the freedom of expression, analyze, abstract generalizations, create imagination explore sexual activity to experience the fun of discussion, tasting innovation of happiness.

5, information to deliver the three-dimensional - multiple senses involved in learning strategies.

In the teaching, through the integration of information technology with curriculum, and strive to provide students with various senses to participate in the study atmosphere, fully lets students move the eye, dynamic ears, move the brain, to begin, saying, and through the hands-on laboratory, operating study, while thinking and doing, edge to practice perceiving, realizing concept, master principle. In the teaching to create a good atmosphere of group communication. Variable information exchanges for information exchange and the diversity of unidirectional.

(3) under the information technology environment of students' autonomous learning strategies

Learning is in teaching contexts through self-study, active explore process. Constructing information technology environment of autonomous learning mode is the sequence: teachers creating context - students about the learning task question assumptions -- the student hypothesized that collect information resources, thus realize learning goals - students through self-evaluation, group evaluation, teacher evaluation system way, to the student's study achievement proper evaluation, causes the student to obtain success experience, fully enjoying under the information technology environment autonomous learning brings joy, improving self-directed learning ability.

(4), information technology environment students inquiry-based learning strategies

Based on information technology environment of explorative study is to use the interactive network platform, using Internet resources or build a topic study site, to construct the virtual reality and combination of inquiry-based learning situations, students and teachers realization based on abundant of network resources interactive learning, so that students in a close to real learning environment, cultivate initiatively explore actively acquire knowledge and apply knowledge and ability to solve problems. Can use the following order: creating context - mission driving - inquiry-based learning - learning outcomes upload.

5   The Integration of Information Technology with Curriculum Core Research

LiKeDong professor said: "the integration of information technology with curriculum, its essence is to let students learn for digital learning."

1. Digital learning concept

American education technology chief executive BBS, namely ET - CEO BBS, in June 2000 held by "digital learning power: the integration of digital content" as the theme of the first three next meeting, digital technology and teaching content conformity way called digital learning, puts forward the concept of digital learning.

2. The integration of information technology with curriculum core - digital learning

With digital learning as the core of information technology and curriculum integration, different from the traditional ways of learning, has the following: (1) clear characteristics study aims at students as the center, the study is personalized, can meet the individual needs, (2) study aims at questions or theme as the center, 3 the learning process is for communication between the learner is consultation and cooperation; (4) learning is creative and reproducibility, 5 everywhere lifelong learning.

3. Digital learning elements

Digital learning has three elements. One is a digital learning environment, so-called information technology learning environment. It passes through digital information processing with information display let alone these flatteries, information transmission network, intelligent information processing and teaching environment virtualization characteristics. It includes facilities, resources, platform, communication and tools. 2 it is digital learning resources. It is to point to by digital processing, can be in multimedia computer or under the network environment, operation of multimedia material. Include digital video, digital audio, multimedia software, cd-rom, website, E-mail, on-line learning management system, computer simulation, online discussions, data files and database, etc. Digital learning resource is the key to digital learning, it can through teacher development, students creation, market purchase and network downloads way to acquire. Digital learning resources is practical, instant credible; Can be used in the multi-level inquiry; Can manipulating; Creative etc. Characteristics. Three is a digital learning style. Using digital platform and digital resources, carry out negotiations between teachers and students discuss, cooperative learning, and through the collection of resources utilization, and explore knowledge, find knowledge, create knowledge and display knowledge study conducted, has the resources utilization, independent discoveries, through consultations and cooperation and practice create several ways.

4. Digital learning mode

"LiKeDong" the professor said digital learning has four modes: "situation - explore" model, "resource utilization - theme quest - cooperative learning" model, "intercollegiate cooperation - remote consultation" model, "project to explore - web development" mode.

Below is a brief description

1) "situation - inquiry" mode

Mainly applied in classroom teaching, tuitional this and we mentioned in the students' autonomous learning strategy based on digital resources same: creating studying situation students put forward thinking problem, this information representation tools, such as Word, BBS etc, form opinions of the digital resources and delivered the operating practice, exploratory discover the features of the objects, relationship and rule by information processing tools, such as PowerPoint, FrontPage etc undertake significance construction with assessment tools, self learning evaluation, timely found the problem, and gather feedback information.

2) "resource utilization - theme explore - cooperative learning" mode

Mainly suitable for campus network environment, the process is as follows: to understand the theme choice to learn and determine learning theme, and formulate topic study plan established cooperative learning group provide teachers and learning theme related resources directory, url and data collection methods and pathways, including social resources, school resources and network resources collection guiding student browse relevant web pages and resources, and to the information of the true, optimum choice to inferior analysis according to need to organize relevant collaborative learning activities, such as competition, debates, design and problem solving or role-play as formation works related to the theme, make a research report, can form is text, electronic manuscripts, web, and etc, and to all the students show teachers organize the students by evaluation work, form opinion, reach significance construction purposes.

"College cooperation - remote consultation" pattern mainly applies to the Internet environment in different countries, region or city, each choose a few schools as a regional members experimental school, each component several cooperation study group, centering on the same subject, different regions of the experimental school, through Internet, looking for related to the theme of the web and through the download, get information, mutual communication, organizes the student to carry on summarize the study and evaluate themselves. "Project to explore - website development" pattern mainly applies in the environment of Internet, on a particular project for more extensive, thorough research learning, and to cultivate students' innovative spirit and practice ability, improving students' comprehensive quality. This kind of learning mode require students to construct "project-based learning websites".

6   Conclusion

Information technology to bring you the open mind, the freedom of knowledge and novel way, simultaneously has also brought the ways of learning huge transformation. However, we must be clearly aware that information technology will not automatically create education reform studying mode, it requires us to teachers in information technology education and curriculum integration in-depth study, can bring learning way transformation.

References

1. Li, K.D.: The Integration of information technology with curriculum core
2. Item trillion fly, Using the information technology teaching and cultivating students' information literacy research subject research plan (February 2009)
3. Yao, Y.: Information technology environment students learning mode change strategy research (January 2009)
4. More than the marin, springs overcome the integration of information technology with curriculum hierarchy

Research on the Value Assessment of Wetland Ecosystem in Qilihai

Kehua Wang[1] and Jieping Dong[2]

[1] College of science, Tianjin Polytechnic University, Tianjin, China
[2] Department of Library, Tianjin Polytechnic University, Tianjin, China
{554380457,397638175}@qq.com

Abstract. As one of the prerequisites of human survival, wetland has a high ecological value, which has provided important ecosystem services. This paper mainly conducts a more comprehensive assessment on the main function of wetland ecosystem in Qilihai, applies various methods to carry out assessment on the direct or indirect use value of wetland ecosystem in the Qilihai to get its total value and expect to provide a scientific basis for ecological construction and ecological protection as well as management of resources of Qilihai.

Keywords: Qilihai wetland, wetland ecosystem, ecological value, value assessmen.

1   Introduction

As one of the prerequisites of human survival, wetland owns high ecological value and provides important human ecosystem services. Qilihai wetland in Tianjin has given birth to the unique biological species for its unique wetland environment, which has a high value for both biological research and wetland ecosystem research, and even the research on the impacts of human existence environment.

2   An Overview of Qilihai Wetland Ecosystem

Overview of the Qilihai natural environment. Qilihai locates in the northeast of Tianjin and in the southwest of Ninghe. Its east is Lutai with the distance of 18 km, surrounding by five towns-Panzhuang, Biaokou, Zao jia, Qilihai town, BeiHuaiding. Its total area is 95 square kilometers and 14.25 million mu with a core area of 45.195 square km and 6.78 million mu. There are 3 canals—Chaobai, Ji and Yongding flowing through the middle and east and west sides of Qilihai. In addition, there are also another 4 2-level rivers flowing through it. In October, 1990, the State Council approved to establish the "Tianjin Ancient Coastal and Wetland National Nature Reserve." Qilihai is the typical reed wetland and gave birth to a variety of specifies. Their biological communities and their dependent natural basis constitute the typical marsh wetland of Qilihai.

Overview of social environment in Qilihai. 2005 survey indicates that the scope of protected areas mainly locates in Ninghe, including Qilihai town, Tawara port town,

J. Luo (Ed.): Soft Computing in Information Communication Technology, AISC 158, pp. 299–306.
springerlink.com   © Springer-Verlag Berlin Heidelberg 2012

Pan Town, Zao jia town and Bei Huaiding, a total of 51 administrative villages. It has a total population of 116,397. In this area, the animal husbandry and fishery is more developed, besides, other related industries such as manufacturing, construction, transportation and warehousing, wholesale and retail trade, accommodation and catering industry, the services of residents, culture, hygiene and sports education as well as social welfare all show booming economic strength in recent years, so its economic strength has been enhanced.

The ecological value of Qilihai wetlands. The wetland is multifunctional, for it can be used as a direct source of water or to recharge groundwater and effectively control flooding and prevent soil desertification as well as retain sediments, toxic substances, nutrients, thereby improving the environmental pollution; it can store carbon in the form of organic matter to reduce the greenhouse effect, protect the coast from storm erosion, provide clean and convenient mode of transport, etc. Because there are so many useful features, it is known as the "kidney of the earth" [1]. Wetland is also the paradise for a number of plants, animals, especially for the growth of waterfowl and it also provides human food (fish, livestock products, and cereals), energy (hydropower, peat, and firewood), raw materials (reeds, timber, and medicinal plants) and tourism sites, which is an important foundation for human existence and sustainable development.

2   Theoretical Basis of the Evaluation on the Ecological Value of Wetlands and the Evaluation Method

Related theoretical basis. Wetland researches in China start late and have not formed a complete theoretical system. The main theories are theory of sustainable development, theory of wetland ecosystem, theory of wetland ecosystem service function and theory of resource value. Wetland Value Assessment is a multi-disciplinary and multi-field cross-application, related to the theories and methods of sociology, land resource management, ecological economics, resource economics, environmental economics, micro economics and asset evaluation, etc [2]. The main theoretical foundations include theory of sustainable development, theory of wetland ecosystems, theory of wetland ecosystem service function and theory of the resource value.

Evaluation Method. Ecological service function provided by the ecosystem is an ecological benefit used to support current production and consumption activities, such as ecosystem nutrient cycling, watershed protection, air pollution reduction and micro-climate regulation. These features, though do not directly come into the production and consumption process, but provide necessary conditions for the normal running of production and consumption.

With people's knowledge and further understanding on the value of ecosystem service function, research methods are constantly expanding to the directions of breadth and depth. Due to the using of different assessment methods, the assessment results may be quite different. The international community has not yet formed a unified, standardized, comprehensive assessment standard and assessment methods currently in use are derived from ecological economics, environmental economics and resource economics. The main assessment methods are: direct market approach, including the market value method, expenditure method, opportunity cost, shadow project method and human capital method; alternative market methods, including the travel cost method, protection cost method and hedonic price method ; method of simulating market value, including the valuation method, carbon tax and the reforestation cost method.

Each assessment method of the value of ecosystem service function has advantages and disadvantages and there is the feasibility and limitations. But overall, the credibility of direct market method is higher than the alternative market method, while the credibility of alternative market approach method is higher than the simulation market method. Therefore, when it comes to the selection of assessment method, the following basic principles should be followed: direct marketing method should be the first choice, if the conditions are not met then use the alternative market approach and when the two methods cannot be used, use simulating market method.

3 The Assessment of Ecological Value of Wetlands in Qilihai

The research on assessment of the wetland's ecological value combines sociology, land resources management, ecological economics, resource economics, environmental economics, micro economics, asset evaluation and other related disciplines to analyze wetland functions and the stakeholder, for the purpose of achieving economic values of wetland's ecosystem services. Evaluation of the ecological value of wetlands is a quantitative analysis about the various functions of wetland ecosystem and it is the behavior to conduct monetization on a variety of service functions of wetland ecosystem system.

Direct value in use. The direct value in use of Qilihai includes three elements: the value of material production, value of cultural and scientific research, value of tourism and leisure, among which the value of material production also includes the value of fish and plant resources.

The value of fish resources. The main fish species of Qilihai wetland are grass carp, carp, crucian carp, etc., with the average market price of about 6 yuan / kg; average market price of river crab in Qilihai is 90 / kg. According to the survey, the fishing ground area of Qilihai are 10,000 acres, the average fish yield is 2000 kg / mu, while the crab yield is about 200 kg / mu. The computing formula of fish resource value in Qilihai is shown as follows:

The value of fish yield = fish yield area ×fish yield of per unit area× the price of fish / kg

$$P = \sum_{i=1}^{n} Si \times Vi \times Ni$$

In this formula:P means the total value of fish yield in wetland

Si means fish yield area of type i

Vi means fish yield of per unit area of i

N_i means the average price of fish of i

n means the number of fish species

According to the above formula we can calculate the total value of fish yield of Qilihai in 2008 is 255 million yuan and deduct intermediate consumption expenditure such as consumption of materials and production services net to get net output----124.95 million yuan.

Value of plant resources. The main plant product in Qilihai wetland is reed, which owns high economic value and ecological value, for it is not only the important raw material of paper industry, but also the important means of production of agriculture, salt, fisheries, aquaculture, weaving industry. The Reed distribution area is about 7186 hectares. According to the survey, the average yield of reed is 500 kg / mu in 2008, while the average market price is 0.6 yuan / kg.. The formula of the value of plant product output in Qilihai wetland:

$$P = \sum_{i=1}^{n} Si \times Vi \times Ni$$

In this formula:P means the total value of plant yield in wetland

S_i means plant product yield area of i

V_i means plant product yield of per unit area of i

N_i means the average price of plant product of i

n means the number of plant species

According to the above formula we can calculate the total value of plant yield of Qilihai in 2008 is 3233.7 million yuan.

Scientific and cultural value. Qilihai wetland not only owns rich species resource, but also precious flora and fauna that is on the State protection list: glycine soja, oriental White Stork, leaving gulls, etc. It locates in the dense transposition section of different routes of bird's north-south migratory and it is also the rest place for endangered bird species to migrate, so it owns great scientific and cultural value. In addition, the Shell Bar and oyster beach in the wetland has important scientific value for the study of ancient topography, oceans, change between land and sea.

This paper uses the average scientific value 382 yuan/ hm2 of per unit area of wetland ecosystems in China to get the cultural research value of Qilihai wetland---3.629 million yuan.

The value of tourism and leisure. Qilihai wetland conservation district has rich tourism resources, for its diverse landscape is manifold. It has utilized natural beauty of east Qilihai reservoir and west Qilihai reed reservoir to develop the sightseeing tours of wetland water scenery and folk customs. The recreational value formula of Qilihai Wetlands Ecosystem is shown as follows:

Total value of tourism=direct income of tourism+ travel costs + direct travel time value.

Direct income of tourism, Qilihai receives 31 million tourists in 2008 annual [3], statistics of the ticket, hotel revenue, adding up tourism revenue, fees of goods and parking, we can get the total number--31 million yuan.

Travel expenses include travel expenses and accommodation costs. Visitors of Qilihai generally come from Beijing, Tianjin, Hebei Province, according to the standard of transportation costs 60 yuan (round trip) as well as accommodation costs 70 yuan / day • per capita, we can calculate that travel expenses are 40.3 million yuan.

Travel time value = daily wage × travel days × 40%, the income level of visitors mainly concentrates on 100-150 (yuan / person), while the travel day is regarded as one day, we can calculate the travel time value is 15.5 million yuan.

Therefore, we can obtain that the tourism value of Qilihai wetland is 86.8 million yuan.

Indirect utility value. Qilihai indirect utility value includes contents of four aspects--conserving water, regulating climate, purifying pollutants and providing habitat.

The value of water conservation. The value of water conservation in Qilihai wetland uses the shadow project method. First calculate the total amount of storage of water conservation that is available in Qilihai wetland and then calculate the total volume according to the project price of water per unit volume. The total amount of water resources in wetland actually includes water content within plant, water content in soil horizon and water accumulating volume. Here, we mainly use the water accumulating volume to replace the total amount of water resources in wetland.

Total value of water conservation = total water conservation storage capacity ×cost of storage capacity in unit storage capacity.

The total surface water accumulating volume of Qilihai wetland is 80 million m3. According to that the service life of for reservoir is 50 years, the cost of unit capacity of reservoir uses the constant price at 2000 and for building every m3 of storage capacity, we need to invest 5.714 yuan, then the total value of water conservation in Qilihai wetland is 457.12 million yuan.

Adjusting the climate value (carbon - fixation and oxygen production). We use the value of fixing CO_2 and releasing O_2 by the plants in wetland to calculate the value of fixing carbon and releasing O_2 in Qilihai wetland. First, calculate the annual net production of wetland plants. The biomass of wetland plants is divided into the two parts of above and below ground. According to the survey, average annual growth of wetland plants capacity in Qilihai wetland is 53895 t. According to photosynthesis equation,6 CO_2 (264 g) + 6 H_2O (108 g) → $C_6H_{12}O_6$ (108 g) 6 O_2 (193 g) → polysaccharide (162 g)

Producing 162 kg of dry matter, plants can absorb 264 kg CO_2 and emit 194 kg of O_2. 1.63 t of CO_2 is needed to generate 1 t of dry matter with a releasing of 1.2 t of O_2. According to that CO_2 formula and the atomic weight CO_2 C/CO_2 = 0.2729, we use carbon tax to evaluate and adopt the Swedish tax rates and the parity between dollar and RMB is 1:6.8, then:= The fixed deposit amount of carbon =fixed amount of CO_2× 0.2729 ×carbon tax rate(here, we use the average value of the afforestation cost of 250 RMB / t in China and international carbon tax standard $ 150 / t -635 yuan / t as a standard of carbon tax)

Thus, carbon fixation value of Qilihai wetlands is 15,223,500 yuan. Volume of releasing O2 = annual release volume of O2 by plants × 1.2 t The average price of industrial production of oxygen in Tianjin is 400 yuan / t, then the value of producing oxygen in Qilihai wetland is 25,869,600 yuan.

As a result, the total value of carbon fixation and oxygen production in Qilihai wetland is 41,093,100 yuan.

Purification value. There are some aquatic plants in Qilihai wetland including emergent, floating and submerging plants with strong ability to remove poisons, and meanwhile, the wetland has the function of adsorption, desorption, oxidation reduction on soil. The function of degradation on pollution of wetland uses the research result of Robert Costanza, that is, the value of per unit area for the function of degradation on pollution of wetland is $ 4,177 / hm2 • year (28,403.6 yuan / hm2 • year). The exchange rate of RMB against the U.S. dollar is 6.8.

The purification value of Qilihai = 28403.6 yuan / hm2 • in 1 year × 9500 hm2 = 269834200 yuan.

Wetland area is 95 square kilometers and we can calculate the purification value of Qilihai is 269.834 million yuan.

Function value of providing animal habitat. Qilihai wetland is a regional habitat for rich species and the main migration route and food supply of migratory birds in the midway. The rich biodiversity of the wetland makes Qilihai the biological nursery, refuge and gene pool of species. The value of biodiversity in Qilihai wetland ecosystem is estimated according to that the value of wetland that can be used as a wildlife refuge when conducting assessment on the global ecosystem and the wetland area of Qilihai that can be used for animal habitat.

In accordance with the price method of alternative market and each protection of a species can obtain a benefit of 1×10^4 yuan, there are a total of 184 kinds of birds in Qilihai wetland protected areas [4], with a value of 1.84 million, which is used to represent the value of habitat for important species.

The function value of Qilihai wetland in providing animal habitat is 1.84 million yuan.

4   Results and Analysis of Assessment

Through the assessment of several ecological functions of Qilihai wetland, if ignore the time difference, we can add the ecological value of each item to get the total value of wetland ecosystem in Qilihai wetland—1017.6031 million yuan / year. Among them, the value of fish resources is 124.95 million yuan / year, the value of plant resources is 32,337,000 yuan / year, educational and scientific research value is 3.629 million yuan / year, tourism and leisure value is 86.8 million yuan / year, value of climate regulation is 41,093,100 yuan / year, the value of water conservation is 457.12 million yuan / year, purification value is 269,834,000 yuan / year and the value provided to habitat is 184 million.

Evaluation results of total value of Qilihai ecological functions are shown in Table 1 and Figure 1.

Table 1. List of total value of Qilihai wetland ecosystem

Value Mode	Economic benefits of wetland ecosystem	evaluation method	monetary value (ten thousand Yuan /a)	percent of all group (100%)
Direct use value	Value of fish resource	Market valuation method	12495.00	12.28
	Value of plant resource	Market valuation method	3233.70	3.18
	Educational and scientific Research value	Travel cost method	362.90	0.36
	Tourism and leisure value	Travel cost method	8680.00	8.53
summary of direct use value		—	24771.6	24.34
Indirect Use Value	value of climate regulation	Carbon tax method and Af-forestation cost method	4109.31	4.04
	value of water conservation	Shadow en-gineering method	45712.00	44.92
	value of purification	Replacement Cost Approach	26983.40	26.52
	value provided to habitat	Market valuation method	184.00	0.18
summary of indirect use value		—	76988.71	75.66
Total value		—	101760.31	100.00

ten thousand/a

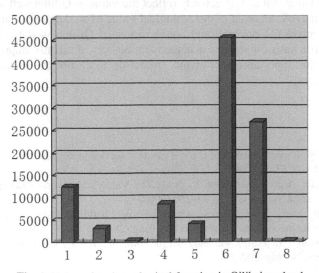

Fig. 1. Value of each ecological function in Qilihai wetland

1 Value of fish resource
2 Value of plant resource
3 Educational and scientific Research value
4 Tourism and leisure value
5 Value of climate regulation
6 Value of water conservation
7 Value of purification
8 Value provided to habitat

This paper calculates the direct use value and indirect use value of wetland ecological functions in Qilihai, a total of 7 types and ignoring the time difference, and then it gets the result that the total value of ecosystem functions of wetlands in Qilihai in 2008 is 1,017,603,100 yuan;

Ecological value in per unit area of Qilihai wetland is 107,100 yuan / (hm2 • a) and it is greater than the national ecosystem service value in per unit area, which indicates that the value of wetland ecosystem service is high;

According to the results, direct use value of Qilihai wetland is 247.716 million yuan / a, accounting 24.34% of total value, while the indirect use value is 769,887,100 yuan / a, accounting 75.66% of the total volume, which indicates that the indirect use value of ecological system is much larger than its direct use value, however, the indirect use value is often neglected by people. The results in this paper can prove that the economic interests created by wetlands virtually are quite huge in order to make people recognize the significance of ecosystem services.

5   Conclusions

By analyzing the ecological value of Qilihai wetlands, constructing the wetland value assessment methodology, and synthetically evaluating the value of Qilihai wetland, we can obtain the value that can objectively reflect the value of Qilihai wetland. Scientific and comprehensive evaluation of the ecological value of wetlands Qilihai can not only help us to improve the research and conservation of Qilihai wetland, but also can provide scientific basis for wetland management, besides, it can also improve the social awareness of the importance of wetland protection to ensure the sustainable use of wetland resources.

References

1. Cui, L.: Study on value assessment of the ecosystem service in Poyang Lake wetland. Journal of Ecology 23(4), 47–51 (2004)
2. Costanza, R.D., Arge, R., de Groot, R., et al.: The Value of the world's Ecosystem services and natural capital. Nature (387), 253–260 (2002)
3. Xu, X., Li, F., Meng, B.: Resources asset management and its sustainable development, pp. 254–268. Social Sciences Academic Press, Beijing (2003)
4. Zhang, G., Wang, S.: Tianjin wetlands and the ruins of ancient coastal, pp. 104–144. China Forestry Press, Beijing (2008)

Study on the Rural Social Endowment Insurance System in Tianjin

Kehua Wang[1] and Jieping Dong[2]

[1] College of science, Tianjin Polytechnic University, Tianjin, China
[2] Department of Library,Tianjin Polytechnic University, Tianjin, China
{554380457,397638175}@qq.com

Abstract. Tianjin rural social endowment insurance system is a part of the social insurance system in Tianjin which plays an important role in ensuring the lives of elderly rural residents. This paper introduces the present status of rural social endowment insurance in Tianjin, and elaborates the necessity for the establishment of rural social endowment insurance. It also puts forward a series of countermeasures to improve the rural social endowment insurance system, and creates both realistic and forward-looking institutional measures meanwhile.

Keywords: social insurance, rural endowment, probability level, aging.

1 Introduction

With rapid development of socio-economy and urbanization in Tianjin, the level of economic development increases quickly, and social life changes dramatically. The rural socio-economy in Tianjin has developed rapidly and the rural social endowment insurance system has been gradually established and made certain achievements. However, there are still some problems of the social insurance system in rural areas, improving the rural social endowment insurance system by exploration is not only the urgent need for a social security system of the rural-urban conformity according to the socialist market economy, but also an inherent requirement of establishing socialist new countryside.

2 The Present Situation of the Rural Social Endowment Insurance System in Tianjin

2.1 The Pension Is Relatively Low on Average

The rural pension per capita level was 22.63yuan per month in Tianjin in 2005, the workers and farmers of each town both have a relatively low level of pension on average. The average level of pension to workers is mainly between 30-40yuan per person, farmers with about 10yuan. Since 2004, Tianjin government implanted the endowment subsidies for old-age famers. Those who are above 65 in rural areas could gain the top-up subsidies. And the standard of the top-up subsidies is 75yuan, namely, when the old aged 65 and above get subsidies lower than 75, it should be added to

75yuan by the financial subsidy. Even if the implement of the top-up subsidies, the average level of rural endowment insurance pension in Tianjin is only 67.5yuan, which is clearly far below the level of basic living standards of farmers.

2.2 Funds Have Realized Socialization Collection and Distribution

Owing to the cooperation between governments and banks for giving farmers free bank cards, which make it possible to realize the socialization of rural social fund payment and distribution. Besides, the funds are directly managed by the county and township levels without having to set the lower agency to manage it. Therefore, it brings convenience to farmers, simplifies the levels of authorities, and significantly costs savings. In 2005, the amount of rural endowment insurance fund collection totaled 256 million yuan. Since 2003, a total of 103.7 thousand farmers drawn the insurance pension, the cumulative total reached to 205 million yuan.

2.3 Government Financial Input Increased Every Year

Since 2004 the government funding input has gradually increased in rural endowment insurance, undertaking significantly increased responsibility. In 2005, the government implemented pension subsidy regulations, which assured 75yuan of subsidies per month to farmers over the age of 65. Furthermore, the government implemented another subsidy benefits system in 2006, stating that those who aged 60 to 64 year-old men could draw 40yuan per month as subsidies. The top-up subsidies to elderly retirees are entirely managed by the municipal, district and township finances, of which 80% funds are undertaken by municipal finance and the district and township finances assumed 20% fund.

3   Necessary Analysis on the Establishment and Improvement of the Rural Social Endowment Insurance System in Tianjin

3.1   Analysis on the Social Aspects

There are mainly two issues in social aspects, namely, the weakness of rural family endowment and the accelerated aging of population. With economic development and social changes, family planning, urbanization and modernization result in the core of family structure, small family size, mobile labor and population migration. All of these aspects not only weaken the rural family endowment function, and make it hard to take life care, spirits comfort into consideration but also lead to the changes of intergenerational relations in the way of support, living, care, communication and so on. In addition, as the special structure of the population, our city takes a hard endowment burden, especially in rural areas. The aging of population in rural areas is faster than that in urban areas. Currently, our city facing the issue of family endowment for the existed elderly (the stock problem), but also confronting with the issue of family endowment for the newly-increased elderly (the incremental problem).

4   Analysis on Economic Aspects

4.1   On the Economic Structures

It has been a long time since the co-existence of two independent but interrelated levels in our social security system. Under the planned economy system, urban residents could enjoy employment placement by the government, access to medical care, retirement pensions, housing and other benefits. Briefly, they were entitled with a much better social benefits. By contrast, the elderly in rural areas could only rely on their sons or daughters to support them. They didn't have work, let alone the social security benefits. However, it cannot be comparable with the securities access to urban residents by "Unit Security System" in terms of security projects, contents and security levels. After the implementation of contract responsibility system in rural areas, the traditional communal accounting system was completed broken. In addition, famers became the independent economic unit and were in a single traditional family security dilemma. [1]

4.2   On the Marketing Views

The current rural social endowment insurance system with higher operating costs, instability of interest rates, difficulty of maintaining and increasing of the value is not attractive to famers in that they could not bear the economic pressure and risk while they could only enjoy the limited amount of pensions. Most of the insured people have been insured a relatively short duration or low standard. Thus, they enjoyed or will enjoy the pension benefits which could not meet the daily needs of their life. Rural endowment insurance practice shows that the payment standard of the insured is not in general terms with the local rural economy, the level of income and the expectation of need for pension, there is still a long way to go in that the current insurance system could not play an effective role on insuring the life of elderly.

4.3   Institutional Aspects

The city-oriented political trend leads to the lack of rural social insurance. Both on the period of revolution and war and the period of socialist construction and development, famers have made a great contribution to our country. They have accumulated a large number of funds for industrialization and urban development. Besides, they have provided a wealth of raw materials for urban residents with a steady flow of agricultural and by-products. However, the benefits of industrialization are almost monopolized by urban residents, even if the terms of investment in agriculture and irrigation are at a significant point to solve the problem of urban drinking and industrial water. The accumulation of a considerable portion of fund by farmers is directly transferred to the urban resident's living benefits. Due to the unequal distribution of national resources, the agriculture loses opportunities for development when it provides our country with accumulation and the same time takes a heavy burden. Only a small number of the national fiscal expenditure on agriculture is for poverty alleviation and social assistance for agriculture. The country takes little responsibility on rural social endowment, which is unfair to farmers.

5   Countermeasures on the Improvement for the Rural Social Insurance System in Tianjin

5.1   The Premise of Improving the Rural Social Insurance System

5.2   To Speed Up Economic Development

Endowment insurance is an important part of social security which is involved with a large number of people and large scale. Besides, the institutional arrangement and capital accumulation need a long cycle time. As to the current actual situation, since all levels of governments take responsibility on famers. From the macro level, improving the endowment system depends on the economic development. However, as to the difference of local economy development, the establishment and improvement of endowment insurance system is relatively involved in the local economic development.

5.3   To Strengthening the Legal System Construction

In addition to social and economic development, the legal system construction is equally important. Although the government has attached great attention to farmers' endowment insurance from its birth till now, it should also depends on the legal response. As to the rural insurance system, only when it is response to laws and some regulations, could it provides farmers with standardized and effective support, and at the same time avoids the probably arbitrary behavior of government.

5.4   To Increasing the Responsibility of the Government

It is certainly that the establishment and improvement of Tianjin rural endowment insurance system could realize with government's support and norms. As the nature of pension, it is not simply the public goods. In Economic Theory of Club, Buchanan stated that the public goods are things or services decided and provided by some social groups for some purposes. Thus, as a social insurance, the endowment insurance is a government-provided consumption and it is endowed with characters of both private goods and public ones. It is needy to put special emphasis on the role of the government and its responsibility, especially the fiscal input of the government. To an individual, the government's support could give people the basic rights to life, and meet the basic survival needs; to the whole society, the social insurance could make it stable and harmonious. Therefore, it is an important responsibility for government to establish and improve the rural social endowment insurance system. [2]

6   Basic Principles to Improve the Rural Social Endowment Insurance System of Tianjin

6.1   The Principles of People-Oriented and Some Basic Insurance Principles

The social- economic development and establishment of a moderately prosperous society should be people-oriented, and it should have the ultimate goal of achieving comprehensive development of human freedom. When it comes to the establishment, development and improvement of the rural social endowment insurance system, at its starting point and destination must be to insure the fundamental interests of the majority of famers. The object of rural old-age insurance is the elderly in rural areas. Therefore, the development of rural social endowment insurance should cover all the famers as much as possible, especially those in poverty. It should make it center to meet the practical needs of farmers for pension and solve effectively the most pressing various problems for farmers. It is necessary to respect the wishes of farmers, but also to ensure that farmers benefit from the insurance system. Only in this way could it get the positive response from the peasants, mobilize the enthusiasm of farmers, furthermore, contribute to the expansion of old-age security coverage and eventually take all farmers into the rural endowment system scope.

6.2   The Principle of Pursuit of Justice

"In order to use the concept of pure procedural justice in distribution share, it is necessary to actually build and manage a fair system of justice system." As a kind of resource, the rural old-age insurance has its scarcity, which determines the principle of justice on allocation. And it is also the appropriate scarcity that constitutes the objective premise of fair value. At the first place, according to the current actual situation of Tianjin economic development, the phenomenon of "guaranteeing the rich but not the poor" should be replaced by the measure of "low level, wide coverage", which insures more elderly get old-age security.

6.3   The Principle of Dynamic Consistency between Social Security Level and Economic Development Level

The rural endowment insurance development could not be achieved without the support of the real economic resources. Besides, though different countries have different theory studies on social security and different practices, there are still some identical roles to follow in the process of the establishment of the social security system. Among them, an important basic requirement is that the social security should be consistent to social and economic development level. Because of rigidity character, "social security often can be increased but not reduced, social security benefits tend to rise but not fall. Otherwise, it would be strongly opposed by the benefit levels, or even lead to the great social crisis [3]."

6.4   The Detailed Countermeasures to Improve the Rural Endowment Insurance System

In order to fully promote the society construction and new socialist countryside construction in Tianjin and further improve the rural social security system, this paper puts forward the following advices on the establishment and improvement of the rural social endowment insurance system in Tianjin:

6.5   The Coverage of the System

Farmers who are 18 to 60 years old in Tianjin should be covered in this system. Those who seek jobs in cities, namely, migrant workers, but could not be covered in the workplace insurance system in urban enterprise workers should be given the right of endowment insurance. Or those local farmers who are engaged in non-agricultural industries but do not have the conditions of urban enterprises and workers to participate in those pension and whose land has been taken has been engaged in non-agricultural industries but they have not participated in the basic living security, if there are some proofs that could prove the truth of the situation, should be included in the rural old-age insurance range.

7   Fund-Raising

(1) The fund-raising of rural endowment insurance take the raising model of individual contributions, collective benefits and government subsidies. Specific ratio: the amount of individual contributions should be higher than 8% of the basic pension, the ratio of collective subsidies and government subsidies should be higher than 7% of basis pension in principle. But make appropriate adjustments should be made as to groups of different income levels.

(2) Payment of the local farmers should be based on a per capita income of the passed year as the base calculation. In the starting of the implementation of the system, the payment should be controlled no less than 60% of this base, but cannot exceed 300% of the base, and according to the local sub-gradual transition to the base of 100% within three years.

7.1   Enjoy Treatment

(1) The insured person could draw a monthly basic pension, but they should meet the following requirements:

Aged 60.To pay the full old-age insurance as required.

Total payment period reaches to 15 years (180 months)

(2) The insured person who reaches pension age, but less than the provisions of contribution years, and can continue to pay pension premiums, also access to basic pension but the time should be delayed. Another approach could be taken, that is, to make up the required fund in the required payment years, and then enjoy the old-age insurance benefits. Besides, there is also another choice. The insured person who is over 60 could delay the date of enjoying the benefits in 5years later. In the5 years, he could continue to pay for the fund. Then, if he could work it out in the five years, he should make it up once only.

(3) The distribution of rural old-age insurance pension has been socialization. The pension receptors could draw pensions on a monthly basis from the designated bank or other financial institution until death. Their individual account balances, according to their heirs or designated beneficiary will be drawn once only.

(4) Exploration of the implementation of old-age pension subsidies. Man who is over 59 and woman who is over 54 in areas where conditions permit may draw the basic old-age subsidies before the implementation of the rural endowment system. What's

more, the elderly farmers over 70 years old and whose family direct relatives have been required to participate in the new rural endowment insurance or urban basic endowment insurance for enterprise employees, should be given the basic old-age endowment subsidies.

(5) Establish a pension adjustment mechanism and adjust the standards of the pension benefits according to the local economic and social development.

7.2 Funds Management

(1) Establish the rural endowment insurance fund supervision and management committee, including financial, labor security, taxation, auditing, statistics, monitoring and other departments, and the committee is responsible for the supervision, inspection and auditing of the operation of rural social endowment insurance fund.

(2) All the rural old-age insurance fund should be put in special financial accounts, and carry out two lines expenditure management. By agreement of the financial departments at all levels and labor security departments to establish "special financial accounts of rural endowment insurance fund." Rural pension insurance agencies at all levels should open the "rural endowment insurance fund expenditure accounts" correspondingly in the banks and take the special management and earmarking.

(3) Enforce the financial discipline and do a good job to rural old-age insurance fund supervision. Benefits resulting from the preserved value and increased value of the Fund should be added into the rural old-age insurance funds. The unit or individual must not seize and misappropriate of it. For corruption, misappropriation of the funds or loss due to a serious loss of work time, the blamed staff, should be punished severely according to relevant management systems. And the stuff in violation law must be disposed to the judicial authority for disposal.

(4) Improve the informationization management level of the rural old-age insurance fund. In accordance with the requirements in payment of endowment insurance project construction, the payment, the business accounting, payment of banks or other financial institutions and inquire services of the insured person should be put into the computer system management. The most advanced management software should be used to achieve standardization of business processes and operation of the service [4].

Solve the problems of endowment insurance for the migrant workers.Migrant and agricultural population in our city has become a large scale with the very special characters which cannot be ignored. For the endowment pension of this group, there is necessary to make more than two options of pension insurance, one of the design is for the stable career working farmers (with a longer period of stability in labor relations and work) and the other option for those migrant workers who have no stable jobs (often in mobile state) to choose, and as a policy introduced. Before the implementation of such policies, the government could make some classifications to the migration workers. To those who have the required residence years, and a relatively fixed residence and units, the government should give them the right to enjoy the city qualifications of the residents' interests and officially into the urban pension insurance system. By contrast, to farmers who do not meet the conditions of migrant workers, the adjustment program should be come up with, and may be gradually incorporated into the insurance system as the case. Specifically, migrant workers with employers could enjoy the benefits in accordance with the urban workers payment policies, and the proportion of individual contributions must be 8% in one step. Those workers without the employers could gain the benefits in accordance to the implementation of individual accounts cumulative storage system, it can be uniformly applied to the employees who pay on an average salary and on the proportion of the salary, that is to say, individuals could pay for the future pension, they can pay more than the prescribed proportion, but they cannot pay less than that. Finally, when the payments reach the required age, the insured who are retired, or aged 65 (employees without employers), can draw the basic pension [5]. Workers with employers can draw a basic pension and individual pension accounts, workers without employers can only draw the individual accounts pension. Two type's migration workers, whose payment period does not meet the required, could be drawn capitals with interests from the individual account storage once only. But they could not draw the basic old-age pension, ever if they have their employer.

8   Conclusions

In summary, through the development and reform of the existing rural endowment insurance system, making a strong safety net for rural farmers is the government and the national social concerns, which attract much social attention to work actively to realize it. The establishment and improvement of the rural old-age insurance system in Tianjin will certainly provide a strong support to the comprehensive, coordinated and sustainable implementation of the "scientific thought of development", but also promote the sustained, rapid, harmonious and healthy development of economy in Tianjin.

References

1. Cai, J.: Legal Research on China's Rural Social Security System. U.S.-China Economic Review 4(7), 80–81 (2004)
2. Li, X.: Question and Answer by Li Bengong – the Deputy Managing Director of Office of National Work Councils on Aging Issues. People's Daily (August 11, 2006)
3. Li, Y.: On the Basic Objectives and Policy Options at Current Stage on Farmers Endowment Security System Reform. Sociological Research (5), 105–116 (2001)
4. Liu, C.: The Establishment of Farmers Social Security System Is the Fundamental Policy to Solve the Issues of Farmers, Countryside and Agriculture. Red Flag Presentation (21), 11–14 (2004)
5. Lu, H.: Theory Thinking and Policy Recommendations on Establishment and Improvement of the Social Security System for Landless Farmers. Economic Dynamics (10), 52–56 (2004)

Comparative Analysis of the Economic Contribution of Sino-US Agricultural Economic System

Qin Ruiqi and Xu Le

College of Economy, Tianjin Polytechnic University,
NO.399 Binshuixi Road, Xiqing District, Tianjin, China
40953175@qq.com

Abstract. This paper will analyze and compare the status of agricultural economic system in the U.S. and China, construct the model on the economic contribution of system to agriculture to quantify the comparison, to find out the shortage of agricultural economic system in China and put forward proposals for reform.

Keywords: China, the U.S, agricultural economic system, comparative analysis, contribution, reform.

1 The Status and Compare of Sino-US Agricultural Economic System

1.2 Status of the American Agricultural Economic System

Land system. Private ownership of land is typical in the U.S., most of which is own by private enterprises and individuals. The proportion of land allocated is roughly as follows: more than 50% is private land, more than 30% own by the federal government. While in federal and state government-owned land, most of them are forest, grassland, marsh and other non-arable land. Most of the rural land is occupied by farms and plays an important role in the development of agricultural production in the United States.

Rural land management system in the United States is based on the family private farms. From the view of the development of agricultural production in the United States, it went from the a small family running extensively to the moderate scale of home, and finally developed into a modern professional integrated large-scale operation of the family, forming a new pattern of land management system in rural America, in order to achieve the integration of production, supply and operation of the enterprise.

Agricultural production system. In the mid of 1980s', the United States firstly proposed the development of sustainable agriculture. The core is to improve the system of crops and farming systems: on the one hand, emphasis on traditional technology and modern biological, ecological technology, research and development, full use of modern agricultural technology development to increase agricultural productivity and the level of production. The other hand, focus on agriculture resources and environmental protection strategies to maintain the development and application of technology development, doing efforts to reduce chemical fertilizers, pesticides and additives, in order to make full use of the resources, improve the quality standards of agricultural products, soil fertility and ecological environment and achieve harmonization of economical and social feasibility.

System related to agricultural management. The financing structure of the U.S. agriculture. In the United States, sectors providing capital for the farms are commercial banks, cooperative agricultural credit system, insurance companies, government agencies, individuals and dealers.

Table 1. Balance sheet of U.S. agriculture (2004-2007)

	2004	2005	2006	2007
Asset	1617582	1835464	2047439	2209924
Real Estate	1340582	1549227	1755794	1912194
Movable Property	277000	286237	291645	297730
Debt	182965	193230	196392	211520
Real Estate	96872	101518	101475	107778
Agricultural Credit Cooperative system	37723	40125	40881	45356
Agricultural Service	2222	2050	2107	2054
Commercial Bank	35233	36939	37777	40598
Insurance Company	10912	11019	11292	11152
Individual and Others	10782	11384	9212	8391
Movable Property	86093	91712	94917	103742
Agricultural Credit Cooperative system	21896	24218	27540	32252
Agriculture Service	3242	3015	2722	2878
Commercial Bank	45830	48523	50995	55475
Individual and Others	15125	15956	13660	13138
Ownership Interest	1434617	1642234	1851047	1998404

UNITE: MILLION U.S. DOLLARS.

From Table 1, the U.S. agricultural balance sheet, it can be seen the U.S. agricultural financing presents the following features: First, the asymmetry of agricultural liabilities and assets. Second, commercial banks and cooperative agricultural credit system are the main funders. Third, financing from cooperative agricultural credit system is rising.

Business enterprise perspective of the U.S. agriculture. First, the United States issues legislation, support policy and a series of measures to establish family farms as the main organizational form of agricultural micro-business organization system, which forms a proper organizational foundation for agricultural enterprises. Second, by expanding the demand of agricultural products in domestic and international market and integrated management of agriculture, it promotes the specialization and commercialization of agricultural enterprises. Again, with the development of commercial agriculture in the U.S., agricultural social service system gradually developed and perfected. Finally, the U.S. government provides the basic condition for agricultural enterprises and system security.

2   Status of the Chinese Agricultural Economic System

Land system. The subject of agricultural production in China changed a lot since the early of 1980s' after the rural reform. Households under the system of contracted responsibilities on the household basis with remuneration linked to output became the subject of the agricultural activities. Under the socialism system in China, we implement public ownership of means of production and land is own by the nation and part of the collective. Farmers are the contractor, having the right to using the land, that is the separation of land ownership and right of use.

Agricultural production systems. Farming system is the factors that affect the environment and its dynamic evolution. In recent years, in order to adapt to modern agricultural development, promoting regional agricultural development cycle that adjust measures to local conditions becomes a key element in the construction of new countryside in China.

3   System Related to Agricultural Management: The Financing Structure of China's Agriculture

First of all, the scale of farmers' financing is small. The main source of household investment is self-financing and bank loans. In 2007, the average household funding was 2135 RMB in China, 1953 RMB of which was self-financing and domestic loans of which was only 67 RMB. However, the financing of each farm was about $ 7,000 in the same period in the United States.

Second, no loan is secured by real estate. In the United States, about half of the liabilities in agriculture are issued secured by real estate. But in China, only a few rural areas begin to try to loans secured by real estate and in its initial state.

Third, source of resources is single. The main sectors that provide household financing currently are financial institutions, commercial financial institutions in rural areas, rural cooperative financial institutions and other financial organizations. More than half of the loans are provided by rural credit cooperatives financial institutions.

The two-tier management organization in China. The two-tier management system divides economic organizations into two levels: one is the unified management of collective economic organizations; the other is the dispersion for the family business. The two tiers are linked by contract. It should be said, the two-tier management system is an institutional innovation, which divides the cooperative economic organizations and functions into two parts linked. The system brings collective land and the key production tasks to farm households under contract, while retaining unified economic functions of the collective, which gives farmers a certain of autonomy to production and management.

4   The Defects of Chinese Agricultural Economic System

The defects of land system. There are more and more obvious deficiencies in the system of contracted responsibilities on the household basis with remuneration linked to output: First, the subdivision of land has led to diseconomies of scale, affecting the further increase of the efficiency of agricultural production. Second, the collective ownership of agriculture is defined unclearly. Third, the right to the contracted management is not complete.

The defects in the system of agricultural production. Family is the basic unit in China's agriculture. The per capita arable land is limited. In order to meet the demand of all family members, it is common to grow several crops on an acre of land. Surveys show that in all farmers doing business in agriculture, forestry, animal husbandry and fisheries (10 thousands households), in the type of production that farmers produce, the proportion of farmers who product 4 and 5 kinds of products were 19% and 19.6%, who only produce one product was 10%. On average, each farmer produced 4.2 kinds of products. Moreover, diversify production is grain-based, accounting for 96.4%, followed by livestock products, accounting for 68.6%; Again, vegetables, accounting for 66.5%. Farmers' generally non-specialized farming systems and management methods limit demand for commodities, including the increase of agricultural products and the expansion of market size. Small-scale product demand and market further impede specialization.

The defects in the way of agricultural operations. China's rural two-tier management system itself lacks a sound system design. On the operating level in the organization, the corporate entity status of such organization, as the market participants, is not clear and their functions are complex. Thus, they look powerless in front of a variety of business organizations and non-business organizations or enterprises. On the level of the family business, the first is that the decentralized management of family exacerbated the limitation of scarce land resources, which affects the scale management of land; second, because of the uncertainty of the harvest of agricultural production and the market, dispersion farmers can not properly use the comparison of price, costs and profits to avoid market risk, leading to a precarious state of agricultural production and management; third, dynamical adjustment of grass-roots organizations make farmers can not achieve the effective input and accumulation.

5   The Model and the Analysis of the Contribution of Sino-US Agricultural Economic System

5.1   Construct the Model

According to the basic principle of AHP, writers constructed index of the contribution of the agricultural system, including rule layer and measures layer. Rule layer is a standard measuring whether target could be achieved, including the two levels: the first level is the principal part of the system and system.

Table 2. Evaluation index of contribution of agricultural economic system

The Contribution of Sino-US Agricultural Economic System A	Principal Part of the System B1 0.33	The Quality of Principal Part of the System C1 0.09	The Quality of Farmers	D1	0.65
			The Quality of Research Institutions 0.12	D2	
			Number of Researchers	D3	0.23
		System Input C2 0.28	Growth Rate of Government Investment in Agriculture D4 0.72		
			Growth Rate of Farmers Investment in Agriculture D5 0.07		
			Growth Rte of Institutions Investment in Agriculture D6 0.21		
		The Initiative of Principal Part C3 0.63	The Initiative of Government	D7	0.26
			The Initiative of Farmers	D8	0.63
			The Initiative of Institutions	D9	0.11
	System Performance B2 0.67	Capacity Performance of the System C4 0.75	Commercial Rate of Agricultural Products D10 0.16		
			Contribution of Agriculture to GDP 0.12	D11	
			Growth Rate of Income per capita 0.36	D12	
			Productivity of Agricultural Labor 0.16	D13	
			Return of Capital	D14	0.16
			Productivity of Land	D15	0.04
		Upgrading of Industrial Structure C4 0.25	Output Ratio of System	D16	0.46
			Comprehensive Utilization Rate of Land D17 0.12		
			Forest Cover	D18	0.21
			Annual Increase of Soil Erosion Control Area D19 0.21		

performance; the second level subdivides the principal part of the system into the quality of principal part of the system, system input and the initiative of the principal part; the upgrading of industrial structure is subdivided into capacity performance of the system and the upgrading of industrial structure. Measures layer refers to the program goals, methods, means, etc., is a further specific refinement layer of the above criteria.

5.2   Parameter Estimation and Model Checking

Make use of the quality of farmers, research institutions and researchers to illustrate processes that determine the weights of AHP in this study. Make use of the method of scoring by expert to get the following matrix:

A	D1	D2	D3
D1	1	5	3
D2	1/5	1	1/2
D3	1/3	2	1

5.3 Calculate the Matrix of Each Row to Determine the Geometric Mean of All Elements

$$W_i = \sqrt[n]{\prod_{j=1}^{n} a_{ij}} \quad (i=1, 2,\ldots, n)$$

Using the above formula to get $W = (2.4662, 0.4642, 0.8736)^T$.

Normalize W_i, that is to calculate

$$W_i^* = \frac{W}{\sum_{i=1}^{n} W_i} \quad (i=1, 2,\ldots, n)$$

We would get $W^* = (0.6483, 0.122, 0.2297)^T$, the relative weights of the three.
Calculate the largest eigenvalue of the matrix.

$$\lambda_{max} = \frac{1}{n} \sum_{i=1}^{n} \frac{(AW^*)}{W_i^*} = 3.0037 \quad (i=1, 2,\ldots, n)$$

Therein, $(AW^*)_i$ is the i element of vector AW^*.

Test its consistency.

$$C.I = \frac{\lambda_{max} - n}{n-1} = 0.002$$

$$C.R = \frac{C.I}{R.I} = 0.03 \quad (R.I=0.58, \text{ can be obtained by looking up tables})$$

It is clear that both of C.I and C.R are less 0.1. Therefore, the weight of the three is acceptable. The weight of other indicators can get by the same method.

Measurement of the Economic Performance of the System

Calculate quantitative indicators
For quantitative indicators, use a comprehensive evaluation that improves the efficiency coefficient method. The basic idea is:
Use power function to calculate individual points. The formula is:

$$d_{ij} = (x_{ij} - x_i^{(s)}) / (x_i^{(h)} - x_i^{(s)}) \times 40 + 60$$

In the above equation, d_{ij} is the individual point of the j expectation value of indicator i. x_{ij} is the j expectation value of indicator i. $x_i^{(s)}$ is the value not allow of indicator i and can be determined based on requirements. 40 and 60 are the given parameters.

Calculate the evaluation points using the arithmetic average method. The formula is:

$$Z_j = \sum d_{ij} w_i / \sum w_i$$

Where, z_j is the first comprehensive assessment in period j; d_{ij} is individual score of index No. i in period j; w_i is the weight of index No. i.

When use coefficient method with improved efficacy evaluation, in the evaluation index, you can include both positive indicators and reverse indicators. No matter what kind of indicators for evaluation, individual evaluation points higher, the better the value of that index evaluation.

Calculate qualitative indicators. For qualitative indicators, use a comprehensive evaluation model in fuzzy mathematics. The basic idea is:

Factor set $U = \{u_1, \cdots, u_n\}$ is composed of a collection of various factors of judged object;

Judge set $V = \{v_1, \cdots, v_n\}$ is composed of a collection of reviews.

Judge single factor means evaluation of individual factor, to get the fuzzy set $(r_{i1}, r_{i2}, \cdots, r_{im})$ based on V. Then it can determine a fuzzy relationship $R \in U_{n \times m}$, which is called the evaluation matrix.

$$R = \begin{bmatrix} r_{11} & r_{12} & \cdots & r_{1m} \\ r_{21} & r_{22} & \cdots & r_{2m} \\ \cdots & \cdots & \cdots & \cdots \\ r_{n1} & r_{n2} & \cdots & r_{nm} \end{bmatrix}$$

It is judged by all of the single factors consisting of F-sets.

Model: AoR=B= $(b_1, b_2, \cdots, b_m)$

Where   A= $(a_1, a_2, \cdots, a_n)$ $\sum_{i=1}^{n} a_i = 1, a_i \geq 0$; $R = (r_{ij}) n \times m$

$r_{ij} \in [0,1]$

$$b_j = \sum_{i=1}^{n} a_i r_{ij} \quad (j = 1, 2, \cdots, m)$$

b_j is the function of $r_{1j}, r_{2j}, \cdots, r_{nj}$. This is the evaluation function.

This model uses the real number of addition and multiplication operations. Using this model, after the comprehensive evaluation index, we can get a membership value. Since it is between 0 and 1, in order to compare the value of individual scores, multiplied the membership value by 100 to obtain the single index score. Then calculate the arithmetic average to get composite score.

5.4   Analysis of the Performance of System to Agricultural Economic Growth

Digital data. Taking into account the availability of data, combined with targets set by the model, collected the following Table 3 and Table 4 of the statistical data and handled them properly.

Table 3. Variable value of agricultural economic system in China (1995-1999)

Year	The Quality of Principal Part of the System		System Input	Capacity Performance of the System		Upgrading of Industrial Structure		
	Junior Middle School Education Level (per hundred)	Agricultural Technology Costs (0.1 billion)	Agriculture Input / Total Income %	Average Grain Production of Agricultural Labor per year (Kg)	The Annual Production of Agricultural Labor (U.S. dollar)	Irrigated Area / Arable Land %	Per capita Acreage of Agricultural Labor (Ha)	Chemical Fertilizers (1000 tons of fertilizer)
1994	44.30	20.16	1.76	100296.00	52561.00	52	74	3035.90
1996	46.70	21.10	1.79	100550.00	53200.00	52	75	3120.50
1997	48.40	22.80	1.82	101323.00	53890.00	53	75	3302.70
1998	49.50	24.72	2.10	101330.00	54270.00	54	77	3500.80
1999	50.70	29.20	2.45	102670.00	54330.00	54	79	3712.60

Table 4. Variable value of agricultural economic system in the U.S. (1994-1999)

Year	The Quality of Principal Part of the System		System Input		Capacity Performance of the System		Upgrading of Industrial Structure	
	Junior Middle School Education Level (per hundred)	Agricultural Technology Costs (0.1 billion)	Additional Education Spending for Agriculture(0.1 billion)	Urbanization %	Fair Trade Volume index	Comprehensive Parity Index of Industrial and Agricultural Goods	Irrigated Area Index	Chemical Fertilizers (1000 tons of fertilizer)	Non-food Area / total sown area %
1995	40.11	3.00	112.89	29.00	92.72	95.70	109.60	3593.70	26.60
1996	42.83	4.94	147.41	29.30	117.56	101.90	112.05	3827.90	26.10
1997	44.30	5.48	147.41	29.90	139.40	105.90	113.95	3980.70	26.70
1998	44.99	9.14	165.02	30.40	158.68	106.30	116.30	4083.70	26.90
1999	46.05	9.13	162.46	30.90	173.66	110.80	118.22	4124.30	27.60

Analysis of the conclusion. Through theoretical study and calculation, we can draw the following conclusions: the system to growth of agricultural economy is significant. Among them, the score of China's system contribution to agriculture is A1 = 28.90; while the score of the U.S.'s system contribution to agriculture is A2 = 38.20. It can be seen that the contribution of the land system, agricultural production and management of agricultural in the U.S. are higher than China.

6   Reform Principles and Proposals to China's Agricultural Economic System

6.1   Reform for Land System

Agricultural land issues are actually problems of the land rights to farmers. Reform for agricultural land property rights has lagged behind, which is an important factor that caused land conflicts and rural poverty.

Clear the property rights to agricultural land and contract permanently
Contracting permanently makes the collective give way to farmers. Farmers with land contract directly possess the property right. Of course, this contract is not private ownership. Only when the use of land would not change, the farmers have the right to transfer agricultural land and get corresponding revenue.

Establish transfer mechanism to protect the property rights of agricultural land
Set up two carriers: the intermediary organizations of agricultural land transfer and farmers holding transfer certificates. The intermediary organizations are in form of agricultural cooperatives, enjoy collective property. They purchase right of use of agricultural land, then transfer to farmers willing to do scale management and hold transfer certificates.

6.2   Reform for Agricultural Production and Management System

Optional system of agricultural production and management
At the present stage, mode of agricultural production and management should build system of share-holding private farms. Private farms are economic entity with legal personality, funded by investors to integrate the existing land resources, where farmers become shareholders with funds, technology, machinery, land and so on, by which to get dividends.

Measures to improve the financing structure of agriculture
First, confirm the ownership of agricultural land and housing. Second, build credit guarantee system for agriculture and expand commercial bank lending for agriculture. Finally, give institutional support and promote the establishment of agricultural insurance system.

References

1. Nelson, R.: National Systems of Innovation: A Comparative Study. Oxford University Press (1993)
2. Porter, M.E.: Clusters and the New Economics of Competition. Harvard Business Review (11-12), 30–35 (1998)
3. Venables, A.J.: Cities and Trade, External Trade and Internal Geography in Developing Economies, Wording Paper (November 2000)
4. Cooke, Schienstock: Structural Competitiveness and Learning Region, Enterprise and Innovation Management Studies

5. Kos, et al.: Property rights and institutional change. Shanghai People's Publishing House (1994)
6. Lin, Y.: Economic Theory of Institutional Change: Change and the mandatory nature of induced changes. Property rights and institutional change. Shanghai People's Publishing House (1994)
7. Liu, M., Zhang, J.: The Economic Analysis of the innovation in China's agriculture. Economies of the North (April 2007)
8. Luo, Y.: Economic analysis of the changes in China's rural land system. Xinjiang University Philosophy and Social Science Edition (June 2007)
9. Xu, G.: Ownership structure and property system in China Studies. Inner Mongolia University Press (1996)

The Research and Application of Case-Based Reasoning in Fixture Design

Cui-yu Li and Rui Wang

School of Textiles, Tianjin Polytechnic University, Tianjin, China
Key Laboratory of Advanced Textile Composites, Ministry of Education,
Tianjin Polytechnic University
Lcy_tju@yahoo.com.cn

Abstract. The application of case-based reasoning in Computer-aided Fixture Design is introduced, including the expression of configuration design, searching and matching of case etc. The method of case structure is described on the base of Case-based Reasoning system. The guide style fuzzy searching functions for mold and die case is used to give the searching range and the most similar case is searched from the antecedent cases. The case is modified to satisfy the new situation and then stored in the case base according to the technique. This fixture design system is expected to possess two characteristics, the first is that the limitation of using the man-machine conversation assembling system is avoided and it is more intelligent than the traditional system; the second is that the system can acquire new knowledge from every consultation, and becomes more experienced.

Keywords: modular fixture, computer-aided design, case-based reasoning, expert system.

1 Introduction

Modular fixture of high flexibility of has become essential aid for modern manufacturing systems, but mature computer-aided modular fixture design software has not yet appeared in domestic. With the rapid development of artificial intelligence technology, designers began to introduce it into computer-aided design of modular fixture system, due to the complexity of modular fixture design, the results are not particularly desirable.

Many studies about intelligentialize/ of mold design and knowledge management has been carried out   in research institutions both at home and abroad, but the system of case knowledge management of mould design and data mining is few. With the rapid development of CAD / CAM technology, especially the rapid development of artificial intelligence technology, researchers have solved different problems fixture design with the artificial intelligence techniques combining different technique, these technologies include CBR (Case-Based Reasoning, CBR) technology. Case-based reasoning is the inference mode different from the rule mode emerging from the field of artificial intelligence, which is the method of similar reasoning, solve new problems through the use of past mature experience and knowledge, and overcome the problem of duplicated effort, a long development cycle and unstable design quality, hence the CBR technology is a very effective method to solve the computer-aided design of modular fixture design.

J. Luo (Ed.): Soft Computing in Information Communication Technology, AISC 158, pp. 325–331.
springerlink.com

2   Case-Based Reasoning Technology

CBR technology was first proposed by Professor Roger Shank from Yale University in his book 《Dynamic Memory》 , which was published in his 1982, and was given the construction method in computer system. Nowadays, the method of this reasoning, now CBR as a reasoning technology of artificial intelligence has been successfully applied in many engineering applications [1-2]. Case-based reasoning, also known as case reasoning has been acquired more and more attention by the researchers as a artificial intelligence modus from the machine copy of surface to the development of deep machine thought. Cognitive psychology research shows that human experts use a lot of experience and knowledge in their minds facing the problem, recall similar problems in ways and do the appropriate modifications for new situations. The famous philosopher Kant once said: "Whenever intellect is short of reliable ideas, the analogy can often guide us forward [3-4]."CBR is a manifestation of this idea.

CBR is a reasoning mode that solve the present problem by visiting the past solving process and the results of the same problem. Compared with the production or other rule-based reasoning system, CBR system solve problem with a completely different way that can increase the efficiency of solving new problems with the past results, rather than derived again. In general, CBR reasoning process include the following basic steps:

1) Extraction, description and importation of issue features: extract the features and relationship between characteristics of to be solved problems and input them into the system;

2) Retrieval of corresponding case: According to the characteristics of the problem, retrieve the past similar cases corresponding to the current issue from the case base;

3) Rewritten and adjustment of the case: find the most similar case from the retrieval case, adjust and modify the case to make it suitable for solving current problems;

4) Evaluation of the program: evaluation and test of implementation effect or satisfaction of new program and results;

5) Storage of new cases: the solving process of the current problems and results form the new cases, according to a certain strategy, and the new case is added to the case base.

Solution procedure of CBR is shown in the figure.

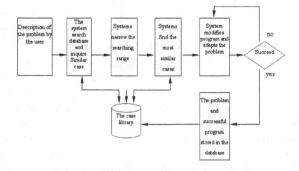

Fig. 1. The process of case-based reasoning

3   The Application of CBR Technology in Computer-Aided Modular Fixture Configuration Design

As the modular fixture CAD system MFS R5.0 simulate the artificial assembly process only with the computer [5], its application is limited owing to its relatively low degree of automation and higher requirements for user technology. Therefore, the paper will introduce case-based reasoning technology to computer-aided modular fixture configuration design, so as to increase the intelligence of modular fixture CAD system [5] to. Overall system model is shown in Figure 2 [6].

3.1   Establishment of Case Library

Case base is the foundation of CBR technology and the core of the knowledge base mainly from the expert knowledge and individual cases. A case library that contains all the past fixture map is the foundation of computer-aided modular fixture configuration design with the CBR technology. Case library include a variety of typical case and information of typical structure. Typical Structure is precedence compound mode of fixture component which workers summarize in the long drawn-assembly process to complete a one or several independent function. Typical structure is regarded as assembly which can be applied in various structure of fixtures that from combination of a number of typical structures. Typical structure is summarized by workers in a long-term and is the crystallization of the wisdom and experience. Therefore, the more typical structure is optional, so it indicates that the intelligence of system is higher, which is a proportional relationship between them. Case library of typical structure in the system is an open independent part, because it is open, users can increase the typical structure of case library and the system's learning ability. Because it is independence, the new cases will not hamper the normal operation of the system, but can increase the system's intelligence [7].

The Microsoft Access database is chosen depending on the application requirements. Compared to other relational database management system, Access is easy to use and powerful and have a good interface with other databases and have some advantages of storing files with a single form, supporting a long file names, the powerful networking capabilities and handling a variety of data, etc.

3.2   Knowledge Representation of Case

Knowledge is the basis of intelligence and is also key considerations in expert system design. Knowledge representation is a description of knowledge and a set of conventions, a data structure used to describe the knowledge that can be accepted by the computer, and it is unity of the data structure and control structure, not only considering the storage of knowledge, but also taking into consideration the use of knowledge. The methods of predicate logic, semantic networks, frames, production rules, object-oriented representation are often used methods of knowledge representation. As the object-oriented representation close to the natural thinking mode of human, superclass, subclasses, and instances of object-oriented approach form a hierarchy structure, subclasses can inherit data and operations of super-class, so it is suitable for representation of classification knowledge. Case knowledge representation of mold design use the method [8-9].

Knowledge library is used to store entities of knowledge and usually is represented with relational database. Table 1 and Table 2 shows the mode of relational database building the case knowledge base respectively.

Table 1. Basic information table of the case

Data type	figure	text	text	text	OLE Object	text	OLE Object	text	OLE Object

Table 2. Basic information table of typical structure

	Numeric field	Type ID	The name of type	Property	Specification	Location of list	Location of structural drawing
Data type	figure	text	text	Text		text	text

In the basic information table of combination fixture (table 1), nine fields clearly expressed the basic features of a set of modular fixture (case) and fixture type, job type, size range, workpiece graphics field as keywords are the basis of search in the next step. Six fields in the basic information table (Table 2) of typical structure expressed basic features of a typical structure and the name, features and specifications as keywords are the base of search in the next step.

3.3   Retrieval and Matching of Case

The purpose of case retrieval is to find one or more "most similar" cases of the current problem from the case library, which seek solutions of current problems.

Establishing the keywords case index:To meet customer needs, retrieve the case to meet user requirements from the case library, and improve the efficiency, four search keywords is used in case tables, respectively fixture type, part type, part size range and the workpiece graphics. The fixture type is selected firstly by the user to determine the type of the fixture, including car fixture, drill fixture or milling fixture, etc.; later, type of workpiece is selected. The system is divided into six major parts: the shaft, plate and shell type parts, fork rod parts, plate parts, shell parts and other parts; after that, the system will display the workpiece graphics of relevant class for the user to further selectting the part of the structure, if the user find a similar group of parts, then the case is used as a similar case.

Searching similar cases: The most similar case is found from the case library through the application of search function of application program. Three types of retrieve is divided by RHStottler [10], that is, qualitative retrieval method, retrieval method quantitative and qualitative and quantitative combined retrieval method. In the method of quantitative retrieval, characteristics of case is expressed by numerical values, the similarity of any two cases is defined as the reciprocal of the distance between the two. In the method of qualitative retrieval, the characteristics value of the case is qualitative value, such as color, etc. At this point, the similarity of any two cases is defined as the number of exact matches between the properties of the case. The method of quantitative and qualitative search is used in the application. Firstly case characteristics of fixture is divided into two groups, that is qualitative and quantitative. Clamp type, job type, size range and workpiece graphics as the qualitative values is used in the process of qualitative searches. When the value of case matched with the qualitative value of case retrieval come out, and then retrieve them quantitatively, in which quantitative values is inverted values of distance of two cases calculated based on experience by fixture assembly workers, the finally similarity value of two is balanced and under the specified sense the highest sense of the case is the best match case. If the similar cases are not found in case library meeting the design requirements, you can search in a case library of typical structure of, a set of integral clamp structure can be expressed with definition of the tree.Search process in the case of the library is based on experience searching the similar case in the tree and can be divided into the following steps:

1) Determine that which type of several typical structural forms fixture, which is the first search process in the tree;

2) Determine that which type of typical structure is used, which is the search process of the second layer in the tree;

3) Determine the specified type of fixture elements in the specific model, which is the search process of the third layer in the tree.

Searing the case library of typical structure according to the thinking of the tree structures has many advantages:

1) Avoid selecting the component without aim, narrow the search range and improve the efficiency and accuracy;

2) It is easy to summarize the assembling experience, acquire knowledge and update case library;

3) Reduce the excessive dependence on the processing parts.

3.4   Case Study and the Maintenance of Case Library

The case in the initial case library is limited, so solutions to new problems is needed to add to the case library in the system running process in order to make the case in case library more enriched and improved, then complete the study of the case. In this system, the case from the case library is modified by the designer to conform to design requirements and added to the case library. But as the cases continue to join, will the growing case base, reducing the efficiency of case retrieval and reasoning, storage strategy of the case is needed to specified.

For the new case, the similarity of all the old cases stored in the case library:

$s = \{s1, \ s2, \ ...sn\}$ (n is the number of old cases, $0 \leq si \leq 1$).

1) If all si=0, Show that the case library does not have the case match with new problems case, then the new case is added into the case library;

2) If there is a case that si = 1, show that the case is exactly matched with the new case, then the new case is discard and do not store it;

3) If all the values of si are less than the value of a given ε ($0 < \varepsilon < 1$, the exact value specified by the experts), then the new cases are joined.

Adopting strategies avoid the endless expansion of case library.

4   Conclusion

The method of CBR technology is more simple than the expert system method in the matter of the simulation of people's thinking. By the index case and design issues, the system can extract a similar case according to functional characteristics of parts of the fixture and can modify to fit the new requirements, this case-based fixture design systems can reduce the defects existing in expert system that is based on rules knowledge. In this way the fixture designer can accumulate design experience by investigations into past cases and the system can learn new knowledge by storing new cases. However, because very mature commercial software has not yet formed, the promotion of new technologies has been limited and this is urgent issues for manufacturing. Future trends should be the set of CAD / CAM technology and CNC technology, case-based reasoning technology in one and the system has the ability of automatically determining the fixture configuration design, automatic detection and integrated intervention, comprehensive automatic fixture design, which is the industry's goal and the needs of the development of modern manufacturing.

Acknowledgment. The project is sponsored by Tianjin Education Commission Research Program Project (20090408).

References

1. Wu, Z., Qin, G., Lu, Y.: A New Algorithm for Locating Error Based on Velocity Composition Law. Journal of Test and Measurement Technology 23, 7–13 (2009)
2. Yuan, D., Ma, J.: Application of CBR in Tube Flexible Welding Fixture CAD System. Science Technology and Engineering 9, 3327–3332 (2009)
3. Robust Ron Sun reasoning, Integrating rule –based and similarity reasoning. Artificial Intelligence 75, 241–295 (1995)
4. Kolodner, J.L.: An Introduction to Case-based Reasoning. Morgan Kaufmann Publisher, San Francisco
5. Li, Y., Liao, W., Li, Y.: Application of CBR Technology in Designing for Aircraft Tools. Jiangsu Machine Building & Automation 36, 64–67 (2007)
6. Yong, Q.: The Study of Case-based Design and its Application in Fixture CAD System. Journal of Yancheng Institute of Technology (Natural Science Edition) 22, 44–46 (2009)
7. Wang, Y., Fan, J.-H., Tian, S.-G.: Method for Case Representation in Expert System Based on Case-based Reasoning. Journal of Shanghai University of Engineering Science 19, 42–46 (2005)

8. Wu, Z., Xiao, J., Wu, T.: Analysis Model of Constraint of Degrees of Freedom of a Workpiece Based on Fixture. Machine Tool & Hydraulics 35, 19–22 (2007)
9. Qin, G., Zhang, W., Li, Y.: New Algorithm of Fixture Locating Scheme Design. Journal of Test and Measurement Technology 22, 236–240 (2008)
10. Stottler, R.H.: Rapid Retrieval Algorithms for Case-Based Reasoning. In: Proc. of the 1 Ith Intel. Con. on Artificial Intelligene, vol. 1, pp. 233–237 (January 1989)

Research on Interaction of Enterprises in the Textile Cluster of China

Gao Jing and Yan Jianping

College of Economic, Tianjin Polytechnic University,
Tianjin Polytechnic University, TJPUTianjin, China
chinagj8647@163.com

Abstract. To find survival conditions of textile enterprises, the paper borrowed the principles of ecology and built Logistic models of textile enterprises. Through analyzing the models' fixed points and stable conditions, the authors conclude that the only way which makes enterprises in the cluster stability is fierce competition between enterprises. If an enterprise evades competition, it will die in complicated and changeable circumstances. Consequently, finding accurately and quickly the vital factors which affect enterprises' success or failure and enhance enterprises' competitiveness is the best way for textile enterprises to survive in complicated circumstances. Using Logistic model of ecology to study textile enterprises is new and original in this paper.

Keywords: industrial cluster, Logistic model, textile enterprise, survival condition.

1 Introduction

Industrial cluster, as an important measure to promote industrial upgrading and improve industrial competitiveness, has been applied in a wide range. Through a large number of enterprises assemble in a particular geographical area and carry out division of labor and network services on the basis of specialization, industrial cluster can improve the technologies and processes, and enhance the competitiveness of the entire cluster. With the development of bionics, a growing number of ecological theories and methods are applied to the study of economics. Industrial cluster, as a kind of economic phenomenon, is surprisingly similar to population of the ecology. Therefore, we can use the theories of studying interaction between species in ecology to study interaction between enterprises in industrial cluster.

Scholar Stephen H. Levine (1999) simulated the pattern of industrial ecology into ecological model of the system. From a perspective of population ecology, he established a model of population ecology on how the products are affected by environmental change and gave kinds of forms. Scholar Qiu Baoxing (2000) raised that each enterprise in industrial cluster must find its own niche to keep the stability of the entire cluster. Based on Logistic model of ecology,scholar Zhou Hao (2003) built symbiotic model of enterprises in satellite-based and network industrial clusters. Scholars He Jishan, Dai Weiming (2005) used the theory of population ecology to study enterprises' relationship in industrial cluster and maintained the conditions of industrial cluster keeping ecological balance are that there are some differences between enterprises in the cluster, and there is perfect network of division and collaboration of work in the cluster, and one cluster keep material and information exchange with other clusters, and the cluster must be an open ecosystem. Scholars Xia Jianming, Nie Qingkai (2005), using bionics principles, divided enterprise relationships in business ecosystem into six relationships and defined them by mathematical language. Some scholars, such as Wu Jin (2007), Fan Haizhou (2010), also used ecological principles to study survival and development of enterprises in the cluster. However, the researches in this academic area, overall, are still in initial stage and need to study further.

2   Logistic Model of Textile Enterprises

Based on the ecological theory, we establish a Logistic growth equation of an enterprise in textile cluster and draw its Logistic growth curve:

$$\frac{dx}{dt} = rx(1 - \frac{x}{N})$$ (1)

In the equation (1), $\frac{dx}{dt}$ is the instantaneous growth rate of enterprise's outputs. r is the growth rate of enterprise's outputs without environmental constraints. $x(t)$ is the enterprise's output, which is a function of time t. N is the maximum output in the natural state. $(1 - \frac{x}{N})$ is retardant coefficient, which shows the scale of natural market negatively affects the growth rate of output.

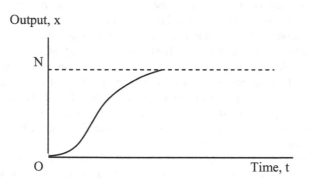

Fig. 1. Logistic growth curve of an enterprise in the cluster

Shown in Shown in the Figure 1, the output of enterprise is S-shaped growth and gradually close to the maximum yield N of the enterprise. The stable equilibrium of this model is $\frac{dx}{dt} = 0$, that is, when $x = N$, the enterprise reaches a steady state.

3 Interaction between Enterprises in Textile Ecosystem

In the textile ecosystem, there are three interaction between enterprises: competition, coexistence and predation.

3.1 Logistic Model of Competition between Enterprises

In order to a simple analysis, we only study relationship between the two enterprises in the ecosystem of textile cluster. And the results In order to a simple analysis, we only study relationship between the two enterprises in the ecosystem of textile cluster. And the results got from it can be easily extended to the case of multiple enterprises.

Suppose enterprise 1 and enterprise 2 survive in the same textile cluster and they are competitive relationship, that one affects the other's output, and changes in their output comply with Logistic Law, we can establish a model, which is:

$$\begin{cases} \dfrac{dx_1}{dt} = r_1 x_1 (1 - \dfrac{x_1}{N_1} - \delta_1 \dfrac{x_2}{N_2}) \\ \dfrac{dx_2}{dt} = r_2 x_2 (1 - \dfrac{x_2}{N_2} - \delta_2 \dfrac{x_1}{N_1}) \end{cases} \tag{2}$$

In the equations (2), the factor $\left(1 - \dfrac{x_1}{N_1}\right)$ reflects that enterprise 1 unceasingly consumes limited resources, which restrains its own output. The factor $\left(-\delta_1 \dfrac{x_2}{N_2}\right)$ reflects that the existence of enterprise 2 restrains the output of enterprise 1 because they are competitive relationship. δ_1 is that the quantities of enterprise 2 consuming enterprise 1's resources are δ_1 times as many as the quantities of enterprise 1 consuming its own resources. Similarly, in the second equation, every factor of enterprise 2 can be interpreted correspondingly. We can calculate this model's fixed points and stable conditions (see Table 1).

Table 1. Fixed points and stable conditions of LOGISTIC model OF COMPETITION

Fixed Point	Stable Condition
$(N_1, 0)$	$\delta_1 < 1, \ \delta_2 > 1$
$(0, N_2)$	$\delta_1 > 1, \ \delta_2 < 1$
$\left(\dfrac{N_1(1-\delta_1)}{1-\delta_1\delta_2}, \dfrac{N_2(1-\delta_2)}{1-\delta_1\delta_2}\right)$	$\delta_1 < 1, \ \delta_2 < 1$
$(0,0)$	Unstable

$\delta_1 < 1$ and $\delta_2 > 1$ mean that in the process of competing enterprise 1's resources, the power of enterprise 2 is weaker than the power of enterprise 1.Meanwhile,in the process of competing enterprise 2's resources, the power of enterprise 2 is weaker than the power of enterprise 1,too.Consequently,enterprise 2 is bound to go bankrupt, and the output of enterprise 1 tends to the maximum yield, that $x_1(t)$ and $x_2(t)$ tend to balance point $(N_1, 0)$.Of course, for $\delta_1 > 1$ and $\delta_2 < 1$,situation is just the opposite with this interpretation. When $\delta_1 < 1$ and $\delta_2 < 1$, it means that in the process of competing enterprise 1's resources, the power of enterprise 2 is weaker. However, in the process of competing enterprise 2's resources, the power of enterprise 1 is weaker. So, two enterprises can be coexistence and reach a stable equilibrium $\left(\dfrac{N_1(1-\delta_1)}{1-\delta_1\delta_2}, \dfrac{N_2(1-\delta_2)}{1-\delta_1\delta_2} \right)$.When $\delta_1 > 1$ and $\delta_2 > 1$, it is impossible to reach a stable condition.

3.2   Logistic Model of Predation between Enterprises

In the textile cluster, the relationship of some enterprises is the relationship of upstream enterprise and downstream enterprise, which is an upstream enterprise supply intermediate product or raw materials to a downstream enterprise. This kind of relationship is similar to the relationship of prey and predator between species in ecology. The upstream enterprise may be considered as a prey and downstream enterprise may be considered as a predator. Consequently, we can use the method of studying predation between species to study the relationship between upstream enterprise and downstream enterprise in the textile cluster.

Suppose the power of enterprise 1 is much larger than the power of enterprise 2, and enterprise 1 can exist independently. Meanwhile, enterprise 2 supplies necessary resources, such as intermediate products or raw materials, to enterprise 1, which is helpful to reducing enterprise 1's cost and promoting enterprise 1's development. If there is no enterprise 1, enterprise 2 will die. Consequently, if enterprise 2 exists independently, the level of its output can be described as:

$$\frac{dx_2}{dt} = -r_2 x_2 \qquad (3)$$

In the equation (3), $(-r_2)$ is the negative growth rate of enterprise 2's output. Considering that enterprise 1 provides a great many orders and market information for enterprise 2, enterprise 2 can enlarge its market scale and gain stable market demand. Consequently, the output level of enterprise 2 can be described as:

$$\frac{dx_2}{dt} = r_2 x_2 \left(-1 + \delta_2 \frac{x_1}{N_1} \right) \qquad (4)$$

In the equation (4), δ_2 represents the contribution of enterprise 1 to the output level of enterprise 2. We can imitate enterprises' competitive model to build Logistic model of predation between upstream enterprise and downstream enterprise, that is:

$$\begin{cases} \dfrac{dx_1}{dt} = r_1 x_1 (1 - \dfrac{x_1}{N_1} + \delta_1 \dfrac{x_2}{N_2}) \\ \dfrac{dx_2}{dt} = r_2 x_2 (-1 - \dfrac{x_2}{N_2} + \delta_2 \dfrac{x_1}{N_1}) \end{cases}$$

(5)

In the equations (5), δ_1 shows the resources supplied to enterprise 1 by enterprise 2 are δ_1 times as many as the resources consumed by enterprise 1 itself. Therefore, from the equations (5),we can get the fixed point and stable conditions of the model (see Table 2).

Table 2. Fixed points and stable conditions of Logistic model of predation

Fixed Point	Stable Condition
$(N_1, 0)$	$\delta_2 < 1, \; \delta_1 \delta_2 < 1$
$\left(\dfrac{N_1(1 - \delta_1)}{1 - \delta_1 \delta_2}, \dfrac{N_2(\delta_2 - 1)}{1 - \delta_1 \delta_2} \right)$	$\delta_1 < 1, \; \delta_2 > 1, \\ \delta_1 \delta_2 < 1$
$(0, 0)$	Unstable

It is easy to get that the realistic stable condition for the predation between enterprise 1 and enterprise 2 should be : $0 < \delta_1 < 1$, $\delta_2 > 1$, and $\delta_1 \delta_2 < 1$. $0 < \delta_1 < 1$ represents the contributions of enterprise 2 to the output level of enterprise 1 should be relatively small. The direct economic interpretation is that enterprise 2 does one or several procedures or provides one or several intermediate goods. Furthermore, there is fierce competition between enterprise 2 and a lot of small enterprises like enterprise 2 around the enterprise 1, leading to accept the choice of enterprise 1. $\delta_2 > 1$ represents the contributions of enterprise 1 to the output level of enterprise 2 are relatively large. Enterprise 1 not only offers the order which is the whole or large part of enterprise 2's output level but also provides the demand structure and the analysis of tendency of the market and sometimes even partly makes direct investment. $\delta_1 \delta_2 < 1$ shows that if enterprise 1 and enterprise 2 are symbiosis, the former enterprise should be large and the later one should be small. Moreover, there are a lot of small enterprises like enterprise 2 around the enterprise 1 and they form fierce competition between each other.

3.3   Logistic Model of Coexistence between Enterprises

On the assumption that there is no disparity in strength between enterprise 1 and enterprise 2 and both of them can prompt the output level of each other, the symbiotic model is:

$$\begin{cases} \dfrac{dx_1}{dt} = r_1 x_1 (1 - \dfrac{x_1}{N_1} + \delta_1 \dfrac{x_2}{N_2}) \\ \dfrac{dx_2}{dt} = r_2 x_2 (1 - \dfrac{x_2}{N_2} + \delta_2 \dfrac{x_1}{N_1}) \end{cases} \tag{6}$$

In the equations (6), the meaning of the symbols is the same as above.

We can obtain that the realistic stable conditions for the symbiosis of enterprise 1 and enterprise 2 should be $0 < \delta_1 < 1$, $0 < \delta_2 < 1$, and the fixed point is $\left(\dfrac{N_1(1+\delta_1)}{1-\delta_1\delta_2}, \dfrac{N_2(1+\delta_2)}{1-\delta_1\delta_2} \right)$. $0 < \delta_1 < 1$ and $0 < \delta_2 < 1$ show the contributions of enterprises 1 and enterprises 2 to each other are not large. The contributions of enterprises 1 and enterprises 2 to each other's output come through in the expansion of the market by division of labor, the imitation of technology and management, the sharing of information of products demand and the channel with trust relationship between the organizations, which is quite different from the predation model above. If many enterprises like enterprise 1 and enterprise 2 want to exist in the same market, they have to meet the conditions that are $0 < \delta_1 < 1$ and $0 < \delta_2 < 1$, which mean that the only way to keep this condition is the fierce competition between each other.

4   Conclusion

Through constructing interactive model of two enterprises in ecological system of textile cluster and using the economics theory to explain, this paper not only theoretically describes the interactive relationship of the enterprises in the ecological system of textile cluster, filling the theoretical blank of this area, but also provides beneficial guidance for enterprises to analyze interactive relationship in the textile ecological system and make enterprise strategy timely in order to get long-term survival and development .

Through the analysis, we can get that the only way to maintain stable state between the enterprises in cluster is the existence of fierce competition for each other. It is meaningless to avoid competition. If an enterprise evades competition, it will die in complicated and changeable circumstances. Therefore, to accurately and quickly find the important factors which have great influence on the success and the improvement of competitiveness of enterprises is the best way for textile enterprises to survive in the complex and changeable environment.

What is more important is that there also exist the choice relation between the enterprise and the ecological system, and the strong enterprise may be strong only in its ecosystem while the weak one may barely survive in its ecosystem. Therefore, the enterprises who barely survive can radiate new in vitality in another business ecosystem, if they can find suitable opportunities to develop in the changeable time and market.

Acknowledgment. In the process of writing the paper, the authors got strong support and assistance from teachers, business people and some of our friends. The authors sincerely appreciate them. In addition, the authors would like to extend heartfelt gratitude to the authors whose words we have cited or quoted.

References

1. Levine, S.H.: Products and Ecological Models—A Population Ecology Perspective. Journal of Industrial Ecology 3, 137–139 (1999)
2. He, J., Dai, W.: Analysis on cluster's the ecological model and balance of the ecology. The Learned Journal of Beijing Normal University, 126–132 (2005)
3. Xia, J., Nie, Q.: Economic Analysis on Enterprises' Interactive Relationship from the Perspective of Business Ecosystem. In: China Economics Annual Conference (2005)
4. Pagie, L., Mitchell, M.: A Comparison of Evolutionary and Co-evolutionary Research. International Journal of Computational Intelligence and Applications, 367–384 (2008)
5. An, C.: Strategy on responding financial crisis for Chinese textile industry. China Textile Leader, 18–20 (2009)

The Mystery of Economy Structural Imbalance between China and America: A New Interpretation of Marshall-Lerner Condition

Han Boyin, Wang Dongping, and Fu Bo

School of Economics
Tianjin Polytechnic University
Tianjin, China
hby188@foxmail.com

Abstract. The economic imbalances between China and America stem from the internal imbalances caused by the inappropriate macroeconomic policies of America over the past years. U.S. loose monetary policy accelerates international capital to flow to emerging market and lead to the formation of a global excess liquidity. But net capital inflows together with trade surplus makes their macroeconomic policies in emerging market countries more complicated. Long-term low interest rates lead to American increasing consumption expenditure and rising double deficit. In order to transfer the conflict, U.S. forces revaluation of RMB frequently. The continued substantial appreciation of RMB challenges the traditional Marshall-Lerner condition. Through formatting an equilibrium model, this paper deeply studies the export elasticity and import elasticity to examine the Marshall-Lerner condition and analysis its adaption and sensitivity, showing that in short term revaluation of RMB can improve U.S. trade deficit, but in long term it is ineffective.

Keywords: component, RMB exchange rate, Marshall-Lerner condition, Balance of payments, Double deficits.

1 Introduction

Recently, the U.S. economy is weak, unemployment remains high, and this is probably a second dip, therefore critics say that the Chinese Yuan is deliberately undervalued, which makes Chinese exports artificially competitive and lead to International Payments Imbalances in China and America. In fact in the past decade, the continuous extremely accommodative monetary policy in developed countries results global excess liquidity and global economic imbalances (especially the United States and China).

In 2008, the Lehman bankruptcy caused the economic bubble burst, we suddenly realized there is still such a big gap.So the governments intervene to provide financial support to financial system to safeguard financial stability. After the crisis balance sheet required to be repaired, person started to pay personal debts to the bank, financial sectors began to return national debt. But real economy was damaged badly, which make it

J. Luo (Ed.): Soft Computing in Information Communication Technology, AISC 158, pp. 341–349.
springerlink.com © Springer-Verlag Berlin Heidelberg 2012

difficult to pay their personal debts for people, meanwhile banks still had a large number of loans, government must ensure that the cost of repayment was not too high, otherwise banks would go bankrupt. However U.S. Economic Policy does not work, America must seek International Co-operation including exchange rate (Il Houng Lee).

If the United States still adopts quantitative easing monetary policy, it will cause negative impact on global economic imbalance. But because the bank balance sheet repair has not been completed and consumption is still low, if it does, it will increase the probability to hit the second dip, this is a dilemma. Meanwhile, it is impossible to turn to fully flexible exchange rate regimes for Asian countries, if so, for some small countries capital inflows lead to exchange rate appreciation and may cause a devastating blow to their economics.

The currency exchange-rate appreciation would significantly increase the price of export commodities, resulting in reduced exports. On the other hand, the currency appreciation will reduce the prices of imported products, resulting in increasing imports and lead to a worsening balance of payments. While in Post-crisis Era, RMB appreciates against the dollar but the trade surplus still increase between China and America.

2   Literature Review

Marshall - Lerner Condition is one of the most important theories which study the effect exchange rate has on a country's balance of payments. The theory suggests that provided that the sum of the price elasticity of demand coefficients for exports and imports is greater than one then a fall in the exchange rate will reduce a deficit and a rise will reduce a surplus.

Among the theories of international payments, the core theories are the Elasticity Approach to the adjustment for the balance of payment proposed by the Marshall, Robinson System, and the Marshall-Lerner Condition , which appeals for depreciating the exchange rate to improve criterions of the balance of payment, proposed by Lerner in 1944. Although Marshall-Lerner Condition has a great of academic and realistic meaning, the depreciation of dollar right after the 1980-1985 continuous advances of it did not improve the balance of payment deficit. Also, after 1985, Japanese Yen continued to rise, increasing the balance of payment deficit in Japan which is also an example of failure to apply the Marshall-Lerner Condition. Rose(1991) checked the date of 1974-1986 in Germany, Canada, the U.S. and the U.K. and found out that there is no co-integration relationship between the import & export and the rate of exchange. After verifying the data of 8 countries, Germany, Canada, the U.S. and the U.K. etc, Boyd (2001) found the theory is not fully validated. How to revise and supplement the model? Dornbusch (1975) proposed the introduction of the noncommercial goods to rebuild the general balance scheme, explaining the effects of the fluctuation of exchange rate to the balance of payment, when the MPC of noncommercial goods is greater than, equal to and less than 1, respectively. And he found that keeping nominal income unchanged is better than maintaining the price of noncommercial goods at enlarging the trade improvement effects of devaluation and that the fluctuation of exchange rate will affect the balance of international payment to its minimal degree when the MPC of noncommercial goods is 1.

Baldwin (1986) , Dixit (1989) took the connection between the entry & exit decision of business and the sunk costs of enterprise under the condition of imperfect competitive market structure into consideration and explored the flexible prolong impacts due to the exchange rate shock on the balance of international payment, namely, when the exchange rate shock is great enough, the rise of local currency will induce foreign products entering the domestic market and as a result the shifted products structure will impose a structural fraction problem to the relationship of exchange rate and import quota, simultaneously the impacts of fluctuation of exchange rate to the balance of international payment will be weaken and become irrelevant to Marshall-Lerner elasticity.

In deed, several domestic scholars have done some positive analysis over whether the Marshall-Lerner Condition is applicable in China or not, and most of their conclusions are positive, for example, Wang Xiangning and Li Xiaofeng (2005) established VAR Error Correction Model to estimate trade elasticity of China and thus proved that Marshall-Lerner Condition is applicable; Lu Xiangqian and Dai Xiangguo (2005) made the empirical test about the chronicle relationship between the weighted real exchange rate fluctuation of RMB to other major currencies in the world and the import & export in China also proved the same. But there is a flaw in these researches that they are all conducted in the term of the entire economic instead of in the term of a specific sector of economic, such as the agriculture sector or the manufacture sector.

Partial equilibrium analysis framework is the flaw in the nature of Marshall-Lerner Condition, while violating the spirit and benefits of free trade, it overlooks dozens of factors: the complex effluences that the changes in the international trade structure will have on the fluctuation of exchange rate and on the balance of international payment, the fact that depreciation will lead to further changes in the price elasticity and affect the import price, the benefits of international cooperation of the division of labor, the impacts that the international cooperation of the division of labor will have on a nation's economic development, and the function of the invisible hands. This article tries to modify and test the Marshall-Lerner Conditions based on those past relating researches and tries to expand the research of the Marshall-Lerner Conditions which explores the relationship between the fluctuation of exchange rate and the balance of international payment from a brand-new aspect.

3   Establishing the Model

3.1   Improved the Marshall-Lerner Condition

The Marshall-Lerner Condition (M-L Condition) is based on the following assumptions: impacts of devaluation to the balance of trade, full employment, unchanged income, full elasticity which is owned by supply of commodities and the balanced trade balance at the initiation of depreciation. Let ex and em be the export elasticity and the import elasticity respectively, here we can arrive at the conclusion that: (1) if ex+em>1, depreciating the local currency will improve the balance of trade; (2) if ex+em=1, the depreciation will have no effluence on the balance of trade; (3) if ex+em<1 the depreciation will deteriorate the balance of trade with negative effects

When the balance of trade is imbalanced, the relation of the import elasticity and the export elasticity will be changed. Using Marshall-Lerner Condition to study the exchange rate issue will be greatly biased. So we here use the improved Marshall-Lerner Condition to diminish the errors. Li Bingyi and Wang Hongguang (1995) extended the M-L Condition, removing the assumption that there would be a balanced trade balance at the initiation of devaluation, and letting Pxb= local currency priced export, Pmb=local currency priced import, $\theta = \dfrac{mP_{mb}}{xP_{xb}}$, to reach the conclusion that:

(1) When ex+em>max (1, θ), depreciation will improve the trade balance both of the local currency and of the foreign currency;

(2)When ex+em<min(1, θ), depreciation will deteriorate the trade balance both of the local currency and of the foreign currency;

(3)When min(1,θ)<ex+em<max(1, θ), the local currency and the foreign currency changes on a opposite direction.

Not only the condition of balanced trade balance at the beginning of the devaluation (here $\theta=1$, and the conclusion is in the content of Marshall-Lerner Condition), but also the imbalanced trade balance under the same situation are included. Both the local currency and the foreign currency can be used to measure the trade balance. This conclusion obviously has a fewer limitations and is the extended version of Marshall-Lerner Condition. According to the import & export trade balance data, we can calculate the price elasticity of and the income elasticity of the import & export in China.

3.2   Introducing the Elastic Model

Goldstei proposed the Imperfect Alternative Theory assuming that the domestic products and the import & export can not substitute one another perfectly, and using the demand function of Demand= Constant×priceα×incomeβ and the assumption that the quantity of export demand is decided by the foreign income level and the relative income index of the export country jointly, that is:

$$D_x = c(P_x/P_f)^{\alpha}(Y_f/P_f)^{\beta} \tag{1}$$

PX is the index of export price, Pf is the index of weighted foreign price, Yf is the index of weighted foreign income. Thus, the elasticity of export price is $\dfrac{\partial D_x}{\partial (P_x/P_f)}$, the elasticity of   export income is $\dfrac{\partial D_x}{\partial (Y_f/P_f)}$, after taking logarithm of both sides of the equation (6) and arranging, we can have the regression equation:

$$\ln D_x = c_0 + \alpha \ln(P_x/P_f) + \beta \ln(Y_f/P_f) \tag{2}$$

It is hard to detect the total export demand, resulting to the obstacle to conduct regression analysis directly. Accordingly, we assume that the fluctuation of export demand is subject to the Inventory Adjustment Hypothesis, namely the demand and the inventory follow the rules that:

$$x_t / x_{t-1} = (D_x / x_{t-1})^k \tag{3}$$

Therefore, we can remove the export demand and introduce the lagged export inventory in the model. After arranging, here we have:

$$\ln x_t = a_0 + a_1 \ln(P_x / P_f) + a_2 \ln(Y_f / P_f) + a_3 \ln x_{t-1} \tag{4}$$

Similarly, we can have the import mode:

$$\ln m_t = b_0 + b_1 \ln(P_m / P_b) + b_2 \ln(Y_b / P_b) + b_3 \ln m_{t-1} \tag{5}$$

According to the above mentioned analysis, in the equation(9) and (10), the price elasticity of the total import and export demand are a1/(1-a3),b1/(1-b3), respectively, the income elasticity of them are a2/(1-a3),b2/(1-b3), respectively.

The above analysis of trade applies to nations indiscriminately, including all the import & export transactions China have done with its trade partners. Further more; the bilateral trade equation has taken cross-price elasticity into consideration. The derivation process is similar to that of the multilateral trade equation. Here are the results:

$$\ln x_t = c_0 + c_1 \ln(P_x / P_f) + c_2 \ln(Y_f / P_f) + c_3 \ln x_{t-1} + \sum c_i \ln(P_{it}) \tag{6}$$

$$\ln m_t = d_0 + d_1 \ln(P_m / P_b) + d_2 \ln(Y_b / P_b) + d_3 \ln m_{t-1} + \sum d_i \ln(P_{it}) \tag{7}$$

xt and mt represents the import and export of China to one certain trade partner respectively in a period t, xt-1 and mt-1 is the corresponding lagged variable, Px is the index of export price, Pf is the price index of the certain trade partner, Yf is the income index of that trade partner, Pm is the index of import price, Pb is the price index of China, Yb is the income index of China, Pu is the commodities' price of the other trade partners, i is a nature number, starts with 4, 5, 6, and the the number of i depends on the number of the rest of these trade partners. Thus, C1/(1-C3)and d1/(1-d3) are the bilateral import price elasticity of and the bilateral export price elasticity of China to that certain trade partner, C2/(1-C3)and d2/(1-d3) are the bilateral income elasticity, Ci/(1-C3)and di/(1-d3)are the cross-price elasticity of that certain partner and another partner.

4   Verifying the Model

Take the data of 2000-2009 as our data base. Using the import & export data of China and of its trade partners in that period to conduct the OLS regression on the Sino-U.S. Bilateral trade equation. The total import & export index of China, the index of the import & export price, PPI of the U.S. and Japan in the model are all from UN Data; The GDPs and PPIs of each year of the P.R.C. and the imports & exports between China and Japan, the U.S. or other major trade partners are from the Statistical Yearbooks of China and the Annual Bulletins of China of each year. The GDPs of the U.S. and Japan of each year are from the International Financial Statistics of IMF. Table 1 and Table 2 present the statistic data of relating variables in the form of index and all the relating variables have a base year of 2000.

Table 1. Data of Sino-U.S. bilateral export model

Year	Export Volume index	Export Price Index	Foreign Price Index	Foreign Income Index	Export Volume Index (pervious year)
2000	100.0	100.00	100.0	100.00	87.9
2001	94.4	96.48	101.2	103.36	100.0
2002	107.9	94.06	98.8	106.94	94.4
2003	120.8	96.92	104.2	111.96	107.9
2004	139.0	102.31	110.6	119.26	120.8
2005	157.2	110.01	118.6	127.00	139.0
2006	170.1	115.40	124.2	134.64	157.2
2007	167.5	120.24	130.1	141.30	170.1
2008	159.4	134.10	142.9	144.39	167.5
2009	145.8	118.70	130.4	141.88	159.4

Sources: calculated according to the China Statistical Yearbook (2001-2010) and the International Financial Statistics from IMF.

Table 2. Data relating to Sino-U.S. Bilateral Import Model

Year	Import Volume index	Import Price Index	Domestic Price Index	Domestic Income Index (In dollar)	Import Volume Index (pervious year)
2000	100.0	100	100.0	100.0	100.2
2001	106.0	99.15	98.7	110.5	100.0
2002	98.2	98.17	96.5	121.3	106.0
2003	98.4	99.73	98.7	136.9	98.2
2004	108.9	103.57	104.8	161.2	98.4
2005	109.5	106.87	109.9	186.5	108.9
2006	122.4	110.68	113.2	221.8	109.5
2007	131.1	116.03	116.7	282.2	122.4
2008	133.0	123.08	124.8	361.0	131.1
2009	129.1	117.31	118.1	410.3	133.0

Sources: calculated according to the China Statistical Yearbook (2001-2010) and the International Financial Statistics from IMF.

After conducting regression on the data of trade between China and the U.S. with the bilateral trade model mentioned above, we can arrive at the conclusion that:

Table 3. Results of the Regression on the Bilateral Trade between China and the U.S. in 2000-2009

	Regression Equation
U.S.	$LnX_t=0.416-0.488ln(Px/P_f)+0.322(Y_f/P_f)+0.5987lnX_{t-1}$ Export Regression Equation
	$LnM_t=2.556-1.547ln(P_m/P_b)+0139(Y_b/P_b)+0.305lnM_{t-1}$ Import Regression Equation

5   Conclusion and Suggestions

Based on the results of regression, we can have such conclusions that:
The Price Elasticity of Import & Export Demand of Sino-U.S. Bilateral Trade Model Complys With the Improved Marshall-Lerner Condition.

From the term of price elasticity: The price elasticitys of import & export are -1.217 and -2.227.with em=1.217, ex=2.227, ex+θem=2.227+0.35*1.22=2.653>1.Using the improved M-L Condition mentioned above, the devaluation of RMB will improve the balance of trade in China. Then, the rise of RMB against the dollar should be an effective method to solve the issue of the trade imbalance between China and the U.S. Thus, we should take the policy of exchange rate into consideration, when establishing policies to improve the Sino-U.S. trade balance.

Since 2005, China established the policy of limited floating exchange rate, devaluation of the RMB has appreciated from 1:8.1917 in 2005 to 1:6.83 in the end of 2009. RMB appreciation rates from 2005 to 2009 were 1.03%, 2.68%, 4.61%, 8.67 and 1.66%. Growth rates of China's trade surplus with the United States in 2005 -2009 were 42.39%, 26.36%, 13.17%, 4.65%, -16.09%. From 2005 to 2009 the economic growth rates of China were 14.60%, 15.67%, 21.41%, 16.85%, 8.70%, respectively. United States economic growths from 2005 to 2009 were 6.49%, 6.02%, 4.95%, 2.19%, -1.74%. Since the devaluation of the RMB, the growth of China's trade surplus with the United States started to slow down and in 2009, a negative growth of trade surplus happened. It is clear the impacts that RMB appreciation has imposed on China's trade are emerging. Let's take conditions and data of 2009 for example to research the Sino-U.S. trade with the M-L Conditions. U.S. economic growth in 2009 was -1.74%, while China's economic growth rate was 8.7%, meanwhile, the income elasticity of China to the U.S. Exports was 0.85, the income elasticity of China to the U.S. imports was 0.2, from the income elasticity we can know that China's trade surplus with the U.S. will reduce for about 0.35%. From the appreciation of the RMB exchange rate in 2008 for 8.67% which will affect the international trade in 2009 and the appreciation of the RMB exchange rate of 1.33%, we can know that China's trade surplus with the U.S. Will decrease 16.53% with the price elasticity of import & export between China and the U.S. taken into consideration. We can estimate that the trade surplus of China with the U.S. will decrease for about 16.88% using the expanded M-L Conditions and integrating the price and income elasticity of import & export between China and the U.S. In fact, the trade surplus decreased for about 16.09%. Then we can draw the conclusion that the expanded M-L Conditions could perfectly explain the issue of trade between China and the U.S.

Based on the results, appreciation of the RMB seems be a "recipe for a crowd" to amend the trade imbalance between China and the U.S. But the RMB has risen steadily and the trade surplus of China with the U.S. have decreased insignificantly, which even showed the tend of increase, since we complement the floating adjustment of the RMB referring to a basket of currency in July 21, 2005. Theory and practice have validated that the moneytary appreciation or depreciation is limited playing the role of regulating the trade balance. In the 1970's and 1980's, the United States had managed to force Germany and Japan to devaluate their currency due to the trade issue, but by the time of 2008, the the United States still had a trade imbalance with Germany, a deficit of $ 42,900,000,000, and with Japan, a deficit of 72.6 billion U.S. Dollars. 2005-2008, the appreciation of the RMB against the U.S. dollar accelerated, over the same period the average annual U.S. trade deficit with China increased 21.6%, which is the fastest growing period in history. 2009, the RMB exchange rate against the U.S. dollar remained relatively stable, while the U.S. trade deficit with China fell 16.1%. Thus, the RMB exchange rate is not the main reason triggering the U.S. trade deficit.

Income Elasticity of the Bilateral Ttrade Models between China and the U.S. Shows a Certain Degree of Asymmetry

From the aspect of income elasticity: the income elasticity of products exported to U.S. From China (0.85) is greater than that of products imported to China from U.S. (0.2). Therefore, the fact that the exports of China to the U.S. are greatly influenced by the income of the U.S., while imports of China from the United States are affected relatively small by earnings in China, indicating that bilateral trade has the character of heterogeneity. High technology products and service consist the majority of U.S. export, while labor-intensive products with low added value consist the majority of the export of China. This phenomenon is subject to the Law of IIM (International Industry Migration). The United States must increase their export of high-tech products rather than rapping the exchange rate of RMB.

References

1. Li, Y.: Research about the Foreign Economic of China and the balance of international Payment. International Culture Press, Beijing (1991)
2. Chen, B.: Research about the Exchange Rate of RMB. East China Normal University Press, Shanghai (1992)
3. Dai, Z.: The elasticity analysis of China's trade balance. Economic Research (7), 55–62 (1997)
4. Zhu, Z., Ning, N.: The elasticity analysis of China's trade balance. World Economic (11), 26–31 (2002)
5. Johansen, S.: Estimation and hypothesis testing of cointegration vectors in Gaussian vector auto regressive models. Econometrica 59, 1551–1580 (1991)
6. Hoper, P.: Trade elasticitiesfor the G27 countries. Princeton Studiesin International Economics, vol. (87). Princeton University (2000)

7. Goldstein, M., Khan, M.S.: Income and price effectsinforeign trade. In: Handbook of International Economics, pp. 1041–1105. North Holland, Amsterdam (1985)
8. Caves, R.E., Frankel, J.A., Jones, R.W.: World trade and payments: an introduction, 9th edn. Addison Wesley (2002)
9. Lerner, A.: The economics of control. Macmillan, New York (1944)
10. Sinha, D.: A note on trade elasticities in asian countries. International Trade Journal 15, 221–237 (2001)

Financial Fragility and the Study of KLR Signal Model Critical Value in China

Jin Ling[1], Han Boyin[2], and Cai Miao[3]

[1] College of Economic
Tianjin Polytechnic University
Tianjin, China
lovejinling1987@163.com
[2] Economics and Business Administration
Hei Long Jiang University
Heilongjiang, China
qinaide51888@163.com

Abstract. Subprime crisis, which happened in America, exacerbated the financial fragility all over the world. With the world economic recession, the possibility of the second bottom is higher than expected. Studying financial fragility in China is of great significance at this time. The measurements of the financial fragility have many standards and every standard has advantages and disadvantages. This article chose the most common method of measuring the financial fragility and got the indexes of China's financial fragility in 1998~2009. Then according to fitting the macro-economic indicators and the financial fragility indexes, this paper calculated KLR signal model critical value in China, contrasted with the international standards and analyzed the financial fragility in China. At last, this article provided some decision-makings for China's development.

Keywords: financial fragility, KLR signal model, macroeconomic indicators.

1   Introduction

The research of financial fragility focused on the currency vulnerability at the earliest. In 1880, Marx thought the currency had particular vulnerability as soon as it existed, which related to the basic function of the currency and decided the currency's fragility by birth. This theory laid the foundation for the future researches. In 1941, Fisher, the first economist who found the cause of the financial fragility, considered there were close relationship between the financial fragility and macro- economic. He held that liabilities were the main causes of financial fragility. In 1996, Francel and Rose studied the Mexico financial crisis and thought the foreign exchange risk, the ratio of credit and GDP, external deficit and economic growth were the main factors to measure the financial fragility. In 1998, Krugman believed that the relation of the moral hazard and the excessive investment must be the banking vulnerability, and the protection from the government make the banking vulnerability is more apparent. In 2000, Stiglitz

J. Luo (Ed.): Soft Computing in Information Communication Technology, AISC 158, pp. 351–358.
springerlink.com   © Springer-Verlag Berlin Heidelberg 2012

supposed the finance is a risky industry which could cause financial crisis. This industry's fragility is inherent and it is magnified by the capitalism.

In China, the research about the financial fragility is to measure the banking vulnerability mostly. In 2000, Jun Han analyzed the cause of banking vulnerability from the macro economy and the micro economy. He indicated that the biggest risk in the banking system is failed to solve the bad debts. In 2000, Zhonghua Shen studied the banking vulnerability from the capital inflows. If the macro economy is not stable, the capital inflows will strengthen the banking vulnerability significantly. On the contrary, it will not affect the banking vulnerability. In 2005 and 2008, Yanhua Zhang and Youfeng Zhang used EVIEWS and SPSS to make models. They thought there were a lot of indicators which had influence to the banking vulnerability, such as the fiscal deficit, the inflation rate, the interest rate and so on.

There were many studies about the financial fragility at home and abroad, but the results were different because of the divergence of the research objects, methods and data. People didn't reach an agreement in the study of the financial fragility. Just like Santos pointed out that the reason was the intrinsic mechanism of the financial fragility, which lacked of understanding. This article used KLR signal model to find out the intrinsic mechanism of the financial fragility though the data from 1998 to 2009.

2   The Fitting of the Model

There is not clear definition to indicators of the financial fragility, the most common method of measuring the financial fragility includes the vulnerability indicators from the banks, the financial markets, the institutions of the financial supervision and the macro economy. Thinking of China's practical conditions, China is good at the financial supervision and the policy implementations, and this article's aim is to research the financial fragility and the macroeconomic critical value of KLR signal model. This paper chose the vulnerability indicators from the banks and the financial markets. The basic content including three parts:

The definition of the fragility critical value was seen in the table1 and table 2 and referred to the international measurement criteria.

Data handling took the interval computing, for example, the bad debts rate in bank was 16.6836% in China and the state of the fragility was normal. Putting it to the region of [12,17], (16.6836-12)/(17-12)=93.67% and mapping the fragility region of [20,50], this article got the bad debts rate's index of 2004 in China, it was 93.67%*(50-20)+20=48.1.

When calculating the vulnerability indexes from the banks and the financial markets, this paper adopted the arithmetic mean. For example, the vulnerability index from the banks in 2004 equaled to the totaling of the fragility of the bad debts, capital adequacy, return   on   assets   and   balance   sheet,   then   divided   4.   It   was (48.1+47.4+16.15+93.225)/4=51.22. When computing the index of the financial fragility, this paper took the Weighted Average Algorithm, the banking system weighted 70% and the financial market system was 30%. So the indexes of the financial fragility were the vulnerability indexes of the banking system multiplying by 70%, then added the vulnerability indexes of the financial market system multiplying by 30%.

2.1   The Measure of the Financial Fragility

The Vulnerability of the Banking System: Banking system is the pricipal part of the financial system in China. The steady the progression of the banking system is the improtant factor to assure China's financial development. To choose the indicators of the banking system, all of the economists from many countries thought the bad debts rate A_1, the capital adequacy rate A_2, the return on assets A_3 and the balance sheet rate A_4 can reflect the actual situations. This paper choiced these indicators and their critical values were listed in the table 1 and their vulnerability was listed in the figure 1 and the vulnerability indexes of the banking system were listed in figure 2.

Table 1. The indicators and critical values of the banking system

the Indicators	the International Critical Value			
the Bad Debts Rate (A_1)	<12	12~17	17~22	>22
the Capital Adequacy Rate (A_2)	>12	8~12	4~8	<4
the Return on Assets (A_3)	>0.4	0.2~0.4	0~0.2	<0
the Balance Sheet Rate (A_4)	<45	45~65	65~85	>85
the Indexes' Mapping Interval Value	0~20	20~50	50~80	80~100
the Extent of the Banking Vulnerability	Safety	Normal	Attention	Danger

(Data source: IMF)

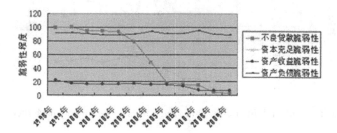

Fig. 1. The Vulnerability of the Four Indicators in the Banking System

The Vulnerability of the Financial Market System: Except the banking system, the other important system is the financial market in the finance. According to the international standards and the pritical conditions in China, this article selected the rate of the stock-market capitalisation and GDP B_1, the stock P/E ratio B_2 and the volatility Shanghai composite index B_3. This paper chose these indicators and their critical values were listed in the table 2 and their vulnerability was listed in the figure 3 and the vulnerability indexes of the financial market system were listed in figure 4.

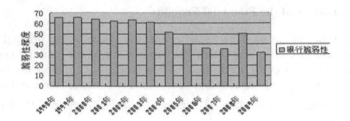

Fig. 2. The Vulnerability of the Banking System

Table 2. The indicators and critical value of the financial market system

the Indicators	the International Critical Value			
the Rate of the Stock-market Capitalisation and GDP (B_1)	<20	20~60	60~90	>90
Stock P/E Ratio (B_2)	<30	30~60	60~90	>90
the Volatility Shanghai Composite Index (B_3)	<100	100~400	400~700	>700
the Indexes' Mapping Interval Value	0~20	20~50	50~80	80~100
the Extent of the Financial Market Vulnerability	Safety	Normal	Attention	Danger

(Data source: IMF)

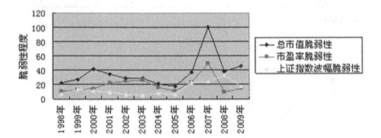

Fig. 3. The Vulnerability of the Three Indicators in the Financial Market System

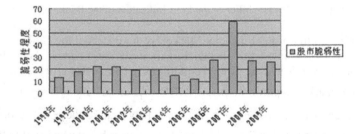

Fig. 4. The Vulnerability of in the Financial Market System

The Indexes of the Financial Fragility: Though weighted mean of the vulnerability of the banking system and the financial market system, this thesis got the the indexes of the financial fragility(figure 5).

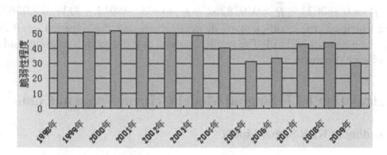

Fig. 5. The Indexes of the Financial Fragility

From the figure 5, the indexes of the financial fragility were high before 2004 and descended from 2004 to 2006, the indexes were lower than 40 in the three years. It was related to the stable development of the economy, the fast growth of the national economy, the reform of the wholly state-owned commercial banks in the three years. But from 2007 to 2008, the indexes rose because of the international financial crisis. In 2009, the index decreased again, this meant that the policy played a part in coping with the financial crisis.

2.2   The Fitting of the Macroeconomy

According to the computing of the indexes of the financial fragility, this thesis got the situations of the financial fragility from 1998 to 2009. It is no doubt that the macro economy has important influence on the financial fragility. So the relation between the financial fragility and the macro economy can calculate though fitting.

This article selected 10 macro-economic indicators, which can reflect the development situations of China. They are the growth rate of GDP X_1, the interest margin of China and America X_2, the appreciation of the exchange rate between RMB and USD X_3, the growth rate of M2 X_4, the growth rate of the foreign exchange reserve X_5, the growth rate of the short-temp capital X_6, the growth rate of the CPI X_7, the growth rate of the fixed assets X_8, the growth rate of the fiscal revenue X_9 and the profit margin of the construction output value X_{10}.

This article used Eviews 3.1 and the OLS in order to fit the indexes of the financial fragility with the 10 macroeconomic indicators. The indexes of the financial fragility were Y, which is called the explained variable. And the 10 macroeconomic indicators were called explanatory variable, which were fitted the explained variable. Then this paper got the equation:

$$Y = 87.86 - 7.02 \times X_1 + 2.79 \times X_2 - 0.74 \times X_3 + 14.89 \times X_4$$
$$-39.51 \times X_5 + 0.21 \times X_6 + 2.59 \times X_7 + 0.44 \times X_8 + 0.81 \times X_9$$
$$-0.71 \times X_{10}$$

And $R^2 = 0.980023$, $\overline{R}^2 = 0.978262$, $F = 4905.738012$, $D.W. = 2.056324$.

In the equation, the coefficient of determination and the adjusted coefficient of determination were closed to 1. The results showed that the model generally had a good effect and high precision.

3   Calculating KLR Singnal Model Critical Value in China

3.1   Leading in KLR Signal Model

KLR signal model was found by Kaminsky, Lizondo and Reinhart in 1999. The main idea included two parts: Firstly, a series of economic indicators, which can reflect the financial fragility, were selected. Then these indicators compared with the international critical value. If the indicators are larger than the critical value, it meant the indicator was a crisis sign. The more crisis signs, the possibility of happening financial crisis will be high. The biggest advantage of this model is that it could relate the economic data with the financial crisis.

KLR signal model's critical value is built though the international standards, so it doesn't reflect the situation in China. This paper calculated the critical value to suit the conditions in China though the fitting equation.

3.2   Calculating KLR Singnal Model's Critical Value

According to the fitting equation, this thesis supposed the indexes of the financial fragility reached the minimum of the danger 80 and other indicators were zero at the same time. Then this paper got this desired indicator's absolute value and this value was the indicator's critical value in China. The critical value was can be seen in the table 3. N meant that the relationship between this indicator and the financial fragility was negative and P meant it was proportional relation.

From the table, the critical values of the growth rate of M2, the growth rate of the foreign exchange reserve and the growth rate of the fiscal revenue are great differences of the international standards'. This means that the macroeconomic situations in China have difference from the developed countries', especially in the growth rate of the foreign exchange reserve and the growth rate of the fiscal revenue. So China should pay attention to the three indicators and prevents from the financial crisis.

The growth rate of GDP has inverse relation to the financial fragility, and its critical value is 1.2%. The growth rate of GDP in China exceeds 7% in recent years. So this indicator is safe now. The interest margin of China and America has proportionality relation to the financial fragility. The critical value is 2.82% and the interest margin is positive in value recent years. So this indicator has big influence to the financial fragility. The appreciation of the exchange rate between RMB and USD also has proportionality relation to the financial fragility and its critical value is 10.58%. The

Table 3. The international critical value and this value in China

Indicators	X_1	X_2	X_3	X_4	X_5	X_6	X_7	X_8	X_9	X_{10}
the International Critical Value	No	No	8	10	6	25	2	16	3	No
the Critical Value in China	1.2N	2.82P	10.58N	0.53P	0.2N	37.7P	3.30P	17.94P	9.63P	11.01N

progress of appreciation of the exchange rate between RMB and USD is very long. In the near future it is safe. It is no doubt that the influence of the short-temp capital to the financial fragility. But the critical value of 37.73% is too high and it means that the short-term capital is flooding in China now. The growth rate of the CPI and the growth rate of the fixed assets have proportionality relation to the financial fragility. They illustrate the inflation is important and the real economy can keep the economy away from the financial fragility. It is the first time to propose the relation of the real estate industry and the financial fragility. But the number is not agreeable with the practical situation. So the real estate industry isn't healthy in China.

3.3   Conclusions and Recommendations

Though the calculating, the states of the financial fragility are attention in most of the years. The others are normal. So the financial fragility is not safe. The vulnerability of the banking system is higher than the financial market system's, this has relationship with the information asymmetry and the underdeveloped domestic financial markets.

This paper gets the KLR signal model critical value in China and compares with the international standards. Firstly, property bubble and the financial risk exist. The profit of the real estate is unreasonable and roots in the thrust of the bank credit funds. If the bubble break, the large bad debts in banks will come true because of Dominos. This will be increase the social contradictions and trigger the financial crisis. Secondly, the big differences of the growth rate of M2, the growth rate of the foreign exchange reserve and the growth rate of the fiscal revenue reflect that the macro economic impact of the financial fragility should be focused. Finally, The growth rate of GDP, the interest margin of China and America and the appreciation of the exchange rate between RMB and USD, the short-temp capital, the growth rate of the CPI and the growth rate of the fixed assets are near to the international critical, it means that China's economy gets in line with the developed country in these parts.

References

1. Stiglitz, J., Weiss, A.: Credit rationing in markets with imperfectinformation. The American Economic Review, 393–410 (June 1981)
2. Huang, J.: Concerning the financial fragility. Financial Reserch, 41–49 (March 2001)
3. Wu, Z.: Financial integration and fragility: international coparison research. Economic and Scientific, 25–31 (June 2008)
4. Yan, T., Li, N.: China's stste-owned commercial bank system vulnerability analysis. The Contemporary Economic Research, 25–26 (January 2009)
5. Liu, M., Luo, J.: Theory of the financial crisis. Economic and Scientific, 72–80 (April 2000)
6. Tan, F.: The development and construction of the banking crisis's KLR early warning system. Reserch on Development, 33–38 (June 2009)
7. Xi, L., Ouyang, S.: Construction and analysis of the Financial risk evaluation index system. Hunan Deministration Institute, 55–59 (April 2010)

The Research and Countermeasures on the College Students' Psychological Stress and Coping Styles

Yang Haicui

School of Information and Communication Engineering,
Tianjin Polytechnic University, Tianjin, China
948042285@qq.com

Abstract. This article analyzes the college students' psychological conflicts on the aspect of adapting the environment, interpersonal relationships, life setbacks, self-awareness, job choosing and so on. It emphasizes the importance of strengthening college students' psychological health education, which is imperative. At the same time, it also gives practical measures for the college students' psychological health education in order to play a positive role for the students forming the good psychological quality.

Keywords: Mental health, Psychological crisis, Coping styles.

1 Introduction

With the social development faster and faster, the psychological pressure of contemporary college students is growing and the students always become extremer and extremer when facing the stress, which have to draw the educators' attention. College students' pressure comes from internal and external. The internal pressure comes from the students themselves due to lack of correct self-awareness, too high learning aim, overpowered motivation which surpass their own abilities and make themselves always feel failure threat. The external pressure comes from the students' heavy learning tasks being difficult to withstand, college competitive environment and social pressure.

A college student in Beijing was sharp chop down dead. And after that a student of the dead was taken by police. It was said that the both had pursued for the same girl and had stricken violently for some feast-------

As the campus violence appeared again and again, we have to be worried. The speeding up social rhythms but the limited person's energy is easy to make people feel blundering. Relatively speaking, social moral restraint strength is weak. When the stress from all aspects accumulating, the damaging force is very big. So when feeling they were damaged, they will instinctively to be anger and attacked without a social acceptance way to abreact, and to seek a kind of extreme behavior to vent which is probably violent to people who hurt him. Therefore for teachers and students, they need to strengthen the psychological supervision, make the energy conversion through normal route and alleviate the problem, which is possible to avoid extreme behavior occurring.

J. Luo (Ed.): Soft Computing in Information Communication Technology, AISC 158, pp. 359–364.
springerlink.com   © Springer-Verlag Berlin Heidelberg 2012

The examples of some unbearable psychological load causing pessimistic depression are also common occurrences. A graduate student in directing department is handsome and tall, but he has a very pessimistic thought: to be a director need to be famous, but there are few of the really famous directors. Besides, the director has to coordinate various aspects of the relationship. This kind of pressure has made him not to endure and he has dealt with dropping out of school in the end. His teachers and classmates all feel sorry for his leaving. What caused his upcoming bilingually, yet choke?

College students now facing stress, causing the larger psychological gap that is inseparable from the whole social development situation and family influence. College students' employment problem, university of enrollment expansion and setting too high target for the future, can cause a lot of psychological gap because of the final undesirable results. So that it requires students to find a reasonable position, make an objective evaluation, know who I am and what I can do, and known that formula "enough is as good as feast". In addition, the students need to be not aiming too high, to be down-to-earth and go their own way.

It is not common to find that there are many college students cohabiting. Xiao Fan and Xiao Lu, a pair of the husband-and-wife tribe, who rent the house around campus for cohabitation. The two people when accepting a reporter to interview, both sides admit that they haven' really clear ideas for the future and they live together currently just for the "mutual heating".

And a recent survey indicates that 13percent people of the surveyed students admitting they had sex during the university.

Why should the now college students be in learning knowledge precious period, but suffer from "sex" problem? In the modern education system, the young man can't make a balance among the "knowing, feeling, and understanding". Its IQ may be higher, but EQ is lower, which causes the weaker willpower and can't control its own life. So the college students when in certain aspects without physical gratification and resulting in a really sexual need will through various channels to obtain satisfy. The students are mostly sexual desire, sexual anxiety, etc, which just show that they lack certain guidance in sex education issue. For students, it is a very important part of the college students mental health to treat sexual problems healthily and scientifically, to learn about sex problems and to be more rational thinking and restrain own behavior.

There are many criminal cases because of the poverty which makes the poor students have the unbearable psychological stress and they eventually go on the criminal road. A college student who was going to graduate only for 17days was theft sentenced. The student in a poor family defaulted on thousands of Yuan tuition for the four years, was going to graduate but found no job. The teachers and classmates analyze that it is the poverty and employment pressures making him be on the criminal road.

Now it has become common problems that the college students can't handle interpersonal relationship. Many college students make a friendly relationship with another dormitory classmates but a very nervous relationship with their own roommates. A university psychological counseling center teacher said that every day he will receive several consulting about how to make a good relationship with the roommates. Sometimes contradictions and friction only due to some trifle, for example, some roommates are rough-up just because of the television remote control. The students who can't handle trivial affairs reasonably lead to the roommates' relation being more strained.

Good capability to deal with affairs and harmonious interpersonal relationship is an important criterion for the psychological health. The pressure of the life will often make some students mood-swings, bad-tempered, small-minded, poor self-control ability, being not friendly with others or getting disharmonious interpersonal relationship.

This fully explained the college students' psychological health suffered great destruction due to lacking the right way to cope with the pressure. In order to help them find the right way to release and solve the psychological pressure, it needs to strengthen the exchanges and communication with them and to strengthen the coaching and education for them, which can make them set up the good life attitude and can cultivate their health and hopeful personality.

It is full of fierce competition in this era which has diversified choices with the coexistence of dream and tribulations, challenge and opportunity, hope and despair, joy and pain, happiness and distress and various psychological disorders occur unexpected. It will affect the college students' healthy growing. Mental health has become the important condition of choosing talents in 21 century. Therefore, we should know the metal crisis the students faced and the ways the students' dealing with.

2   The College Students' Psychological Problems

2.1   Increasing Competition Brings the Psychological Problems: Fidgety and Fragile

In the present, it is an important manifestation that the social competition becomes increasingly fierce. In this social background, some people may feel confused and disoriented, especially for those college students who will go out of school and enter the society. In the face of the fierce social competition, some students are bewildered about what kind of career they want to choose. We can image that how big the students' psychological pressure is.

2.2   Interpersonal Communication Brings the Psychological Problems: Loneliness, Melancholy

With the advancement of science and technology and the development of Internet technology, networking, such as online dialogue, online communication, online making friends, etc, has become a prevalent phenomenon. Because of the appearance of many new technologies, such as office automation and online teaching ,which makes people spend more and more time in interacting with computer and long-term man-machine interaction dilute the interpersonal contact. So people, the teachers and students, the classmates, the old and the young, are lacking in communicating. Whether happiness or sadness, anxiety, worries or uneven, which they can't speak to others just bottled inside alone. While the computer is only a machine, it can't give a person with emotional comfort and spiritual enlightenment except bring the convenient for working and studying. In the long run, negative emotions such as loneliness and unhappiness have aroused spontaneously. If this kind of lonely and melancholy psychology cannot be dispelled and persuaded in time, it is very easy to cause mental illness.

2.3 Rapid Growth Brings the Psychological Problems: Tension and Stress

Along with the social economy high-speed development, the pace of life is accelerating which is inevitable to bring mental tension and pressure to university students. They don't know whether they can adapt to the high-speed development of the society or they can be in the high-speed development of the society. At present, this problem is becoming more and more severe and we have to face it.

2.4 Information Bombardment Brings Psychological Problems: Lost and No Main Force

It is one of the world's big trend that different thoughts and cultures influent each others with the world pattern diversification, economic globalization and intellectualization, and the surging wave of scientific and technological revolution. As new knowledge and new technology continuously emerging, how to choose and use it and do people lose themselves and independent in numerous information when they face so vast information? It is also a very urgent problem in the face of the college students needing to be researched and solved.

3 Countermeasures of Getting a thorough Self-understanding and Cultivating the Sound Personality

In the college students' growing process, it is inevitable to face this or that kind of psychological problems. The problem is not terrible but the important is how to understand and solve problems. I think they can take efforts from the following aspects.

3.1 To Foster Independent Self-awareness

Ancient Greek philosopher Socrates had proposed the slogan of "know yourself", which symbolized the human beings begin to be self-awareness consciously very early. As the successful learning founder Hill said: "success is produced in those with the success consciousness. Failure is rooted in those who let themselves produce failure consciousness of them unconsciously." Only by knowing self-awareness, controlling the self-awareness and improving self-awareness, can they be invincible and from success to success. All outstanding figures get the success and excellence beginning from positive pursuing success self-awareness. Therefore, only by gradually setting up self-reliance, being independent, being self-confidence, being self-improvement and being sense of self-mastery consciousness, will they eliminate the psychological disorders gradually and success.

3.2 To Pay Attention to the Cultivation of Their Elastic Personality

Elastic personality is a kind of healthy personality which includes there respects:
   Firstly, harmonious interpersonal relationship. Interpersonal relationship can best embody one's heath level of personality. People with healthy personality enjoy interaction with others and build up good relationship with others. Healthy people often

respect others with cordial, fairness, humility, lenient attitude, and also receive other people's respect and acceptance.

Secondly, good mood regulation and social-adaptability. People with healthy personality, whose emotional responses are moderate. They have the ability to adjust and control mood and often maintain happy, satisfy, cheerful mood as well as sense of humor. They can vent, dispel, transfer and sublimate reasonably when negative emotions appear. The ability to adapt to society reflects the coordination degree between the human beings and society. They can keep a good intimate contact with the society and can be initiative to care about the society, understand and contact with the society with an open attitude. At the same time, they can make their own thoughts and behavior keep pace with the era's development and comply with the social demands under the knowing of the society.

Thirdly, optimistic attitude towards life. Optimists always see the bright side of life, are full of hope and confidence for the future, are highly interested in the work or study they are doing and play their wisdom and strength in it. They also have the bull by the horns and are brave to win even in difficulties and frustrations.

3.3   To Cultivate Self-adjustment Ability

Cultivating self-adjust ability can gradually eliminate one's tension and pressure caused by the fast-paced life of society. Adjusting ability is that people can change themselves with the changing environment to adapt to the environment. People must learn to update ourselves constantly as circumstances changing and learn to adjust themselves constantly along with rapid development of society to adapt to the changed situation.

3.4   To Cultivate Peaceful State of Mind

In the contemporary society, division of professional labor is finer and finer, there is nobody may or have any need to master all the knowledge and receive all information resources. Although it needs compound talents for the development of society in the future, there is no a top scientist can master all the information. So, people just need to master well the information about their own works and interests, browse the knowledge of other aspects as widely as possible and keep the heart usual to meet changes with constancy.

With the progress of the society, people need to update and expand our knowledge constantly to enrich and perfect themselves in order to adapt the continuous development of social needs. Learning and learning continuously is the eternal pursuit of life. Only being in the learning process, laying a solid foundation, improving skills, can they do a job with skill and ease and live happily strong in the complexity of the modern society.

4   Conclusion

As for the college students' mental health, it is the bounden duty for the educators and an important respect needing the social support. This article gets a further investigation and study on the problem caused by students' psychological pressure and gives some countermeasures, such as raising the independent self-awareness, paying attention to cultivate elastic personality, cultivating flexibility self-regulation, cultivating peaceful state of mind and so on. We being care about the students' psychological health with the improvement and unremitting efforts, the students can devote to the society with all their capability and make due contributions to the social development.

References

1. Li, B., Li, G.: The reasons and countermeasures of college students' mental setbacks. China Youth Research (July 2003)
2. Meng, Z. (ed.): Common psychology. Peking University press (1994)
3. Stailman, K.T. (writting), Zhang, Y. (trans.): Emotional psychology. Liaoning People Press (1987)
4. Zhao, X. (ed.): The university students' psychological health tutorial. Modern Knowledge Press (2003)

The Application Research on Cotton Fabric of Natural Tea Pigments

Zhi-peng He and Zan-min Wu

College of Textiles Tianjin Polytechnic University
penpen966922@163.com
Key Laboratory of Advanced Textile Composites (TJPU),
Ministry of Education, Tianjin 300160, P.R. China
wuzanmin54@tjpu.edu.cn

Abstract. The paper by leaching tea pigments primarily researched the chemical structure of tea pigments which were gained by infrared spectral analysis and then did ultraviolet-visible spectrum analysis. Using natural tea pigments dyeing cotton fabrics, discussed the influence of the factors of dyeing. Results showed that:to tea pigments the best dyeing temperature, time and pH value were 80 degrees, 50min, pH = 4.0; NaCl dosage for 5g/L the promote effect was obvious; Fe^{2+} had the most obvious effect of dyeing, compared with other mordant on its color. Using the simple method can obtain natural tea pigments, accord with current "low carbon", "environmental protection" trend. And fabric after dyeing fastness to washing soaps resistance, resistance to sweat fastness, rubbing fastness were good concurrently had good antibacterial, ultraviolet performance.

Keywords: tea pigments, dyeing, natural pigments, low carbon, environmental protection.

1 Introduction

With the worsening of current environment, humans begin to realize and have been going into action to protect our survival and development space, along with the environmental protection wave, will bring unprecedented challenges to the textile and dyeing industries. In the production process of synthetic dyes can consume large amounts of energy and raw materials and have a lot of toxic wastewater which is tremendously harm to the environmental, this is the problems which must be solved.

As a water-soluble pigments tea pigments is extracted from low-grade green tea , have good stability, colorbility and multiple health protection function, also have antibacterial and uvioresistant function. Because of its unique physiological function and edibility, tea pigments mainly used in medicine and food science domain now, its application research in dyeing is still in the stage of exploration.

J. Luo (Ed.): Soft Computing in Information Communication Technology, AISC 158, pp. 365–372.
springerlink.com

2   Materlals and Instruments

2.1   Materials

Low-grade green tea(From Fujian sanming area)
Pure cotton bleach fabric C40×40,133×72
Instruments
DataColor SF 600 Plus (America DataColor company); Camspec M350 UV/Visible (Britain SDL company).

3   Test Methods

3.1   Dry/Wet FastnessTests

Testing standard: GB/T 3920-1997«Textile·Color fastness test·Rub resistance test»,GB/T 3921.1～5-1997«Textile ·Color fastness test·Washing fastness test», GB/T 3922-1995«Textile ·Perspiration resistance test».

3.2   UPF Value Tests

Testing standard: USA:AATCC Test Method 183-1998.
Testing three times to each sample,the angle are 0°,45°and 90°,after that take average.

3.3   Antibacterial Property Tests

Testing standard: GB/T 20944.2-2007(absorption method).

4   Experiments

Extraction of Ttea Pigments[1].green tea→leaching→filtering→filter liquor→mixing the pH value→adding precipitant→suction filter→filter liquor→reduce pressure distillation→vacuum drying→tea pigments
　　Dyeing process of pure cotton fabric Analysis of dyeing property Influence of Temperature and Time Selecting temperature of 343.15K,353.15K,363.15K,373.15K, respectively dyeing under the time of 30min,40min,50min,60min,70min. Influence of pH Value Mixing the pH value to 4,5,6,7,8,9, respectively dyeing under the temperature of 363.15K. Influence of Salt DosageRespectively selecting salt dosage 5g/L,10g/L,15g/L,20g/L,25g/L, according to the dyeing curve (Fig.1) finish dyeing.Influence of Tea Pigments DosageConfigurating dye liquor(pigment dosage is owf 1%,2%,3%,4%,5%)→fabricenter(363.15K)→dyeing(60min)→finishing dye process(according with Fig.1)→taking out fabric→aftertreatment

Influence of Mordants. Configurating dye liquor(respectively adding 1ml 0.1M AlCl3,MgSO4,CaCl2,FeSO4·7H2O,CuSO4)→fabric
enter(363.15K)→dyeing(60min)→taking out fabric→aftertreament

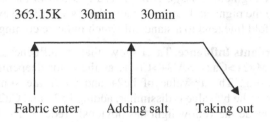

Fig. 1. The dyeing curve

5 Results and Analysis

5.1 Analysis of Dyeing Property

Analysis of Temperature and Time Influence. Fig.2 shows the influence of temperature and time,with the increasing of temperature the k/s value is decreasing after dyeing 50min;the results of 353.15K and 363.15K are analogous;comprehensive consider various kinds of factors ascertainning the optimum dyeing temperature and time:353.15K,50min.

Analysis of pH Value Influence. Fig.3 shows the influence of pH value,the optimum pH value is 4.

Analysis of Salt Dosage Influence. Fig.4 shows the influence of salt dosage,the adding of salt has directly influence,when the dosage is 5g/L the k/s value reach to the biggest and obtain the deepest shade.Similarity as the the direct dye,the adding of salt has promoting effect to dyeing.

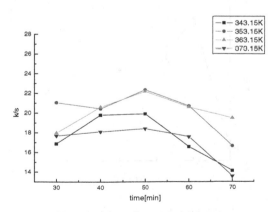

Fig. 2. k/s value of different temperature and time

Analysis of Tea Pigment Dosage Influence. Fig.5 shows the influence of pigment dosage, the more the pigment dosage the more the k/s value, however when the dosage reach to 4%,the fold line tend to a standstill which indicate coming to dyeing balance.

Analysis of Mordants Influence. Tab.1 shows that the influence of metal ion on color depth is $Fe^{2+}> Mg^{2+}>Ca^{2+} >Al^{3+}>Cu^{2+}$, as the color deepening the brightness is gradually decreasing. The a* value of Fe^{2+} and Cu^{2+} are similar, the others' are different; moreover the b* value is dissimilar, adding Fe^{2+} and Cu^{2+} have great impact on shade which made the yellow light has obviously reduce. The C* value shows the bright--colored color degree, Fe^{2+} has great influence, which make the color change from orange to kelly.

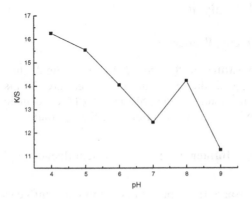

Fig. 3. k/s value of different dyeing pH value

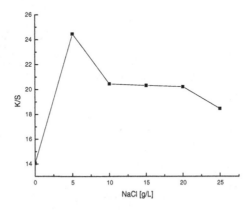

Fig. 4. k/s value of different salt dosage

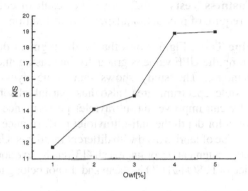

Fig. 5. k/s value of different tea pigments dosage

Table 1. The influence of mordants

	k/s	L*	a*	b*	C*	h
Al^{3+}	12.22	76.82	5.31	8.75	10.24	58.78
Fe^{2+}	25.45	64.70	2.52	1.84	3.12	36.15
Mg^{2+}	17.54	70.72	4.82	4.86	6.84	45.26
Ca^{2+}	17.12	71.88	6.14	6.86	9.21	48.19
Cu^{2+}	10.81	77.82	3.11	6.50	7.20	64.43

Table 2. Color fastness

Temperature (K)	Rub resistance		Washing fastness		Perspiration resistance	
	dry	wet	fading	staining	fading	staining
243.25K	4~5	4	3	4~5	4	5
253.15K	5	4~5	4	5	5	4~5
263.15K	4~5	4~5	4	4~5	4	5
273.15K	4~5	5	4	4~5	5	4~5

Results of Color Fastness Tests. Tab.2 shows the result of color fastness tests,the color fastness fit the require of production,dispense with fixation.

Results of UPF Value Tests. Fig.6 shows that as the pigment dosage increasing the color depth is deepening,the UPF value is gradually increasing and the anti-ultraviolet performance is enhancing. The study shows that some dyes besides has strong absorption in the visible spectrum area, also has partial absorption in ultraviolet spectrum area, thereby can improve anti-ultraviolet performance of fabric, generally with the increasing of color depth the anti-ultraviolet performance is improved[2].

Tab. 3 shows UPF value of fabrics dyed with different mordants. Using $Fe2+$ as mordant UPF value is greatly augment,accordingtoTab.4[3]reach extraordinary, outstanding safeguard, and the price of $FeSO4 \cdot 7H2O$ is cheap and do not belong to heavy metal ion.

Results of Antibiotic Property Tests. Fig.7 shows the bacteriostatic rate of fabric dyed using different tea pigment dosage. With the dosage increasing the bacteriostatic rate is augment, when dosage is 5% reach the best ideal bacteriostatic rate. As polyphenol, the molecular structure of tea pigment have many oxygen atoms for the coordination complexation, which can combine with protein inside the microbial body, make them degenerate, destroy protein structure, make inactivation of various metabolic enzymes inside the bacteria cell, and achieve sterilization effect.

Table 3. UPF value of fabric dyed with different mordants

	Al^{3+}	Fe^{2+}	Mg^{2+}	Ca^{2+}	Cu^{2+}
UVA%	18.2	5.3	18.4	18.1	19.0
UVB%	5.2	2.0	6.2	5.5	5.6
UPF	14.0	40.2	12.4	13.5	13.1

Table 4. Evaluation criteria of UPF[3]

Range of UPF value	The protective effect	Ultraviolet Transmitttance	UPF rank
15~24	better	6.7~4.2	15,20
25~39	superduper	4.1~2.6	25,30,35
40~50,50+	best	≤2.5	40,45,50,50+

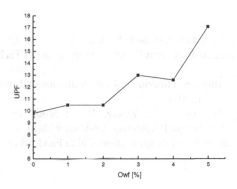

Fig. 6. UPF value of fabric dyed with different tea pigments dosage

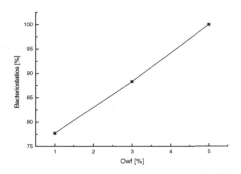

Fig. 7. The bacteriostatic rate

6   Conclusion

Through dyeing experiments of tea pigment and each test results show: tea pigment have good dyeing affinity to pure cotton fabric, cotton fabric color is deep,every fastness are preferable, enhance the anti-ultraviolet performance and have anti-bacterial property.

Use simple extraction method can obtain tea pigment, greatly reducing the consumption of energy and chemical raw materials, according with current green textile, ecological textile trend.

Tea pigment as tea by-products with wide source,extraction method is simple, and its use is gradually popularity and widened, so tea pigment is one kind potential product[4].

References

1. He, Z.-P., Wu, Z.-M.: Extraction and Separation of Tea Effective Ingredients. In: International Conference on Intelligent Control and Information Technology, Tianjin (2011) (in press)
2. Wu, J., Gao, Z.-F.: Textile anti-ultraviolet effect evaluation and influencing factors. China Fiber Inspection 12, 31–32 (2004)
3. Han, W.-W., Deng, H., Xie, Y.-J., Zhang, B.-R.: Study on fabric performance of anti-ultraviolet. Journal of Tianjin Polytechnic University 28(6), 69–72 (2009)
4. Ye, Y.: Tea pigment research and its application. China Food Additives 4, 23–24 (1997)

Extraction and Separation of Tea Effective Ingredients

Zhi-peng He and Zan-min Wu

College of Textiles Tianjin Polytechnic University
penpen966922@163.com
Key Laboratory of Advanced Textile Composites (TJPU),
Ministry of Education, Tianjin 300160, P.R. China
wuzanmin54@tjpu.edu.cn

Abstract. Using hot water extraction method conducts the single factor and orthogonal experiments and explores the best condition to extract three effective ingredients: theaflavin, tea polyphenol, theabrownine from green tea. Through a series of experiments gets the best extraction process: 368.15K, 35 min, ratio 1:40. Using simple method can obtain higher yield, accord with the current trend of low carbon and environmental protection. And the tea effective ingredients have immense application value on textile industry.

Keywords: green tea, theaflavins, theabrownine, tea polyphenol, environmental protection, low carbon.

1 Introduction

Green tea is without fermentation, retained more natural ingredients within fresh leaves, which has special effects to prevent anile, cancer prevention, anti-cancer, sterilizing, anti-inflammatory and others, and is incomparable for other tea. In 2008 China tea-garden area reaches 160 million hectares, 124 million tons of tea production accounting for the world total 1/3. Because domestic and international market demand steady growth tea industry growth potential is tremendous in our country. Besides large low-grade tea, the quantity of tea stem and old tea after pruning is more tremendous, so explore new ways to the development and utilization of the waste tea and low-grade tea can bring huge economic benefits to tea farmers, on the other hand can reduce synthetic chemicals production enterprises and related industries' environmental pollution, true realizing "low carbon and environmental protection".

2 Materials and Instruments

Materials. Low-grade green tea(From Fujian sanming area)
zinc chloride,sodium hydrogen carbonate,ethyl acetate(From the third chemical agent factory of Tianjin)

Instruments. 723-type visible spectrophotometer VIS—723
(Shanghai precision &scientific instrument limited liability company), Electric vacuum drying oven DZF-6030A(Shanghai YiHui instruments limited company)

J. Luo (Ed.): Soft Computing in Information Communication Technology, AISC 158, pp. 373–379.
springerlink.com   © Springer-Verlag Berlin Heidelberg 2012

3   Test Methods

According to the TP content to make sure the best extraction processes,so before that no ingredient is separated.Test standard:GB/T 8313-2002.

4   Experiments

Single   Factor Experiments(the dosage of tea is 0.3g in each experiment)
    the effect of temperature: fixing the time, ratio and extraction times (45min, 1:30, 1time), extract separately under the temperature of 353.15K, 358.15K, 363.16K, 368.15K, 373.15K, and dilute the extract solution to 50ml,then test the Abs value.
    The effect of time: fixing the temperature, ratio and extraction times(368.15K, 1:30,1time),extract separately under the time of 30min,35min,40min,45min,50min,and dilute the the extract solution to 50ml,then test the Abs value.
    The effect of ratio: fixing the temperature, time, extraction times(368.15K, 35min, 1time),extract separately under the ratio of 1:30, 1:35, 1:40, 1:45, 1:50, and dilute the extract solution to 50ml,then test the Abs value.
    the effect of extraction times: fixing temperature, time, ratio(368.15K, 35min, 1:40),extract three times, and dilute the extract solution of each time to 50ml,then test the Abs value.
    Orthogonal Experiments
    According to the results of single factor experiment
    to determine the orthogonal factor table(Tab. 1).

The Separation of Effective Ingredients[1]. On the basis of orthogonal experiment results sepatate the effective ingredients.Extracting route: Gree tea→hot water leaching→filtering→filter liquor→mix the pH value to 6~8→adding precipitant zinc chloride [2]→leaching→(1), (2)

    (1)filter liquor→reduced pressure distillation→
vacuum drying→theaflavin
    (2)filter cake→dissovle in acid→extract with equivalent ethyl acetate→separating funnel dividing→①, ②
    ①organic horizon→reduced pressure distillation→
vacuum drying→tea polyphenol
    ②water layer→reduced pressure distillation→
vacuum drying→theabrownine

5   Results and Analysis

5.1   Results of Single Factor Experiments

Fig. 1-4 shows that the best extraction condition is 368.15K, 35min, 1:40, extract 1 time, the fold lines of Fig.1-3are first increased then decreased.

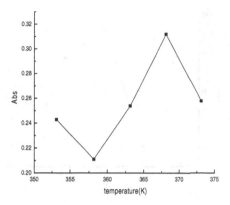

Fig. 1. The Abs value under different extraction temperature

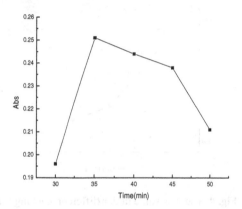

Fig. 2. The Abs value under different extraction time

Results of Orthogonal Experiments. Based on the results analysis of single factor experiments the leaching times has less influence on yield, can be ignored. Tab. 1 is factor table,and Tab. 2 is the result of orthogonal experiments.

The range analysis of Tab. 2 shows that the influence level on yield of each factor in turn: ratio, time, temperature, which indicates that ratio has the biggest influence, moreover temperature is the least. Analysis by synthesis of Tab. 1 and 2 and other factors make sure the best process conditions are: ratio 1:40, 35min, 368.15K. The experiments also demonstrate the process of the paper is better than related document's yield of 10.5%[3] when convert to the yield.

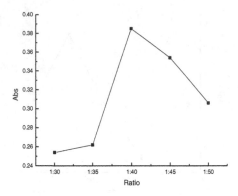

Fig. 3. The Abs value under different ratio

Fig. 4. The Abs value under different leaching times

Table 1. The orthogonal factor table

factors	Temperature(K)	Ratio	Time(min)
Level 1	363.15	1:35	35
Level 2	368.15	1:40	40
Level 3	373.15	1:45	45

6   The Infrared Spectrum Diagram Analysis of Effective Ingredients

Fig. 5-7 are the infrared spectrum diagram of theaflavin, theabrownine and tea polyphenol.

Fig. 5 shows that the peak in 3384cm-1 is hydroxyl, the peak in 2958cm-1 is hydrogen of benzene,the peak in 1699cm-1 is carbonyl, the peak in 1232cm-1 is stretching vibration of ether bond.

Fig. 6 shows that the peak in 3429cm-1 is hydroxyl, the peak in 2960 cm-1, 3089 cm-1, 3147 cm-1 is stretching vibration of hydrogen of benzene, the peak in 1625 cm-1 is the double bond of benzene,the peak in 1168cm-1 is ether bond, the peak in 1031cm-1 is flexural vibrations of hydrogen of benzene, by analysis we can see the structure of theaflavin and theabrownine are similar.

Fig. 7 shows that the peak in 3795cm-1 is hydroxyl, the peak in 2920cm-1 is hydrogen of benzene, the peak in 1714 cm-1 is carbonyl, the peak in 1622cm-1 is the double bond of benzene.

Table 2. The results of orthogonal experiments

factors	temperature	ratio	time	yield
Exp.1	1	1	1	20.5485
Exp.2	1	2	2	16.9607
Exp.3	1	3	3	15.6560
Exp.4	2	1	2	21.3965
Exp.5	2	2	3	19.3743
Exp.6	2	3	1	12.6553
Exp.7	3	1	3	22.4403
Exp.8	3	2	1	13.1771
Exp.9	3	3	2	17.7435
K_1	53.165	64.385	46.381	
K_2	53.426	49.512	56.101	
K_3	53.361	46.055	57.471	
k_1	17.722	21.462	15.460	
k_2	17.809	16.504	18.700	
k_3	17.787	15.352	19.157	
R	0.087	6.110	3.697	

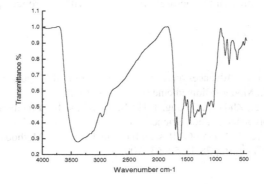

Fig. 5. The infrared spectrum diagram of theaflavin

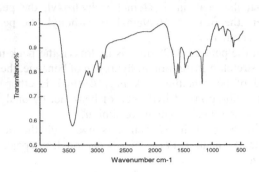

Fig. 6. The infrared spectrum diagram of theabrownine

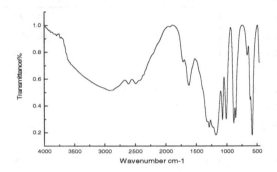

Fig. 7. The infrared spectrum diagram of tea polyphenol

7   Conclusion

The best extraction condition is: ratio 1:40, temperature 368.15K, time 35min, extraction times 1 time.

Through infrared spectrum diagram analysis to identify various ingredients' functional groups, their main structure are coincide with references [2,4,5].

References

1. Liu, M.-J., Chen, J.-P.: Advances on reaearch of extraction of tea polyphenols from green tea leaves. Progress in Modern Biomedicine 6(7), 70–72 (2006)
2. Ong, W.-B., Hu, Y., Zhou, L., Zhang, J.-H.: Study on the Ion Precipitation Technology for Preparing Tea Polyphenols. Food Science and Technology 9, 45–47 (2002)
3. Li, Y.-M., Wang, X.-S., Yu, Y.-C.: Research on the Extraction Method of Tea Polyphenol in Tea. Guangzhou Chemistry 28(1), 59–63 (2003)

4. Li, D.-X., Wan, X.-C., Xia, T.: Preparation and Component Analysis of Tea Pigments. Journal of Hygiene Research 33(6), 698–700 (2004)
5. Qin, Y., Gong, J.-S., Zhang, H.-F., He, J
6. Zhang, S.-H., Zhou, H.-J., Li, B.-C.: Extraction Technology of Theabrownine from Pu-erh Tea and Its Physico-chenical Properties. Chemistry and Industry of Forest Products 29(5), 95–98 (2009)

The Basic Elements of Modeling – The Use of Lines in the Phoenix Vein

Using Mid-Western Zhou Dynasty Phoenix Vein Cup for a Case

Wenli Shao and Li Zhang

Institute of Art and Fashion, Tianjin Polytechnic University, Tianjin, China
shaowenli1314@yahoo.com.cn

Abstract. Based on the analysis of phoenix vein, the paper explains that different lines have different personalities which give the audience different visual psychology, and explains the relationship between the lines in phoenix vein and the overall modeling of the vein. It comes to a conclusion that phoenix vein is not a pure visual style, and phoenix vein can be extracted and symbolized to be applied to wider areas of design.

Keywords: Line, Phoenix vein, Phoenix.

1   Introduction

The lines are the important part of composition of a picture; lines are the brim of the shape. The definition of the lines depends on the compare between the shape and the background. The lines not only decide the shape, but also transfer the sight of the audience from one location to another. The sight of people usually move along with the lines, not only in spite of they are bright lines which are composed of roads, straight trunks in rows and telegraph poles but also in spite of they are indirect lines which are hidden in the contour of body, hue and color.

The lines can be divided into materiality and immateriality. Lines of Materiality are pointed to contour of the sculpture and brim of the sculpture, the form of the sculpture, and the boundary between sculpture and sculpture. While the lines of immateriality are pointed to the lines that can be experienced and felt. Lines can generally be divided into beeline and curve in despite of materiality or immateriality. Beeline can also be divided into perpendicular, horizontal line and diagonal. While the curve can be divided into free curve, undulance curve and geometry curve.

1.1   Perpendicular's Vision Psychology

Perpendicular can show the rise, erect, dignity to people. It represents life, dignity, forever, power and ability of resisting change. Perpendicular can make the view line move up and down. It shows height, so can produce the effect of rise, tall and up.

1.2   Horizontal line's Vision Psychology

The longer horizontal line makes people feel quiet, peaceful, steady, open, extendable and stretchable. The use of horizontal line can bring the feeling of permanence, peace,

J. Luo (Ed.): Soft Computing in Information Communication Technology, AISC 158, pp. 381–387.

quiet and steadiness. Horizontal line can make the view line move left and right, so it can produce the effect of wide, extend and stretch.

1.3   Diagonal's Vision Psychology

Diagonal makes people feel pulsatile, dangerous and strange. Diagonal looks like vivid, lively, profound and pulsatile. Diagonal makes people feel the stretch or shrink from one side to another, and produce the effect of uncertainty, so it makes pulsatile.

1.4   Curve's Vision Psychology

Curve alters people's vision ever and again, leading view line to barycenter. Although the forms of Curve are abundant, they are basically the transmutation of undulance lines. Not only the different direction of the curve makes people feel different, but also the change of itself has the abundant expressive force. For example, bold curve can stand for adamancy, massiness, stabilization and stupidness, while the thin curve stands for smartness, acuity and sensitivity.

2   Phoenix Vein's Artistic Characteristics

As shown in the Figure 1 : Phoenix vein has very dignified, elegant exterior modeling, which is constituted by curve and beeline. The grain is balanced, left and right sides is symmetrical, overall stability, in most cases it uses the combination of single-phase artistic image. Which makes the vein look like a whole and a independent part. So, it uses the typical procedure exaggeration method.

Fig. 1. Phoenix vein cup in Mid-Western Zhou Dynasty

Using individual patterns to emphasize the Phoenix veins and square continuous veins to surround bronze wares. The purpose is to gain more complete and overall impression to give people a kind of circulation, perennial emotional appeal. An shown in Figure 2.

Using the compare between positive and negative, virtual and actual, bold and thin, big and small, the method of heave in plane, concave in interspaces to draw cloud and thunder veins. That makes the whole sculpture produce a rich and changeable arrangement of positive and negative.

Fig. 2. Phoenix vein

Shutting up and staring, long crown and bend tail, bridle and gaze, regular and strict, to show a power, purpose and a savage beauty, in the power and purpose, dynamic can be seen in the static, a inside mysterious power will extend as the time goes. As shown in Figure 3.

Fig. 3. Phoenix vein

3   The Application of Lines in Phoenix Veins

Line is a means of modeling. Wireless is non-painting. Lines of phoenixes in the main effects: first, a sense of the performance of the screen: second, the creation of a visual image of the screen. In this bronze of phoenix vein, the ideal combine of line and pictures makes people focus on the subject matter .The screen lining the main lines and complementary patterns present us the overall concept of imagination the image of the Phoenix, also showed a feeling of depth and movement.

3.1   Line Perspective

Due to the bronze special design structure, the design in the portrait of time itself should be considered with the perspective, for example, the line with the break, the curvature of the transition, conversion of square. Only the full consideration of the space can express the deepness of it. In order to make the expression in the space of consciousness comforts to man's experience and feel more truthful .In addition, a compare of far and near makes the primary stand out.

3.2   Practical Application of the Line

Line is an objective things, which restricts the external form surface shape, each existing objects are outside the outline, presents certain lines. People have a deep impression on the contour of the object outside and the changes of moving objects' line during a long time. So, in turn, through some combination of lines, one can think of some kind of object shape and movement. Therefore, all the plastic arts have attached great importance to the broad lines of force and expression; it is an important language of plastic arts.

Phoenix vein's success is the emphasis on the extraction and the use of lines .What's more, we highlight and emphasize the image of the expressive use of the outer edge of profile to reflect the main object of the texture, feel and sense of space volume, reproduce accurate, clear and vivid visual images and in order to express their emotions.Thus, Line makes phoenix vein living, which embodies the ancient superb styling ability and excellent use of practical ability. Specific use as follows.

Using simple lines of generalization to express the main characteristics of the phoenix vein: For simple and summarize the graphics is peculiar visual psychological tendency. From art performance, it is to simplify the first line of outstanding features, main purpose, subtract needless little detail. In the simplified form processing line is not for simplification, but to choose some fairly obvious candidates and refine the contents to stand out the primary one's expression and to appeal to audience. A lot of simple and general lines are used to shape modeling in phoenix vein decorations. Simplify phoenix's shape. The main line is sturdy and strong, outline is also clear. A crown, a foot and a tail are enough to show phoenix's posture and expression. It is concise general and charming in the characterization of the image. It exaggerates some parts of the body and bypasses the rest parts of it in order to achieve the goal of highlight the main features. It brings people the beauty of momentum, power, a kind of good wishes' realization, a savage beauty and so on.

Using line's structure beauty to convey feeling: In the form of art, the line is an important lyric means, in other words it has a strong emotion.

The symbolism of traditional phoenix pattern is related to religion and religious worship. The general patterns with imagination express some kinds of thoughts and ideas. So these graphics for people are not for appreciation, but as a sign or symbol to the family or the human.The lines of phoenixes are fluent, rounded, and lively. The connection between lines is natural and ingenious. The handled structure is programmed, and the technique of expression is flexible. No showing of any traces of rational thinking. It advocates strong, exaggerated strength, neat, and has a strong spirit that people can not bear.

People have accumulated the deep impression and feeling to the line structure in long-term real life, so they could develop association to some types of lines structure when they talked about it. When it comes to the phoenix patterns, it is natural to think of bird body, head with a specialty long beak, large round eyes, huge wings and other structural features. Then it will inspire the corresponding emotion. For example, the phoenix is the king of birds in the Chinese legend; it represents the luck, peace and clarity of politics.The phoenix, soaring ambitiously, just looks like the Chinese Kunpeng Bird spreading the wings to pass 90,000 miles; Phoenix is the embodiment of wisdom and courage, and has unlimited capacity; Phoenix symbolizes strong-will, brawniness, peace, rebirth; Phoenix symbolizes dignity, the holy man; Phoenix is noble and pure which eats pearl and drinks dew, so it is the world's holiest thing, the chief representation of the best things. Phoenix also represents the wishes of freedom and marriage, etc.

Using the line structure of rich character shown in a moment to express act and scene: A moving object, it is difficult to catch the accurate its basic features. Because it is a moving target object, its shape, gesture and contour lines are changing constantly. Therefore, extracting the line structure of a moving object is important. In addition to other means of modeling, catching moment is also very important. The line structure of the moving object is different at different moment, and its modeling expression power is different too. Lines of moving object are generally elegant, stretchable, smooth, vigorous and effective, while the shape of the lines is beautiful, graceful. so it can express movement, strength, skill and unrestrained feelings. As shown in Figure 4 and Figure 5.

Fig. 4. Dynamic phoenix vein

Fig. 5. Dynamic phoenix vein

The artistic process of Phoenix vein is the model of combining visual thinking, abstract thinking, imaginative thinking and creative thinking. Even if the designer did not really see the movement of the phoenix, at least he observed the movement of other kinds of bird carefully, and caught the dynamic moment of the bird perching, proudly standing posture and so on. The designer made each of the dynamic moment static, after the brain processing, refining, streamlining, creating, with a moment of points, lines, or even faces, the designer can show the characteristics of the ideal phoenix. At the moment, the gesture of the lines in the picture varied, and became graceful and restrained, which made the Phoenix vein contain dynamic moment and static moment.

in the sculpture, the Phoenix seemed to fly high because of tension shown in the picture. The accurate capture of object's dynamic images reflects that aesthetics plays an important part in the Shang and Zhou dynasty, and people consciously combine the aesthetics means with the strong religious myth organically.

Using the beauty of lines to form rhythm: We are quite familiar with rhythm. Rhythm comes from the movement, using the arrangement and combination of lines can create the tempo and rhythm. The arrangement of duplicate lines forms the visual rhythm, the changes and differences of similar lines can product rhythm. Lines of different shapes and different arrangement density in the picture will lead to different rhythm. Because these rhythms, by different visual rhythm, some are crisp, some soft, some sharp and some slow.

The formation of the rhythm of a picture comes from life and nature. But to produce rhythm does not mean to copy the formation of images of life or to imitate nature,but to deal with the rhythmic lines which are formed in life and nature in ways of art,making it stand for a specific meaning to express feelings,convey ideas and reflect theme and content.

The process of form and rhythm in phoenix vein modeling and the treatment of relation among point, line and face gain much attention, so does the order of lines and modeling arrangement.

The picture shows the two sides' continuous pattern, using line-type composition, phoenix using deformation techniques, black and white ash handle properly, which reflects a rational, masculine beauty. Designers successfully grasp the primary part ,secondary part and screen density, full of rhythm. As shown in Figure 6.

Fig. 6. Two sides' continuous phoenix vein

Using compare among different lines and different individuality to figure whole sculpture: Using compare among different lines and different individuality to figure whole sculpture: The character of the layout of lines such as stiff or agility, orderly or disorderly, symmetric or balanced, steady or deft is reflected directly in the whole sculpture of the Phoenix vein. Facing these different character and speciality of the lines, the visual feeling such as positive and negative, big and small, bold and thin, static and dynamic, curve and straight is coming directly. The phoenix vein use character compare among different lines well to emphasize the main character of the phoenix sculpture. The main part of the phoenix vein and the secondary part supplement each other. Which produces dynamic illusion in the whole stiffness. This is the master art produced by static and dynamic. At the same time, separating the main part of the phoenix vein from the secondary part independently, then fit them together makes the sculpture has a art of not only integrity but also individual.

4   Phoenix Vein's Meaning in Modern Times

The art of Phoenix vein in Shang and Zhou Dynasty's bronze is charming in art, it is the product of Chinese labor people. Otherwise, it is the valuable fortune of our offspring and can be used for reference. So, only should we band the Chinese traditional design and modern innovation and image together, can we create the modern design rich in Chinese nation style.

In modern design, such as advertising packaging design industry, Chinese aesthetic psychology is taken into consideration by numerous designers .Art in the traditional Chinese, dragon, phoenix, carp and other symbolic graphics are used in design. With this image in popular culture to reach psychological emotional appeal. Let the Chinese feel warm and accessible in the level of cultural psychology.

It is widely used and it always works. The diversity of phoenix vein visual arts also provides rich visual language for modern design. It becomes a kind of specific symbol factor. It doesn't only reflect human consciousness but also embodies humanism. This feature makes it become a kind of indispensable and specific design language in modern design.

References

1. Jianqun Lin, G.: Fundamentals of Modeling, pp. 50–56. Higher Education Press, Beijing (2001)
2. Lin, J.: Design innovation and education, pp. 22–26. SDX Joint Publishing Company, Beijing (2002)
3. Garmaabazar, K., Maeda, A.: Retrieval Technique with the Modern Mongolian Query on Traditional Mongolian Text. In: Proc. 9th International Conference on Asian Digital Libraries (ICADL 2006), pp. 478–481 (November 2006)
4. Garner, S.: The undervalued role of drawing in design. In: Thistle Wood, D. (ed.) Drawing Research and Development, pp. 98–110. Longman Group, London (1992)
5. Laseau, P.: Graphic thinking for architects and designers. Van Nostrand Reinhold, New York (1989)
6. Schenk, P.: The role of drawing in the graphic design process. Design Studies 12(3), 168–181 (1991)
7. Won, P.H.: The comparison between visual thinking using computer and conventional media in the concept generation stages in design. Automation in Construction 10, 319–325 (2001)

On the Exploration and Research of Reinforcement on the Construction of University Students' Learning Atmosphere

Yang Haicui

School of Information and Communication Engineering,
Tianjin Polytechnic University, Tianjin, China
948042285@qq.com

Abstract. It's urging to construct the university students' learning atmosphere, which is aimed at implementing the arrangement of National Ministry of Education, accelerating teaching reform, and cultivating talents with high qualities for the society. Based on the important ideas of "The Three Represents" which directs the construction, this article puts forward the goals of every phase, obtains the measures and manipulating points, and lastly points out how to make sure the realization of the goals of the learning atmosphere's construction. The exploration and research on these aspects are an eternal theme of university's learning atmosphere construction, and also a talents' project with strategic meaning.

Keywords: construction of learning atmosphere, goals of every phase, education of learning atmosphere.

1 Introduction

The construction of learning atmosphere is an important component of university style's building, and also a key to drive quality education comprehensively and cultivate talents with high qualities. Good learning atmosphere is an important condition to improve the quality of education and teaching. The good or the bad learning atmosphere directly affects the key of teaching reform, the improvement of teaching quality, the shaping of university's atmosphere, the realization of talents cultivating, the social prestige and image of university. We should put importance on the construction with high social responsibility, and take "The Three Represents" and the advanced ideology as guiding to carry out the party and the country's educational policy, updating educational concepts constantly, deepening educational reform, emancipating thoughts and requiring true action. Also we should make a positive and aggressive learning atmosphere in the students' groups, build the adaptive cultivating environment for this university's development and harmony, comprehensively enhance the multiple qualities of students, and make efforts to educate qualified talents.

2 Clear Objectives

Considering the specific situation of nowadays students, the construction of learning atmosphere should be carried out phase by phase.

J. Luo (Ed.): Soft Computing in Information Communication Technology, AISC 158, pp. 389–395.
springerlink.com   © Springer-Verlag Berlin Heidelberg 2012

The long-term objective. Through a period's effort, the students can understand the conception, "happy to learn, diligent to learn, capable to learn", and they will further their study attitude, explicit study objective, strengthen study regulation, enhance study quality, and consolidate professional thoughts. The students can also shape the rigorous, strict, realistic, and true learning atmosphere with their promising and pioneering spirits put in their study life.

Phase Objectives. Make use of a period to create new atmosphere—let the students obtain explicit learning goals and good professional ideology, change their study attitude, and achieve the "Six Obviously", which are: learning enthusiasm increases obviously, the performances of absence, being late, leaving early, playing computer games and chatting decrease obviously, atmosphere and regulation of examinations become better obviously, the learning atmosphere of classes improves obviously with the improvement of study results, the ideological state of students become more positive with an obviously improvement in comprehensive qualities and creative abilities.

Based with that, it takes another period to make bigger change—let the students transform from "make me study" to "I want to study" under the dominance of subjective initiative, and master some professional applied skills under the premise of studying the professional basic theories well.

At last, another period is taken to get the result of the learning atmosphere of our college—to obtain the long-term objective. The students possess good moral quality, innovation consciousness, necessary theoretical knowledge and application skills connected with the inquiry of society. And we should make sure the comprehensive qualities of students are generally praised by the society.[1]

3   Specific Measures

To strengthen students' ideological and moral cultivation, and correct learning motivation, we should consider the students' practical ideology under the new situation. We should strengthen college students' political work and cultural education; scientifically guide them to build up the correct learning, successful, and working conceptions; constantly enhance student's social responsibility, stimulate their self-consciousness of learning for national rejuvenation and a prosperous motherland.

Strengthen students' party construction work to promote learning atmosphere. Further improve students' party branch work rules and regulations, focus on the development, training and education of student party members. The learning condition will become a prerequisite for developing student party members and also a standard of evaluation. The learning atmosphere construction will be seen as student party members' duty and obligation. These students will display model leading role positively in the construction of learning atmosphere.

Enforce the propaganda to promote the students' study and lead the students to put their energy and attention to study. Make full use of various forms, such as class meeting, blackboard paper and briefings, to discuss the problem of the atmosphere's construction, and propaganda good people and good deeds. Carrying out an activity with the theme of setting up "class with good learning atmosphere" in the groups of students, create a good environment of "everyone studies, every class has good learning atmosphere", then drive the deepening promotion of the construction and gradually eliminate bad habits relatively affecting the building.[2]

Deepen teaching reform and stress the teaching style to promote the learning atmosphere. According to the requirements of quality education, we should actively promote curriculum system, teaching content, the reform of teaching method and improve the attraction of classroom teaching. We should also deeply carry out the education of "ethics of teachers", and encourage them to take an active part in teaching reform, track frontier disciplines, update their knowledge structure, and improve the teaching level. If a teacher wants to teach well, besides being an exemplary virtue for the students, he or she must become a leading role in the construction, combine the teaching contents and characteristics of this course and introduce students to the latest research trends, achievements and research methods. They also should introduce the function of the knowledge in this course which can improve students' personal knowledge structure and quality development in their career life in the future.

Strengthen the management of class, stress class style and build learning atmosphere. Instructors and teachers should strengthen the guidance of the work of class atmosphere, and also give full play to the class committees, youth corps, student cadres, student party members and active students for entering the party organization in the atmosphere construction. They should establish perfect class rules and regulations; strengthen the management of classrooms, dormitories and study lounges; increase inspecting efforts of the students' attending class, self-study, and morning exercise. Besides, they can organize abundant and colorful, sound and beneficial class activities, create good class atmosphere with the spirit of seeking knowledge, studying lively and solidarity and striving to be the best.

Make the examination disciplines more severe, stress and correct the examination atmosphere. Positively reform the examination contents and methods, emphasize inspection on the student's innovative spirit and practice ability. Strengthen the standardization and scientifically management of examination, improve examination quality, make the examination discipline more severe, and make the rating criteria stricter. Resolutely stop cheating in the exams, and severely deal with the students who violate the examination rules according to related regulations.

Strengthen the reform of students' rewards and punishment system, and take the models to construct leaning atmosphere. Further strengthen the reform of students' rewards and punishment system, implement two levels of rewards—one of the university, and one of the colleges, and make learning atmosphere as an important index when the students are evaluated. Praise the collectives and individuals if they act outstanding and carry out comparison activity of fighting for the honorary titles of "top ten classes in constructing learning atmosphere", "top ten learning pace-setter" to set up models.[3]

Give full play to our college's features—most students have computers—to establish students' work website and BBS for the construction of learning atmosphere, then guide students to develop big discussion on the construction, exchange learning experiences and solve professional doubts online, make forum between teachers and students, etc, lastly   make the construction work swiftly into the students.

Hire experienced teachers to hold the post of "supervision expert of the construction on learning atmosphere", and select students acting well in every aspects to assume the office of "informant of the construction on learning atmosphere" (study committee could also take this job) . Establish a perfect feedback system of the construction's information, and more effectively to inspect and supervise the construction.

Summarize and honoring at the end of the year, to praise the classes, dormitories and individuals which act outstanding in the work of the construction of leaning atmosphere, deepen and consolidate the achievements of the construction.

4   Key Operations

Carry out education of learning atmosphere separately in different grades. The psychological condition, thought condition and learning situation of the students differentiate from the first year to the fourth year, thus the education should be aimed at the differences of their psychology, thought and learning features with main points.

The freshman year is the transformation and adapting period from middle school to university of the students. In this phase, the students' autonomous learning desire is intense, the learning attitude is earnest, but they don't deeply understand professional knowledge or master flexible skills and methods of university's study. They are also lack of self-manipulating ability and have inadequate mental preparation for the difficulties in study. In this phase, we should lay particular emphasis on the education of following aspects:

Education on professional confidence: invite expert professors to introduce the function and market demands in the development of social economics of the major to the students, thereby enhancing students' social responsibility consciousness, removing the doubts in their thoughts, setting up the students' confidence of learning the major well and strengthen its professional thoughts.

Education on the importance of basic curriculums: invite famous teachers to explain the importance of basic curriculums on the study of professional curriculums and further education in postgraduate phase. Strengthen the initiative and autonomy of the students in learning basic curriculums.

Education on learning methods and skills: invite excellent senior students to exchange experiences on learning methods and skills with freshmen, and help them adapt university's learning life as soon as possible.

The second year is the period for students to stably develop and gradually shape themselves. In this phase, they basically adapt to university life and master the study links. Their stress on study is eased but learning attitudes appear differentiations—students who are enthusiastic on the study are earnest, but some may relax themselves. Several students' grades tend to decline. In this phase, we should lay particular emphasis on the education of the following aspects:

Through introducing the more and more severe employment situation of recent years, we should point out the importance of knowledge when finding a job to cause the classmates' attention and set up the crisis awareness.

Hold exchanging meeting of English learning experience and improve the learning interests and skills.

In the class, establish groups of various interests, actively develop all kinds of learning competitions and create a good learning atmosphere.

Students in the third year have their own active thoughts and certain independent ability to analyze problems. They highly accept all kinds of education and influence, and their self-care ability and self-discipline ability are improved obviously. But parts of students diminish the collective concept gradually, so the centripetal force of class begins to abate. Part of the students who want to continue postgraduate study is trying hard, but a small part of the students relax the study of major course. Several students' disgusted emotion to study is increasing. In this phase, we should mainly focuses on the following aspects:

Guiding education for pursuing a higher degree: invite outstanding postgraduates to analyze social economic development trend and introduce more and more severe employment situation in forms of report, discussion and exchanging meeting; then tell the students to see the success of getting a postgraduate enrollment as their goals; inspire more students to take part in the examination of postgraduates' enrollment; improve the participating rate and the acceptance rate; thus facilitate the positive development of the university's learning atmosphere and form good learning atmosphere.

Education on collectivism: strengthen class contribution, cultivate collective consciousness, and improve students' collective concept with taking the dormitory as center.

The fourth year is the time of a psychological preparation for the students to move towards society. Students can be psychologically mature in this phase and concern personal prospective extremely. They abandon collective concept and independently act more. So we should lay special emphasis on the education of following aspects:

Guide students to be patriotic and love university. Actively find a job is both the demand of society and a performance of loving country and school. We should strengthen their ideology on employment's significance, guide the students to correctly understand the contradiction between ideal and reality, and set up the correct employment conception--first obtain a job; second choose a job; and third create a career.

The cultivating of social responsibility: extensively carry out the education on concerning the society and others and abiding the professional ethics education to make the graduate smoothly move towards on their jobs.

Civil education on leaving university: through developing colorful activities for school leavers, we should guide graduates to form host consciousness, maintain the campus orders, and walk towards society with full spirit and good image.[4]

Conduct learning atmosphere education in different stratifies. In order to enhance the effectiveness and efficiency of education on learning atmosphere, we should conduct the education in different stratifies and continue to highly care about students who have difficulties in study, in economy, and in psychology, carry out our work aiming at students' characteristics on study, life and psychology.

For students who have learning difficulties, we should organize the older students and excellent ones to coach them, launch activities such as "one helps another", "one helps several others" and "several students help one", take effective measures to help them to establish learning confidence and improve the learning results.

For those who have economic difficulties, we should specially pay attention to their life and study. We should also make full use of national student loan and work-study approaches, mine the available resources to solve their practical difficulties and lastly let them devote themselves into study.

For the students who are overwhelmed by pressure of economy, study, and life, we should pay special attention on them, conscientiously master basic situation, and let them devote themselves into study to full efforts through psychological counseling, heart-to-heart talk to release their psychological pressure.

5   Responsibility Implementation

In order to ensure the realization of the goal of the construction, we must strengthen management, implement the responsibility system, enhance the inspection supervision and trace the person with principal responsibility if he or she doesn't have earnest attitude, work efficiently, take measures effectively, or giving rise to bad impact.

The Leaders should inspect and supervise the teacher's teaching contents, teaching process and teaching effect, executing checking classes by listening classes.

Based on the ideas of being responsible for the students, the teacher should teach and learn rigorously and earnestly, avoid accidents and well play the part in "preach, impart knowledge and solve problem".

Instructors and teachers should actively guide students, strengthen the class management and care for students' life. They should also conscientiously carry out students' ideological education work, try to cultivate students' innovative spirit, and attract the students to professional study and comprehensive quality training.

The class cadres must enhance awareness of responsibility, seriously pay special attention to the daily classroom administration work, strengthen the checking system of learning and actively organize students to participate in the construction activities on learning atmosphere.

The Youth Committee and Students' Union should propaganda the construction well, earnestly organize the activities in each class, increase the checking system of classroom attendance and atmosphere in students' dormitories including sanitary inspection, and promptly inform the situation of atmosphere construction activities.

The construction supervisor and informant are responsible for the inspection and supervision on the situation of the construction and promptly inform whenever they discover a problem.

6   Conclusion

As an important aspect of the university's construction, the construction of learning atmosphere directly affects students' knowledge and moral quality, affects the talents' quality cultivated by the university, and lastly affects the school competition in the market economy. Therefore, the strengthening of the construction of learning atmosphere is not only an eternal theme in the reform and development of higher education, but also a strategic talents project.

References

1. Sheng, D.-S.: Construction of Style of Study in Colleges and Universities. Journal of Xiaogan University 2(25), 94–98 (2005)
2. He, J.-M., Huang, J., Jiang, X.-Y.: On the Construction of study style in Colleges and Universities in the New Situation. Journal of Chongqing Institute of Technology 3(17), 98–100 (2003)
3. Yang, C.-Q.: Thinking of University's Student's Community Building and Construction of Style of Study. Journal of Xiangtan Normal College 1(27), 35–37 (2005)
4. Cheng, Y.-G., Shu, T., Han, Y.: Thinking of Strengthening and Improving Construction of Study Style of University. Journal of Nanchang University(Social Science) 3(36), 163–168 (2005)

The Analysis of Mental Health Education among Impoverished College Students

Xiao Yue

School of Economics, Tianjin Polytechnic University, Tianjin, China
494092703@qq.com

Abstract. The impoverished students is a special group, whose mental development is drawing social attention increasingly. The research combining economic aid with mental health education from the aspects of the status on the part of mental health among impoverished college students and relevant educational measures conduces to the perfection of management strategies and financing system.

Keywords: Impoverished college students, mental health, measures.

1 Introduction

As China's economy and development of higher education, there is a special group - impoverished students arising among colleges and universities in recent years. According to the statistics from Soong Ching Ling Foundation in 2006, there is about 700 million impoverished students in China, accounting for 20% of the students at school, the most needy students ratio is about 8% -8.5%.

In fact, the problem of poor students is not merely family problem, not just an educational issue, but a widespread social problem that requires attention of the whole society. Poor College Students suffering from economic pressures result in greater psychological pressure than the other students, therefore relatively more prone to psychological problems. Thus, the understanding and mastering of impoverished college students on the part of the psychological and personality development has become an essential part of the management of university education.

2 The Significance of Mental Health Education among the Impoverished College Students

With the worsening mental health status after larger enrollment, the Ministry of Education issued in 2001 a "Circular on Strengthening Mental Health Education Among Colleges and Universities " which required that local education departments and universities should treat the work of strengthening mental health education among college students as the work of further strengthening and improving moral education in universities, and promote quality education with favorable measures.

J. Luo (Ed.): Soft Computing in Information Communication Technology, AISC 158, pp. 397–402.
springerlink.com   © Springer-Verlag Berlin Heidelberg 2012

To help alleviate the economic pressure on poor students, our country continuously improves the student aid system, including a series of positive measures such as scholarships, grants, grants for the poor, tuition remission and free social assistance, but compared with the large groups of poor students, these are still is a drop in the bucket, only a small part of the mitigation of economic pressure of poor students, on the other hand, the problems of the privacy of poor students, depending on the social and psychological pressure generated by the funding occur countless. Therefore, the essence of university education requires us to improve the process of national student aid system, not only "give him fish", but also to "teach him how to catch fish." While helping poor students to solve their economic difficulties, universities should carry on mental health education among them, form a good social personality and psychological quality, and enhance their capability to adapt to the environment in times of frustration with a positive attitude to overcome the difficulties and complete their studies successfully.

3   The Status of Mental Health among Impoverished College Students

During the period of life and study in universities, the impoverished students are under greater economic pressure, the psychological shadow is also hanging over them, the psychological "poverty" caused by great economic burden directly affects their quality of life and academic progress.

3.1   Inferiority Complex

Low self-esteem is one's own dissatisfaction, contempt and other negative feelings; it is a extremely strong psychological experience of the pros and cons of individual honor and shame. Because of family poverty and so caused low self-esteem, the impoverished college students have a strong requirement of self-esteem, their heart is extremely sensitive, so they prone to strong emotional fluctuations about the matters concerning themselves. In addition, their tuitions are relied on the borrowed money from their parents or the loans which will be repaid by their future salaries, thus in face of employment pressure, they may be worried about or have no confidence in the future of their own, the performance of their external depression tends to be depressed, and their conversation and behavior are of instability.

3.2   Tension Anxiety

Students from poor families often carry more and higher expectations of the future from families and individuals, they hope to completely change the fate of themselves and their families through strenuous study. So some of them set high targets for themselves. However, in real life, the too high expectations and unrealistic goals potentially increase the psychological pressure on himself on the one hand; on the other hand, many practical problems are difficult to resolve by their own efforts. These factors can lead to physical and mental fatigue of the impoverished college students.

3.3 Mentality of Loneliness and Sensitivity

Loneliness refers to a subjective experience of no understanding of emotions and thoughts from others' knowing and respect, expressed as withdrawal and afraid of contact, self pity, and closed mind, the psychologists define it as "psychological closure" phenomenon. Poor students have a strong desire to integrate into the community, eager to gain other people's acceptance and recognition, because of great inferiority complex, too strong self-esteem, over sensitivity, they neither want others to know their difficulties, nor want to accept the sympathy and help of others, in collective activities, they prone to obedience and default behavior, and their self-confidence would be dampened over time, the so caused psychological contradictions of poor students bring difficulties of various degree in interpersonal communication and community life, obstacles in their relationships, therefore it is easy to fall into the position of loneliness over time.

3.4 Extreme Psychology

Some poor students are vulnerable to "harmful" emotional and seek for extremes due to sensitivity, low self-esteem and anxiety, their attitudes on many things are with stereotypes and prejudices, absolute one-sided tendency, lack of a rational opinion and objective criteria and the recognition into a dead end, they often look at the negative aspects of society with a magnifying glass, and bear enormous hatred to the negative phenomena of the society. They like to violently attack against social evils, see few good sides on the work of class leaders and teachers in school and see more inadequacy and shortcomings, they often engage in recklessness and disregard of consequences.

3.5 Lacking of Strength in Study and Mentality of Tiredness

The impoverished students often worry about tuition fees as well as that of food and clothing so that they can not concentrate on their studies. Some students are absent from class and engage in off-campus work and other business ignoring the school rules at random, which as a result directly affects their studies. There are still some poor students who can not face their own poverty, claiming that poverty is a shame. Therefore, they are buried in their own pain, lose hope in life, and lose interest and motivation in learning, finally they are lack of strength in study and even tired of learning.

3.6 Mentality of Belittling Oneself in Capability

They hope to demonstrate their individual abilities, but they are neither able to participate too much in various activities because of the constraints of the economic situation, nor do they want to over-express themselves, so that they suppress their own potential. The feeling of belittling themselves and bitterness overwhelm the impoverished students as they are inferior to others in language expression, social interaction and adaptability.

4   The Measures of Improving Social Personality and Psychological Development among Impoverished College Students

The various manifestations of psychological problems among poor students reflect both the "material poverty" and the "psychological poverty". While solving a variety of real-life problems of poor students, more emphasis should be laid on the "psychological poverty". The study period in universities is just the same period of gradually forming outlook on life, values, and the world, thus, the ideological political work, economic assistance and psychological aids should work hand in hand with reference to the combination of practical life, study and mentality of the impoverished college students.

4.1   Ideological Political Education

According to the specific conditions of poor students, while helping poor students to solve their economic difficulties, universities should help these students to establish a correct outlook on life, values, and the world and lofty ideal with the same sense of responsibility, also encourage them in times of hardship, never bow to adversity, face up to difficulties, be proactive and foster their will of ceaseless self-improvement, and improve their psychological endurance. Furthermore, let the poor students recognize the limited financial resources in our country, the joint efforts of individuals and their families are needed to solve their economic difficulties instead of blindly relying on others' help. Through the ideological political education, we strive to achieve the goal that every poor college students can take the initiative to bear psychological pressure caused by economic distress and avoid the occurrence of mental health problems.

4.2   The Establishment of the Correct Guidance of Public Opinion

The power of role models is endless. Universities should notice the selection of the outstanding representatives of poor students to set a good example. By various means, universities should disseminate a good example of self-sufficiency, establish the incentive programs for poor students in recognition of advanced individuals in work and study, therefore create a "grown up poor excellent students more glorious "atmosphere, so that poor students who get inspiration from role models, could healthy face the challenges of life with positive and healthy attitude.

4.3   The Construction of Harmonious Atmosphere on Campus

A harmonious cultural environment on campus is also fertile ground for growth for each student. Campus should be a n inclusive pure land promoting equality and harmony. Universities should strengthen the construction of educational environment such as campus, class and dormitories, give full play to the importance of students' union and associations, and actively carry out psychological mutual activities, promote mutual care, sincere acceptance and mutual growth between the students, create a favorable environment for growth through the solidarity between students, caring about and helping each other, so that poor students can get help on economy and life, more importantly, get psychological support, and ultimately achieve the goal of solving psychological problems of poor students and cultivating a sound personality. Strongly propose the merit of thrift and oppose extravagance and waste, create hard working and positive campus atmosphere. A "compete for learning, progress and creation" environment is more favorable than a "competing for eating, wearing and enjoying" environment in terms of healthy growth on the part of students.

4.4   A Good Work of Mental Health Education

1） Set up mental lectures focusing on education of health, personality and emotional regulation. Mental lectures should help update the impoverished students' concepts of health, personality and quality, help poor students to form a correct understanding of poverty and healthy mentality, foster their healthy personality, enhance their sense of participation, self-esteem, self respect, self-improvement, and emotional regulation.

Personality education. Personality education reflects a person's soul and spirit. During the education of poor students, we should give them assistance and care in their daily life, and encouragement and aid in their study according to their characteristics instead of bearing any prejudice. Adhere to the principle of "providing aids for will before poverty", give more attention to their psychological world in addition to material help. At the same time, counselors should respect the privacy of poor students, adopt appropriate ways of coping with issues concerning poor students, so that they could live in a relatively natural, relaxed environment and develop comprehensively and healthily.

Mental education in terms of venting. Living in the "stress", the impoverished students are prone to feel stress, anxiety, and other unpleasant emotional response which leads to various psychological maladjustment and psychological disorders. Counselors should give poor students timely psychological counseling, open their heart through talking, discussing and providing advice, help them freely share their trouble, boredom, depression in order to vent the unpleasant emotions; deliberately give guidance and help them pour out bitterness deep in heart to their close friends through the establishment of friendship, and finally relax tension.

2） Establish counseling group for the growth of the impoverished students. The economic situation of members of the group should be roughly equal. The group members share the same feelings with mutual respect, sincere acceptation, open their hearts to talk freely, and invite poor students to share their own experience. Therefore, treat poverty correctly by mutual encouragement and a new cognition of oneself, recalled the experience of growing up and reconstruct objectives of growth.

3） Carry on individual counseling. In the individual counseling, advisory teachers offer the impoverished students understanding, support and comfort through the basic insisting steps and methods of emotional guidance, cognitive restructuring and behavioral guidance, analyze causes and consequences of helplessness, confusion and low self-esteem, help poor students face bravely environment and pressure, learn self-acceptance and self-appreciation to support their mental world and spiritual world, strengthen the subjective consciousness of self-esteem, self-reliance, self-confidence, and actively pursue   knowledge and self development.

5 Conclusion

The ideological and political education of the impoverished college students is in essence on the basis of the living conditions of the poor students to adhere to education as a precursor,   as a goal, to encourage poor students to correctly understand and deal with poverty, to improve the comprehensive quality of poor students, to establish poor students' courage and self-belief in face of life, to help poor students fundamentally get rid of poverty.

References

1. Zhang, Y., Fang, R., Miao, X., Huang, J.: The Mental Health Survey of 760 Newly Enrolled Impoverished Students
2. Jia, H.: Psychological Health problems and Countermeasures of the Impoverished College Students. The Journal of Shangdong Youth Management Cadre Institute (1), 51–52 (2005)
3. Zhang, Y.: Focus on Psychology – the Subject Not Being Ignored in Ideological and Political Education
4. Li, Y.: The Research of Mental Health Status and Measures of the Impoverished Students. Chinese Health Education (9), 566–567 (2002)
5. He, F.: Mind Analysis and Support of the Impoverished College Students. Explorations (December 2006)

Research in Garment Cutting Plan Optimization System

Xiaoyun Wang[1], Zhenzhen Huang[1], Manliang Qiu[2], and Fenglin Zhang[2]

[1] School of Art and Clothing, Tianjin Polytechnic University, Tianjin 300160, China
[2] Inter-China 3502 business wear Co., Ltd., Shijiazhuang, China

Abstract. In clothing customization manufacture, cutting schemes are worked out mostly based on manual experience. it needs lots of time and workforce, especially for various orders in business wear customization industry. In order to solve the overwork problem, in the paper, cutting optimization system is developed based on theory method and thinking. It can generate cutting schemes automatically or interactively by using Visual Basic language combined with the greed algorithm and select sorting algorithm. After generating cutting schemes and marker making, according to the actual marker information, it can distribute the cutting task with the assignment problem algorithm. The software's application would reduce the cost and the sub-bed error, improve efficiency and optimize management. Finally, it promotes the information development of our garment industry.

Keywords: cutting scheme, optimization, business wear, greed algorithm, assignment problem.

1 Introduction

In the actual production of business wear customization, the workers need to work out a reasonable cutting scheme which should be material-saving and convenient for cutting process. The cutting scheme is different according to the various orders. It needs rich experience and records in the paper. So it is a tired, low efficiency work and has no entire database. For most clothing customization industry, it is anxious to use information technology to work out the cutting schemes so that the industry can improve profit and efficiency. Based on the present development in this respect, this thesis develops the cutting optimization system which includes two functions: the cutting scheme optimization and the cutting task allocation. The cutting scheme optimization is in terms of theoretic method and thinking, assume that the production is a cutting scheme at first, and then do the limiting optimization with limited production condition and production task. It not only meets the industry's requirement but also advances the information development of China garment.

2 Factor Analysis for Cutting Schemes

2.1 Conventional Problems

For garment customization corporations, there are various orders and clothing styles; besides, the quantities of size in each order and productions are different. So it must

J. Luo (Ed.): Soft Computing in Information Communication Technology, AISC 158, pp. 403–409.
springerlink.com

combine the sizes in one order or combine the orders to work out a cutting scheme in order to save fabric and pursue profit-maximizing. As seen from the process, the cutting scheme is a key in manufacture. In order to solve the problems fully, we research and analysis relative references about the cutting scheme and the practice in industry. The conventional problems in cutting scheme can be summed up as follows: the improper size combination, significant difference of size and quantities, the time of receiving orders, delivery and incoming material, the accuracy of ration, the remaining size, workers' experience, specialization and cooperation in each department, the fabric breath, the chromatic aberration and the cutting power loss and so on [1].

2.2   Analysis in Practice Rresearch

We have a fully research in a clothing customization industry, the main effective factor analysis is shown in fishbone diagrams in figure 1. The cutting scheme optimization system can distribute the cutting task in the base of optimization results which is generated according to size and quantities in the orders [2].

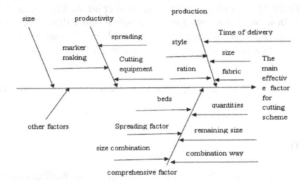

Fig. 1. Effective factory analysis for cutting schemes

The products of each enterprise were various, so the influence of each factor on the production is different, comprehensive analysis is needed, meanwhile the actual production model of the clothing industry should be considered. [3] The optimization of cutting task distribution is mainly concentrated on the full use of factory resources and achieving maximum benefits. Based on the fabric cost which can be got from the layout, the efficiency of the cutting equipment can be analyzed, by using assigning method, the distribution of the cutting task can be realized. [4]

3   Algorithm Analysis

By comprehensive analysis in demanding above, we can see that a cutting scheme is generated mainly by the combination of three factors, that is, orders, size in the order and quantities of each size. In order to fulfill the ration (the limit fabric consume of each size), it usually needs to put different sizes in one segment, at the same time, it must consider the production quantities of sizes in one segment, thereby it can determine how many each size in the segment puts. By this way, we can get a cutting scheme until all sizes in the order are cut. Table 1 is a sketch map about the cutting scheme.

Table 1. Cutting Scheme

bedNo.	layers	beds	Total combined pieces	size	Pieces of each size in one segment	Pieces of each size in a bed
P_j	CP_j	U_j	V_j	a_1	T_1	W_1
				...	...	...
				a_ω	T_ω	W_ω

With the mathematic method, computer application technique, the practice experience and relative reference research, we use the following algorithm to optimize the cutting scheme and distribute cutting tasks.

3.1   Optimization of Cutting Schemes

Greedy algorithm and selection sort algorithm were adopted to optimize the cutting scheme. Selection sort algorithm is used to sort the quantities of sizes because sizes which have large production quantities in one order are produced first. So we need to combine those size first in cutting scheme. The most important point in optimization is the determination of the limit combined sizes' pieces and the limit layers. So greedy algorithm is used which means that the sizes' combination first meets the limit situation to generate cutting schemes, and then, if it can not meet the limit situation, it must generate a local optimization scheme. The character of the greedy algorithm is an overall optimal solution, it can be realized through series of local optimal choice. By using top-to-down method the main problem can be simplified to smaller sub-problems. When the optimal solution of a problem includes the sub-problems' optimal solution, it may be said that the problem has optimal substructure properties.

Based on the actual condition , the limit spreading layers and the limit combined sizes' pieces are selected as the optimization goal. In the case the remain production quantities of sizes and the sizes' number is less than the limit situation, that is, the limit layers and the limit combined sizes' pieces, the optimization is that the combined pieces and layers should be as many as possible while the beds should be as little as possible. Finally, the results should be the best choice in current situation.

The limit combined sizes' pieces can be reached by calculating reasonable cutting length, fabric width, quota and the production quantities. The limit layers also can be got by calculating monolayer fabric thickness and reasonable limit cutting thickness. According to the set limit condition, the optimized scheme can be generated after comprehensively analyzing the sizes and their production quantities in the order. Different methods can be applied to deal with different situations. Here, we can list the different conditions.

Where, t is the amount of size, m is the limit combined sizes' pieces, c is the limit spreading layer, Qi the production quantities of the ith size ($1 \le i \le t$, $t \ge 1$, $m \ge 1$, $c \ge 1$). The conditions are as follows:

In the condition of $m > 1$ and $t \ge m$.

$Q_i \geq c$ when $i = m$; $Q_1 \geq c$ and $Q_m < c$ when $i = m$ and $i = 1$; $Q_i < c$ when $i = 1$

In the condition of $1 < t < m$.

$Q_i \geq c$ when $i = t$; $Q_1 \geq c$ and $Q_t < c$ when $i = t$ and $i = 1$; $Q_i < c$ when $i = 1$

In the condition of $t = 1$ and $i = t$.

$Q_t \leq c$ (2)$Q_t > c$

In the condition of $m = 1$ and $i = 1$.

$Q_i \geq c$ (2)$Q_i < c$

3.2   Optimization of Cutting Task Allocation

Cutting task allocation belongs to assignment problem, so allocation optimization method can be used. Here, each bedNo is a cutting task and different cutting tasks have different time consume in different cutting equipments. According to the actual length of each segment in the cutting scheme, we can get the actual fabric consume. So we can allocate cutting tasks to those cutting equipments by applying the allocation optimization method. Thereby, the industry can make full use of the production resource and realize the most profit and the least work time consume.

Q is the total times of fabric changes, and then the cutting task would be: Q_{1-i}, Q_{i-j}, ..., Q_{j-n}. M is the amount of cutting equipment and N is the amount of manpower cutting. when $Q = M + N$, the situation is shown in Table 2.

Table 2. Cutting Task Allocation

Cutting Task	Amount of Cutting Equipment	
	M_1, M_2, M_i, M_n	N_1, N_2, N_i, N_n
Q_{1-i}	C_{1i} , C_{2i}, C_{ii} , C_{mi}	R_{1i} , R_{2i}, R_{ii}, R_{ni}
Q_{i-j}	C_{1j} , C_{2j} , C_{ij} , C_{mj}	R_{1j}, R_{2j}, R_{ij}, R_{nj}
...	...	...
Q_{j-n+m}	C_{1m+n} C_{2m+n} C_{im+n} C_{mm+n}	R_{1m+n}, R_{2m+n}, R_{im+n}, R_{nm+n}

C and R were the time consume used for mechanical and manpower respectively. Introduce the variable X_{ij} , and then the mathematical model for tasks Q finished by all production facility in the shortest time can be established. Let i, j = 1,2, ..., n.

If the ith facility have completed the jth task, then $X_{ij} = 1$, otherwise $X_{ij} = 0$, and the time consume is $C_{ij} \geq 0$. In order to realize the least time consuming, the following model can be established:

$$(AP)\begin{cases} M \inf = \sum_{i=1}^{n}\sum_{j=1}^{n} c_{ij}x_{ij} \\ s.t. \sum_{i=1}^{n} x_{ij} = 1, j = 1,2,......\tilde{n} \\ x_{ij} = 0 \, or \, 1 \\ \sum_{j=1}^{n} x_{ij} = 1, i = 1,2,......n \end{cases} \quad (1)$$

Coefficient matrix can be established and solution may be got by optimizing the related matrix.

$$\begin{bmatrix} C_{1i} & C_{ii} & C_{mi} & R_{ii} & R_{ni} \\ C_{1j} & C_{2j} & C_{mj} & R_{ii} & R_{ni} \\ \vdots & \vdots & \vdots & \vdots & \vdots \\ \vdots & \vdots & \vdots & \vdots & \vdots \\ C_{1m+n} & C_{2m+n} & C_{mm+n} & R_{im+n} & R_{nm+n} \end{bmatrix} \quad (2)$$

When Q> M + N, there is a hypothetical that Q - (M + N) pieces of equipments should be needed to complete the tasks which are finished by M ~ N equipments in sequence. the least time in the actual time consume when each equipment finish different tasks is the hypothetical time consumed. A new matrix can be created and an optimal solution can be reached. When Q <M+N, the same hypothetical method can be adopted to deal with the problem.

3.3 Development of the Functions

The cutting optimization system can be divided into two parts: the cutting scheme optimization and the cutting task allocation. Each part has several function modules. We takes the cutting scheme optimization for example.

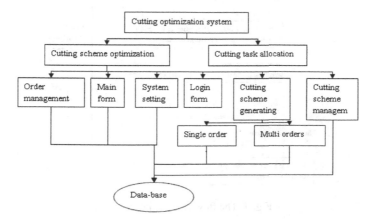

Fig. 2. The function modules

```
If m = 1 Then
    If Q(m - 1) >= c Then
        ReDim H(s, m - 1)
        H(s, m - 1) = a(m - 1)
        ReDim f(s, m - 1)
        ReDim W(s, m - 1)

        For i = 0 To m - 1               '确定各号型套排的件数
            f(s, i) = 1
        Next i

        b(s) = c                         '确定层数

        X = 0
        Do
            X = X + 1
            Y = Q(m - 1) - b(s) * f(s, m - 1) * X
        Loop While Y > b(s) * f(s, m - 1)

        u(s) = X                         '确定床数
        v(s) = 1                         '确定套排总件数

        For i = 0 To m - 1
            W(s, i) = f(s, i) * b(s) * u(s)
            Q(i) = Q(i) - W(s, i)
        Next i                           '确定套排数量和余数

        Call AD03_4ADD1(Q0, b0, u0, v0, H0, f0, W0, s, m)
        GoTo L13
    End If
```

Fig. 3. Partial program of cutting scheme generating

It has Login form module, the Main form module, Order management module, cutting scheme management module, System settings module and Cutting scheme generating module. The models are shown in Figure 2, the flow chart of process is shown in figure 4 and partial program of cutting scheme generating is shown in figure 3.

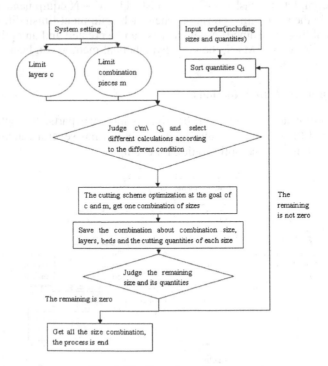

Fig. 4. The flow chart of process

4   Conclusion

The cutting optimization system has a tryout in the garment industry. It shows that materials are saved by 2%, working efficiency improves by 20%, the error descends to least. It proves the system's feasibility and practicability. It can improve work efficiency, optimize management mechanism and lower the cost. It not only has market value but also has significant scientific research meaning. This thesis offers a new theoretic idea to China garment industry's information development, at the same time, it reveals that the combination of theory and practice is important, the practice demands promotes the development of the technology.

Due to not enough time and the ability limit, this research on the garment cutting optimization is only in primary stage and it needs to study further and perfect in system's integrality.

Acknowledgment. Guiding projects of Science and Technology of Textile Association of China, subject No. 2009059.

References

1. Ling, C.: The Application of XML Technology in a Web-based Custom-tailor System
2. Xu, J., Zhang, W.: Theoretical study on optimal cutting plan of custom made garments 23(2), 24–27 (2004)
3. He, J.: The optimization method. Tsinghua University Press, BeiJing (2001)
4. Jiang, X.: The research of garment cutting scheme optimization system. SiChuan Silk (2), 41–44 (2000)

An Econometric Analysis of the Consumption Difference in the Urban Household in China

Fu Xiaohuan and Li Yiwen

College of Economic, Tianjin Polytechnic UniversityTianjin Polytechnic University,
TJPUTianjin, China
fu-xiaohuan@163.com

Abstract. Household consumption plays an important role in the socio-economic development, People's reasonable consumption patterns and moderate consumption size conducive to sustain economic health of the scale of growth. Since the 20th century, 90 years, China's consumer has changed greatly, affect consumer spending differences between regions are many factors, In order to analyze the differences in consumer spending and the quantitative relationship between the levels of consumption and the influence factors, choice is the use of econometric analysis the relationship between consumption and income of function model, with a view to reveal the main factors that affect the per capita consumption expenditure of urban residents in various regions of a significant difference in this stage and the corresponding policy recommendations.

Keywords: household consumption, household income, econometric model.

1 Introduction

Household consumption plays an important role in the socio-economic development. People's reasonable consumption patterns and moderate consumption size conducive to sustain economic health of the scale of growth, and this is the concrete embodiment of people's living standards. Since the reform and opening up, with China's rapid economic development, people's living standards are rising, the levels of consumption are growing too. But at the same time, we should also see the different pace of economic development and the significant differences in consumption levels in various regions. For example, the national per capita consumption expenditure of urban residents in 2009 are RMB12264.55 yuan, the lowest of Qinghai are RMB 8786.52 yuan, the highest of Shanghai are RMB 20992.35 yuan, which are 2.39 times as Qinghai. In order to study the consumption level of the whole country and the reasons why it changes, we need specific analysis. The factors that make household consumption expenditure differences significantly may be many. For example, income level, employment status, the retail price index, interest rates, the residents of property, shopping environment, etc. In order to analyze what are the most important factors that make household consumption expenditure significantly different and analyze the

J. Luo (Ed.): Soft Computing in Information Communication Technology, AISC 158, pp. 411–417.
springerlink.com

quantitative relationship between the levels of consumption and the influence factors, we should create the appropriate econometric model to study.

2 Selection of Variables

This study focuses on the differences in regional household consumption. Household consumption can be divided into urban and rural household consumption. Because of the big differences between the proportion of the population and the economic structures in urban and rural areas, the most directly comparable factor is the urban household consumption. Moreover, since the total population and the economy are different in various regions, we can only use "the annual per capita consumption expenditure of urban residents in various regions" to compare, the data variables of which can be obtained from the Statistical Yearbook. Therefore, we choose "the annual per capita consumption expenditure of urban residents in various regions" as the explained variable Y of the model. Because the purpose of the study is the regional differences in consumption of urban residents, not the changes in consumption of urban residents at different times, so the consumption expenditure of urban residents should be selected during the same period to build the model. Considerations based on the ease of the data collection, we take the cross-sectional data model in 2009 as an example. There are a variety of factors that affect the per capita consumption expenditure of urban residents in various regions to have significant differences, but from the theoretical and empirical analysis, the most important factor should be the people's income. Although other factors also have an impact on the household consumption, it is difficult to get some data, such as "resident property" and "shopping environment", some may be highly correlated with income, such as "employment" and "resident property", also some factors are not much differences between regions when using cross-sectional data, such as "Retail Price Index" and "rate". Therefore, these other factors can not be included in the model, even if they have some impact on household consumption, they can also be classified in the stochastic disturbance. In order to correspond to the annual per capita consumption expenditure of urban residents, we choose "the annual per capita disposable income of urban residents", which can be obtained from the Statistical Yearbook as the explanatory variable X.

3 Selection of Sample Data

The per capita consumption expenditure and disposable income of the urban residents in various regions of China in 2009.

Table 1.

Region	The annual per capita consumption expenditure of urban residents(yuan)Y	The annual per capita disposable income of urban residents(yuan)X
Beijing	17893.3	26738.48
Tianjin	14801.35	21402.01
Hebei	9678.75	14718.25
Shanxi	9355.1	13996.55
Inner Mongolia	12369.87	15849.19
Liaoning	12324.58	15761.38
Jilin	10914.44	14006.27
Heilongjiang	9629.6	12565.98
Shanghai	20992.35	28837.78
Jiang Su	13153	20551.72
Zhejiang	16683.48	24610.81
Anhui	10233.98	14085.74
Fujian	13450.57	19576.83
Jiangxi	9739.99	14021.54
Shandong	12012.73	17811.04
Henan	9566.99	14371.56
Hubei	10294.07	14367.48
Hunan	10828.23	15084.31
Guangdong	16857.5	21574.72
Guangxi	10352.38	15451.48
Hainan	10086.65	13750.85
Chongqing	12144.06	15748.67
Sichuan	10860.2	13839.4
Guizhou	9048.29	12862.53
Yunnan	10201.81	14423.93
Tibet	9034.31	13544.41
Shaanxi	10705.67	14128.76
Gansu	8890.79	11929.78
Qinghai	8786.52	12691.85
Ningxia	10280	14024.7
Xinjiang	9327.55	12257.52

The date from the Statistical Yearbook in 2010.

4   Model Selection and Parameter Estimation

4.1   The Construction of the Mode

According to the relationship between the annual per capita consumption expenditure of urban residents (Y) and the annual per capita disposable income of urban residents (X), We can make the scatter diagram between (Y) and (X).

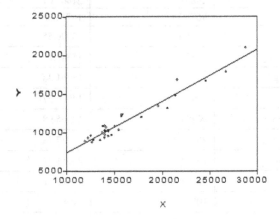

Fig. 1. The scatter diagram between (Y) and (X)

It can be seen from the scatter diagram that the annual per capita consumption expenditure of urban residents (Y) and the annual per capita disposable income of urban residents (X) are generally showed a linear relationship, so the econometric model of establishment is the following linear model:

$$Y_i = \beta_1 + \beta_2 X_i + u_i$$

4.2   Parameter Estimations

Assume that the model and the stochastic disturbance (u_i) satisfy the classical assumptions, we can use OLS method to estimate its parameters. Use computer software Eviews to make econometric analysis, it appears the regression results as in Table 2.

Table 2. The regression results

Dependent Variable: Y

Method: Least Squares

Date: 11/07/10   Time: 15:23

Sample: 1 31

Table 2. (*continued*)

Included observations: 31				
Variable	Coefficient	Std. Error	t-Statistic	Prob.
X	0.668059	0.030926	21.60193	0.0000
C	755.0114	520.3031	1.451099	0.1575

R-squared	0.941490	Mean dependent var	11628.97
Adjusted R-squared	0.939473	S.D. dependent var	2978.791
S.E. of regression	732.8515	Akaike info criterion	16.09410
Sum squared resid	15575067	Schwarz criterion	16.18662
Log likelihood	-247.4586	F-statistic	466.6435
Durbin-Watson stat	1.644234	Prob(F-statistic)	0.000000

Therefore, the estimation results as:

$$\hat{Y}_i = 755.0114 + 0.668059\, X_i$$

$$(1.451099)\ (21.60193)$$

$$R^2 = 0.941490\quad F = 466.6435$$

5   Check the Model

5.1   Economic Sense Test

The estimated parameters $\hat{\beta}_2 = 0.668059$, which is consistented with the absolute income hypothesis that the marginal propensity of consume is between 0 and 1 in the economic theory. We can conclude that a difference of 1 yuan in the annual per capita consumption expenditure of urban residents can lead to a difference of 0.668059 yuan in the annual per capita disposable income of urban residents.

5.2   The Measure of Goodness of Fit

As can be seen from the Table 2, coefficient of determination is 0.941490, indicating that the model fit the sample data on the whole good, It is to say that the explanatory variable "the annual per capita disposable income of urban residents" explains most of the differences of the explained variable "the annual per capita consumption expenditure of urban residents".

5.3   T-Test of Regression Coefficient

Suppose $H_0 : \beta_2 = 0$, Derived from the Table 2, The standard error and t value of the estimated regression coefficie $\hat{\beta}_2$ are: SE($\hat{\beta}_2$)=0.030926, t($\hat{\beta}_2$)=21.60193. Take α=0.05, check the t distribution table under the degrees of freedom as n-2=31-2=29, we can get the critical value t $^{0.025}$ (29)=2.045. Because t($\hat{\beta}_2$)=21.60193 > t $^{0.025}$ (29)=2.045, the suppose should be rejected. This shows that the annual per capita disposable income in urban have a significant impact on the annual per capita consumption expenditure.

6   Suggestions

According to the econometric model above, we can conclude that the annual per capita disposable income in urban have a significant impact on the annual per capita consumption expenditure. Therefore, the most important factor that affect the annual per capita consumption expenditure of urban residents in various regions of a significant difference is the income differences, household consumption plays an important role in the socio-economic sustainable development. In order to promote the development of China's economic rapidly and reasonably, according to the definition of the consumer behavior, we should take an active consumer policy to guide household consumption, form a reasonable income gap and improve the consumer propensity in order to facilitate the upgrade of consumption. Income convergence may easily lead to the concentration of low-level consumption, but if the income gap is widening and the low-income people are increasing, it would result in lower marginal propensity to society in general which is not conducive to the growth of consumption. Therefore, the income distribution system should be reformed speedily. The orientation of the reformation is to mobilize the enthusiasm of workers to create wealth, taking into account the interests of all social relations and social tolerance, and actually increase the real income of low-income people. On the one hand, distribution of national income should be adjusted, greatly improve the distribution of residents in the proportion of national income, expand corporate profit margins, help companies and residents tide over the difficulties. On the other hand, the government should open access to markets or through the split and reorganization of monopolies, introduce market competition mechanism, eliminate the institutional basis that monopoly revenue distort the relationship of the income distribution, at the same time, increase income of the low-income people, especially farmers, increase policies support and protection on agriculture, improve the proportion of the expenditure of public services in agriculture and rural area. and speed up construction of the new socialist countryside.

Acknowledgment. The authors would like to thank you for the great support and assistance obtained from teachers and some friends in the process of writing this paper. In addition, the authors extend sincere gratitude to the authors whose words are cited or quoted. Finally, the authors are profoundly grateful to the continuous support, understanding and selfless dedication of the family.

References

1. Carroll, C.D., Overland, J., Well, D.N.: Saving and Growth with Habit Formation. American Economic Review 90, 341–356 (2005)
2. Michael, K.Y., Zhu, L.: Financial liberalization and economic growth: a theoretial annlysis of the transforming. Pacific Economic Review, 125–158 (2005)
3. Zhang, J., Zhang, J.: The Effect of life Expectancy on Fertility, Saving, Schooling and Economic Growth: Theory and Evidence. Scand. J. of Economics, 45–66 (2005)
4. Yang, W.: The discusstion of househould consumption of china. Economic Theory and Economic Management, 76–80 (2005)
5. Rhodes, M.: Diversification efficiency and deposit rates. Applied Financial Economics, 935–945 (2005)
6. Zhou, Z.: A view to the relationship between savings and investment. Economic Forum, 18–19 (2004)

References

The Application of Simulated Experimental Teaching in International Trade Course

Ma Tao

School of Economics, Tianjin Polytechnic University, Tianjin, China
hebeimt@sohu.com

Abstract. International Trade Practice is a professional basic course for specialty of International Economy and Trade. As the core of International Trade Practice, it is extremely related to foreign affairs and needs much practical experience. This paper used methods of comparative and analyzed the current situation of teaching. Then the paper puts forward some suggestions on Simulated Experimental Teaching in order to educate the outstanding graduates, which are urgently demanded by the foreign trade enterprises.

Keywords: international trade course, simulated experimental teaching, practical experience.

1 Introduction

To create the professional brand of international trade, enhance the competitive strength, it's an inevitable choice for college to strengthen the teaching reform and focused on "High Quality, Sufficient Abilities", to improve the status of professional practice in the education system, so as to adapt the public's new requirement for the international professional. In order to adapt the new position of our entering the WTO and receive new challenge, speed up to cultivate high diathesis professionals with the ability of innovating, reining the market, and managing the international business affairs, who are shortage in the international trade, many domestic colleges adopt the simulated experimental didactics one after another. In theory, they do widely research; while in practice, they summarize a mass of experience. But meanwhile we should know that, the simulated experimental didactics has not been long-playing used in China. So we may have something to be ameliorated and ulterior improved in our education activity, which cause us to go on intensifying education reforms, so as to create conditions for the further application of simulated experimental didactics.

2 The Comparative Advantages of International Trade Simulated Experimental Didactics

The international commerce is not only a subject researching the specific process of international commodity exchange, but also a comprehensive and practical subject which is practice-needed and involving foreign interest. As a practical subject, just by learning teaching material to make students be "High quality, Sufficient Abilities "are never enough. We should also get help from simulated experimental teaching to reach the goal. Compared with traditional teaching method, the simulated experimental didactics undoubtedly have many obvious advantages.

Simulated Experimental Teaching is favorable for training the student's ability of integrating theory with practice, consolidating and deepening their understanding of theory. Theories originate from practice, and are the refinement, abstract and summary of practice. The penetrate of practice will in turn deepen our understanding of theory. In teaching, students get abstract theory, which is hard to make sense of and accept. For example, the teachers will take about half a period to explain the international contracts, which contain contract subject matter with its quality, quantity and packing, mode of transportation, shipment clause, transport documents, form of contracts and its basic components, performance of import and export contracts, main import and export document, claims and settlement of claims and so on. But for students, it is low efficient and difficult to grasp. However, if you use international trade simulated experimental didactics, students with the aid of adviser, fill the transport documents and main import and export document by their own hands and simulate signing import and export contracts. So we can bridge the gap between theory and practice. One more example, to explain the payment in the international trade, agreement and fulfill the Contract of International Goods Sales and international electronic commerce in e-commerce and so on in business of international trade settle accounts, will make the beginning students puzzled and at sea. Yet once you finish a period of simulated experiment and change the abstract theory into tangible data, you may get a thorough comprehension of international trade rudiments.

Simulated Experimental Teaching is favorable for making up the need of practice outside the college to protect their trade secret. Many of them refuse to accept students to practice. They worried that the interns may give out their trade secret. Or perhaps they settle for the interns for some relations, they won't arrange them to do pacific operation commonly. Instead, they may give them some documents to fill, some data to tabulate, or even ask them to do some laughing matter. Besides, because of expand admission, the schools' internship expenditure per student is decreasing year by year, while the cost of internship is increasing gradually. The only way is to practical work placement nearby and as simple as possible. Due to multifarious factors above, students' practicing outside the college is just a form and results in worse effect. The International Trade Simulated Experimental Teaching therefore plays a growing part in education system.

3   Optimization the Positioning and Content of International Trade Simulated Experimental Teaching

Whether the design of this system is scientific has a close bearing on the quality of simulated experimental teaching. It should has a specific aim, which is strengthening the students' understanding and grasp on booklore (the Terms of International Trade, Clause in the Contract for the International Sale of Goods, the Negotiation and Fulfillment of International Sale of Goods, Pattern of International Trade, etc.), setting up the concept of competition and have an all-around idea of what you have learn.

Getting some practice effects, lead the students to have perceptions of the requirement about international business and master the skill of international trade business work.

The international trade instructional system makes international trade business work as the central task. Analog simulating the whole procedure of international trade business. First, preparation work before trade. In order to ensure compliance of import and export contracts, we should keep good preparation before trade. It includes choosing suitable markets and trade partners, application import and export license, etc. Second, consultation of import/export contracts. Business consultation is a procedure that buyers and sellers consult on condition of merchandising, to reach an agreement. It's the basis of entering into a business contract. Whether the business consultation is good or not directly influences the award of contract and has respect to each others benefit. We should attach importance to it. Negotiation covers Name of Article, Quality, Packing, Price, Shipment, Insurance, Payment, commodity inspection, claim, arbitration, force majeure, etc. The procedures of business negotiation can be concluded into four steps, that is Inquiry, Offer, Counter-offer and Accept. We take the business process below into consideration in the education system design. Export company give an offer---Import company counter-offer to bargain---Reach an agreement---Enter into a contract. Third, signing contracts. In the international trade, we need to do business negotiation to conclude business, and the form of law is contract. Business negotiation is a process while award of contract is a result. Fourth, each negotiating party carries on the contract. After signing a contract, both parties should fulfill obligations as the contract tells. The export company's main obligation is delivering the goods, handing over all the relevant documents, and transferring the ownership of the goods to the import company. While that of import company is paying and receiving the goods.

Simulation experiment is good for the development of college and student, and it also cater for the need of first-rate international college in 21st century. It can not only change the adverse consequences of low efficiency and the drudge caused by the traditional teaching of the theory of international trade, which make the students realize the actual hands-operation, familiar with the specific trade procedures and operation process, but also enrich teaching model, enhance the enthusiasm and innovation of college research, expand the impact of college, enhance the school's reputation. So we should vigorously promote the application of simulation teaching in the international trade courses.

References

1. Liu, Y.: Analysis on Teaching Methods of International Trade Practice. Journal of Huazhong Agricultural University (Social Science Edition), 24–25 (January 2008)
2. Zhang, Q.: Probing into Reform of Teaching Methodology for International Ttrade Practices. Journal of Changchun University of Science and Technology (Higher Education Edition), 35–36 (January 2007)
3. Wang, W.-Y.: Analysis of the Teaching Reform of International Trade Business and Researches on the General Teaching Scheme. Journal of Anhui University of Technology (Sociel Sciences), 40–41 (January 2002)
4. Zhu, W.: Thinking and Practice of the International Business Simulation Laboratory. Laboratory Research and Exploration, 17–18 (February 2001)

Analysis on the Design of Intelligent Buildings Based on Low Carbon Ideas

Wang Jin

Institute of Art and Fashion, Tianjin Polytechnic University, Tianjin, China
double_tiger@126.com

Abstract. With the increasing concentration of carbon dioxide, cycling balance of carbon in nature is completely broken. "Low carbon concept" is not a popular vocabulary or slogan but something related to the future direction of all human beings. Intelligent buildings are not the same as energy-saving buildings. There are too many automation equipments for the purpose of maximum comfort and convenience which greatly increase the consumption of energy. Air-conditioning Variable flow technology and Optimal Management of supply and distribution system in the Building's automatic control system can maximize energy savings. This is the most effective solution to the existing buildings for energy saving modification. Designs of new intelligent buildings should take the use of new and forward-looking energy-saving technologies into full account. The principles of sustainable development and harmony of environment, architecture, humanities, science and technology should also be followed.

Keywords: Low Carbon, Intelligent building, Design, Building Automation System, Variable Flow.

1 Introduction

1.1 Low Carbon Concept

Carbon dioxide is not only the carrier and product of the natural activities but also a product of human production activities. Human being's activities like cultivating land, deforesting and burning wood will release carbon dioxide into the atmosphere. Long time before the industrial revolution, the concentration of carbon dioxide in the atmosphere is generally stable in the 270-290ppm which indicates that the influence of human activities to the Earth's atmosphere is very small. But 1800 years later, for the reason of rapid development of modern industry and transportation, steady increase of urbanization level and rapid increase of coal and oil consumption, the concentration of carbon dioxide in the atmosphere continues to increase, and the increase becomes faster and faster. Carbon's cycling balance in nature is completely broken and the earth begins to run a "fever".

After the climate conference in Copenhagen which with the theme "for tomorrow" and aims to control the "greenhouse effect" finished, "low carbon", "energy saving" have become Community-wide attention new keywords. Before this, our government has promised that China's energy consumption of GDP will fall to 80% of the original by 2010 and said that this figure will be even lower by 2020.We can see from this that

J. Luo (Ed.): Soft Computing in Information Communication Technology, AISC 158, pp. 423–430.
springerlink.com   © Springer-Verlag Berlin Heidelberg 2012

"Low carbon concept" is not a popular vocabulary or slogan but something related to the future direction of all human beings.

1.2   Intelligent Buildings

In the 21st century, the building construction industry can provide concrete buildings which no longer are ignorant and cold but replaced by a warm humanistic intelligent buildings. With the development of information technology, modern buildings have been given the thinking ability.

American Intelligent Building Institute (AIBI) defines the intelligent building as: by optimizing the four basic elements of the building, namely, structures, systems, services and management, and the inner link among them to provide a reasonable investment environment which also has high efficiency, comfort and convenience . The so-called intelligent building is to organically combine the high-tech (computers, multimedia, modern communications, smart security, environmental monitoring, etc.) with architecture and design and construct safe, comfortable, efficient, energy saving, convenient and flexible modern buildings. Generally speaking, the building has a 5A system can be called intelligent building. 5A systems include Office Automation System (OAS), Communications Automation System (CAS), Smart Security System (SAS), Fire Alarm System (FAS), and Building Automation System (BAS).

1.3   The Direction of Intelligent Building Design

Building automation system is an important part of intelligent building and its future direction is the refinement and classification of management area. Through a variety of sensors that automatically carries out Variables control technology to the change load of the end bases on the changes of external conditions and to find the appropriate point between comfort and energy efficiency.

An important feature of the information society is the digitization of information and the diversification of communication business. The types of communication services is increasing, such as videotext, data communication, video telephony, video conferencing and high-resolution transmission of still and moving images, even the comprehensive services digital network which integrates voice, data, images and other information transmission services will also become a reality.

The generation and development of office automation systems is not an improvement of individual office tool or the perfection of office ways, but a new way to solve problems and achieve office integration, procedure and intelligence in the support and assistance of intelligent tools.

Intelligent buildings are not the same as energy-saving buildings, some are opposite. Excessive automation equipment in the buildings for the purpose of seeking maximum comfort and convenience makes the energy consumption is much higher than ordinary buildings. Intelligent building design should follow the principles of sustainable development. Through the scientific overall design and integrated energy configuration to use natural ventilation and natural lighting properly. Maintaining the structure with low power consumption and using new energy such as the solar energy and intelligent control of high-tech to fully demonstrate humanities and architecture, environmental and technological harmony.

2   How to Reduce Energy Consumption through the Design of Automation and Control Systems

2.1   Analysis on the Factors That Affect Energy Consumption of Intelligent Building

Take modern buildings for example, the air conditioning and lighting system's energy consumption is greatly, accounting for 70% of total energy consumption of the building approximately, among which air conditioning, refrigeration and heating equipment account for 40%. In addition to air-conditioning systems, the lighting is also consumes energy largely, accounting for 30% of the total energy consumption. Therefore, making use of many kinds of control measures such as variable air volume, variable fresh air, return air ratio, variable water volume, variable air state point and variable control values to air conditioning system on the assumption of meeting the environmental requirements of users to run the air conditioning system in the best energy-saving conditions, which can minimize energy consumption. That would be more than 30% energy saving than the usual constant temperature and humidity control. Another optimal lighting control can save about 30%-50% electricity for lighting.

2.2   Building Automation System

Building Automation System (BAS) is a major system of intelligent building systems, also known as building automation control system. BA system is an important part of 5A intelligent building system which not only provides a comfortable and pleasant living environment but also saves a lot of energy. It uses modern sensor technology, computer technology and communication technology to control all electrical and mechanical facilities in the buildings automatically. The electrical and mechanical facilities are power distribution, water supply and drainage, air conditioning, transportation, fire protection, lighting, security and other systems which fully integrated monitoring and management automatically by computers. BA system adjusts temperature, humidity, fresh air automatically to meet environmental requirements, which can greatly improve efficiency. The development of intelligent artificial environmental technology will further settle the parameters and control of noise, color, odor, air and air components and so on, which will achieve the comprehensive objectives mentioned above with a higher standard.

2.3   Control Function of Air-Conditioning System

1) Automatic Control Mode of Central Air Conditioning Units

In HVAC systems, power consumption of refrigeration equipment and air conditioning units in the whole system occupy a major part of the energy consumption. In order to achieve optimal start and stop and the best operating condition, Building Automation System (BAS) adjusts the heat and cooling output and the run-time of units by calculating the load of the end. With the development of technology, manufacturers of refrigeration equipment and air conditioning unit equipment have used computer control technology commonly, that is, use the own control system to enhance the traditional analog control to the digital element control, dual-rate control to the intelligence control. In which the energy regulation of compressor develops from the traditional two-bit control into the no class unloading control, multiple linkage control, bypass control and variable speed control. The energy regulation is more precise. Load regulation of units develops from single variable supply air temperature control, return air temperature control, return water temperature control, condensation pressure control and evaporation temperature control into multi-variable input fuzzy control, which greatly improves the operating efficiency of units

2) Variable Flow Technology and its Automatic Control Mode

In the buildings, the energy consumption of HVAC system accounts for more than 40% of the total energy consumption of the whole building, in some areas even more than 60%. Most general designs of HVAC systems are designed for all load conditions, but in the actual use, the system is part of the load most of the time. So Variable flow system (VAV, VWV and VRV) is widely applied from the perspective of comfort and energy saving. While the normal operation and the operating characteristics in actual use of variable flow systems depend on the system control totally, for the purpose to achieve the effect of energy conservation in ensuring a comfortable environment.

a) Variable Air Volume System (VAV)

Variable air volume system is that maintaining the same inlet parameters when the room's heat and moisture load is less than the design value, and by reducing the air volume in order to keep the room temperature constant. Compared with fixed air conditioning systems, variable air volume system reduces the mount of the heat and the corresponding cold. Moreover, with the change of the room air amount, the total air volume of system changes accordingly, this can reduce the energy consumption of fan. In addition, according to the operational characteristics of variable air volume air-conditioning system, simultaneous occurrence of each room's load should be taking into consideration when the total load of air-conditioning system is calculated, which can also reduce the fan capacity appropriately. The control of variable air volume system can be divided into two parts, there are VAV control and the control of VAV air-conditioning unit. In addition to the precise design calculations, reasonable system layouts and good construction installations, to choose the best control method is also critical for a good variable air volume air-conditioning system. In the practical use of engineering, the ways adopted more often are Static Pressure Control Method, Variable Static Pressure Control Method, Direct Digital Control Method and Fan Total Air Volume Control Method.

Having a good flexibility, the VAV system can be changed and extended easily, especially for the building with flexible pattern, such as leasing office space and so on. When the indoor parameters change or re-partition, it may be needed to replace the branch pipe and terminal devices, move outlet position or even just to reset the indoor thermostat. The VAV system belongs to all-air system. It has some advantages of the all-air system, so it can eliminate indoor air load by using new wind, without any water condensation problems of fan coil and mold problems. However, the VAV system needs careful design and construction, commissioning and management, otherwise, there may arise a series of problems: the lack of fresh air, poor airflow, the negative or positive pressure of the room is too large and the noise is too large, the unstable running of the system, the unobvious energy-saving effect and so on. The principle of VAV air conditioning system is not complicated and the key of it is required to implement air terminal device (Terminal Box) of the principle of variable air volume, particularly to implement the related terminal units and the whole the automatic control equipments of the VAV system.

Typically, the VAV systems have two kinds: one is VAV air conditioning system (AHU-VAV system) in the AHU duct system; the other is the indoor fan variable air volume system (FCU-VAV system) in FCU the system. AHU-VAV system will supply constant air temperature in the whole duct system, and regulate the air supply blower means to cope with the changes in the indoor load of the air conditioning. FCU-VAV system, which regulates the cold water supply, installs stepless variable power controller in the indoor FCU to change air flow, that is, changes the heat exchange rate of the FCU to adjust the indoor load changes. These two methods reduce power consumption of blower through the adjustment of the air flow, which can also increase the efficiency of the machine then to save the power consumption of the heat source. Therefore, it can gain the efficiency of energy conservation both in the air and the heat energy at the same time.

b) Variable Water Volumes (VWV)
The so-called Variable Water Volume (Variable Water Volume, referred to as VWV), increases the efficiency of the heat source machine by using air conditionings which can supply certain water temperature and changes the water delivery by using special water pumps, so as to achieve the effect of saving the pump power. The Variable Water Volumes can diversify the energy conservation efficiency of pumping systems according to the pump control modes and use ratio of VWV. Generally, control modes of VWV are variable speed control mode (SP) and bidirectional valve control mode. The control of central air conditioning, that is, concentratly set the pipe, pipe, valve or valves, in order to control fluid to provide cold air. As a result, the effective combination control of central air-conditioning, so to speak, it is effective to limit the energy cost, which is a air conditioning system, whose aim designs to energy conservation.

c) Variable Refrigerant Flow Systems (VRV)
V RV air conditioning system's full name is (Variable Refrigerant volume) system, that is, variable refrigerant flow system. The system structure is similar to the split system air conditioning units, which uses a single outdoor unit corresponds to a set of indoor unit (usually up to 16). Application on control technology is variable frequency control method. The inside of the outdoor unit in control room turns on the scroll compressor speed according to the number of opening indoor unit thus to control the refrigerant flow. VRV air conditioning system compares with all air systems, all water systems, air and water systems, which is better to meet the operating requirements of individual users. Its equipment takes up relatively small building space and is more efficient in energy conservation. Just because of these characteristics, it is more suitable to the office construction projects for the people who need to overtime work frequently and independently.

3) Control of the Fan Energy Consumption
Energy consumption in HVAC systems, the fan energy consumption can account for about half. In the unconditional use of technology of variable air volume HVAC system, the full use of control technology and reasonable control the running time of

the fan, both are one of the effective energy saving means. The control of air-conditioning fan coil, mainly through the following ways:

- In summer, air-conditioning fan coil can be changed the temperature setting in public areas according to the principle of the outside to inside, successively lowering the setting temperature of public buildings.
- At night, it can be increased the temperature setting by lowering the room temperature control pointer at night to reduce energy consumption.
- It can be naturally cool, where it is possible, using the outside air anyway.
- The system optimization improve the control accuracy of the indoor temperature and humidity
- It use a special thermostat with remote control, through the determination of the difference between the environmental temperature and setting temperature, automatically control the state of two-way valve and fan speed.

2.4   Optimization Management of Supply and Distribution System and Control Modes

Energy-saving optimization management of supply and distribution system and control modes are mainly in two aspects: one is the lighting in public areas. The lighting systems in public areas always have a serious appearance of energy waste. The following three ways can be effectively put an end to this phenomenon: 1. drawing up the run schedule statement of lighting according to the season of the barrier. 2. Increasing lighting circuit design in public area, in the case of non-essential, especially at night, the lighting system runs only the minimum supply circuit power. 3. Photoelectric sensors control the lighting circuit. On the other hand, another one is management and control of electricity utilization in the office. Because of people's carelessness and neglection for the electricity utilization in the office, they forget to switch off the electricity equipments in the office after work which cause the waste of electric energy and lead to potential safety concerns. It is so common for people to work overtime in the office. In that case, it is effective that the districts increase the strength of monitor and control management. If there is no overtime work after the quitting time, remoting control can turn off the power thus to achieve conservation of the office energy through using technology and management ways.

3   "Low Carbon" Design Applications

The key implementation of energy-efficient intelligent buildings is the application of new environmental technology. While it applies energy saving modification for the old buildings, the new intelligent building design should take full account of the new forward-looking energy-saving technologies. Our university (Tianjin Polytechnic University) has achieved fruitful and highly effective practice. The construction of new campus makes use of ground source heat pump heating and air conditioning system, which is currently the most advanced green air-conditioning system in the world. The heating and cooling and sanitary hot water systems have a good effect. There are more than 200 meters loose settled layers within the plain area in Tianjin province, which have obvious advantages in shallow geothermal energy development, so this new application of technology has a realistic basis.

The ground source heat pump is an energy-efficient air-conditioning system which can supply both heating and cooling by using shallow earth resources (including soil, groundwater, surface water or city water). It uses the heat exchange pipes which are laying in the soil and surface water to achieve the heat transferring in air-conditioned rooms, the soil and surface water, for the purpose of air conditioning effect. The supply of heat in winter mainly extracts the heat of geothermal water from the water source heat pump, supplemented by peak shaving of gas boilers; in summer, the supply of cooling is mainly from the ground source heat pump, water source heat pump units add with outdoor cooling tower peak shaving. Our university has already dug 588 shallow wells in the new campus stadium, whose aim is to achieve hot and cold exchange of underground heat exchanger ground source heat pump system with soil: when there should be supplied with heat in winter, the heat in the ground needs to be "taken out" for the interior heating and it accumulates the cold from the underground at the same time; when the temperature drops in the summer, the heat from the indoor needs to be "taken out" to release to the surface and it stores the heat from underground, in order to prepare for winter heating.

Principally, domestic hot water comes from utilization of solar energy. The roofs of student apartments have installed 1,500 square meters of glass vacuum tube collector system, supplying non-potable water to canteen and shower water to students, as well as providing 24-hour live hot water for university students' activity center and international exchange center. When it is necessary, the university will assistant to offer hot water, heat pump and gas boiler.

In addition, our university has already utilized the LED semiconductor lighting materials technology of independent intellectual property rights. Our university has installed more than 1,500 LED street lights and more than 2,000 indoor lighting LED lamps in the new campus and installed more than 20 LED solar garden lights in the lake region as well. It is the world's largest semiconductor lighting demonstration project, which can save 1.3 million degrees electricity annually.

The energy saving design of the new campus in Tianjin Polytechnic University is not an individual case. Our country has been built up a group of public buildings and residential quarters which are environmentally intelligent types. In the "low carbon" concept, our country will continue to put the most advanced energy saving technology into the intelligent building design, taking the large-scale state investment construction projects as demonstrations to encourage the new energy-saving technologies promotion by using policies.

References

1. Long, W.: Energy Conservation and Energy Efficiency Management of Building. China Building Industry Press (2005)
2. Jia, H.: Human and the Building Environment. Beijing Industry University Press (January 2001)
3. Liu, G.: Building Automation System. Machinery Industry Press (2002)
4. Wu, L.: The Key Technology in Building Automation System. Building Energy Conservation (July 2010)

5. Hong, H.: Energy Saving Building, Green Building and Low Carbon Building. Building (April 2010)
6. Li, G.: Building Energy Conservation and Low Carbon Building. Science and Technology Information (July 2010)
7. Ou, Y.: Building Energy Conservation Is an Important Part of Promoting Sustainable Development. China Engineering Consulting (February 2005)

The Integration of Industry with Finance's International Comparison

Hu Qiangren

School of Economic & Management, Shanxi Normal University, Shanxi Normal University,
SXNU, Linfen, China
huqiangren@163.com

Abstract. The integration of industry with finance is one important way of the national economy development and which direct impact on one country's economic performance. This article describes two typical models, the market-based and bank-based; compare the economy efficiency of the two and what we can learn from the experience.

Keywords: market-based, bank-based, economic efficiency.

1 Introduction

The integration of industry with finance is the industry sector and financial sector through equity relationships interwoven, which made industrial capital and financial capital to achieve the mutual transformation and direct integration. This is an inevitable and very important economic phenomenon. With the economic structure, size and operating system further development, integration of industry with finance became the optimal allocation and effective way to achieve the capital growth and expansion.

Each country taking different models of industry with finance's integration, as law, economic environment, status of bank and capital market development level are different from country to country. In general, we can divide it into two typical models: the market-based model which led by U.S. and Britain, the market-based model led by Japan and Germany. Other countries, on the basis of these two models, combined with their economic development level and characteristics to formation the industry with finance's integration

2 Two Models of Industry with Finance's Integration

2.1 "Japan and Germany's Model"

The mode of Japan and Germany based on "society" or "social" market economy to run, in which enterprise based indirect financing, banking and business in the economy play an important role, while the capital market less important. In this model, banks and enterprises maintain close and continuous contact and communication, both long-term dependencies. Enterprises have stable trade relations with the bank or main bank and cross-shareholdings. The bank have an impact on enterprises decisions, through loan covenants, equity share, personnel and agents of small shareholders with voting rights,

J. Luo (Ed.): Soft Computing in Information Communication Technology, AISC 158, pp. 431–436.
springerlink.com

etc. "Japanese model" and "German model" have similarities to indirect financing, such as main banks in the dominant financial system, capital market development inhibited.

Although Japan and Germany's integration of finance with industry are bank-based models, but they still have two important differences: First, Germany's banking system play a greater role in the integration of finance with industry, as the universal banking *business*, a strong proxy voting rights and the restricted stock market; second, the mode of Germany is more market than the Japanese mode, and finds a reasonable equilibrium between the market efficiency and the government behavior.

2.2   "Anglo-American Model"

The U.S. mode run on a free market which banks and enterprises have weak property rights restricted, mainly depending on short-term debt to contact and by the strict legal system to resolve disputes, so enterprise funds mainly from self-accumulation, direct ownership of business enterprises securities and credit guarantee for enterprises to eliminate possible risks.

Ownership structure of U.S. companies are more decentralize and socialization than Germany and Japan', which determines the *shareholders* do not have enough power in corporate governance, also determines legal ownership have very limited capacity to control and management company. The dispersed ownership structure and investment philosophy determines the U.S. is mainly on the external capital market to achieve integration of finance with industry. This way has high transparency and binding, because the pricing and allocation of financial capital is market determined, not by suggesting the contract to complete.

2.3   Governance Structure of the Two

Anglo-American corporate governance structure's advantages are that the relationships between corporate governance structures are more balanced, business transparency, shareholders' equity usually can get a better protection and private income in each period *relatively* high, due to strong shareholder constraints and market discipline. British and American companies made investment rate of return as criteria to evaluating performance, so if managers want to keep their jobs, they must strive to maintain a high dividend rate of profit. Compared with the Anglo-American enterprise, Japan and Germany's enterprise have high market share, competitive enterprises and small pressure from shareholders dividends. And all these stimulate the continuous, aggressive investment to increase production capacity, improve the productivity of existing business activities. From this high accumulation and low dividends to achieve the development of companies, in long term, while the shareholders will have their interests followed the rapid growths of capital appreciation in the end.

3   Two Modes' Economic Efficiency Analysis

Whether it is market-based or bank-based, have contributed to the development of national economy. Here we analyze the economic efficiency of two models from three levels: sustainable development, international competitiveness and risk prevention ability.

3.1 Sustainable Development

Judged from an economic point, sustainable development refers to the current period of economic growth will not affect economic growth in the next issue and the resource allocation in the future. Dr. Huang Ming document that: the capital as a resource, the use and management concerned with the sustainability of a country which associated with the three factors: First, capital formation ability; second structural adjustment capacity; Third, technological innovation support. The two modes' sustainable development capacity comparisons are in table 1 below:

Table 1. Two modes' sustainable development capacity comparisons

Item		Market-based	Bank-based
Ability of capital formation	Ability of capital gain	Weak	Stronger
	Ability of capital transformation	Strong	Stronger
	Capital productivity	Strong	Uncertain
Ability of structural adjustment	Incremental adjustments to capacity early	Stronger	Stronger
	Ability to adjust the stock	Strong	Stronger
	Ability to adjust post-increment	Strong	Stronger
Technological innovation capability	Innovation investment operations	Good	Common
	Innovation investment	Large scale and obvious effect	Small scale and uncertain effect
	Control of the main body of innovative investment	Enterprise-based, government lead banks activities	Bank + government

Table 1 shows that:

The ability of capital formation, in long term market-based better than bank-based;

The ability of structural adjustment, market-based adjustments in incremental and stock both have advantage, better than the *bank*-based focusing on incremental adjustments;

The ability in technological innovation, the bank-based innovative investment under the potential risks, but the gains *may* not be high, making the market-based investment system superior to bank-based in the environment of innovation and operation. Therefore, in sustainable development, the market-based is slightly stronger than the bank-based.

3.2 International Competitiveness

Refers to the international competitiveness is enterprises have more attractive price and quality to production and sales good and the ability to provide services, competition with the domestic and foreign enterprises, in their current and future environment. As the difference of every country's economic status, the integration of finance with industry's mode cannot be simply horizontal comparison. But we can compare the economic efficiency of financial sector or industry sector with its overall economic strength, which can reflects the ability of potential competition broadly.

Table 2. The 2008's international competitiveness of the financial system rankings

Country	Economic performance	Operational efficiency		World competitiveness ranking
		Ranking	*Index*	
U.S.	1	12	7.34	1
Singapore	3	8	7.63	5
Germany	6	50	4.99	7
Japan	29	44	5.44	9
China	2	43	5.44	30

Source: "Global *Competitiveness* Report 2008" http://www.imd.ch/index.cfm?nav1=true&bhcp=1

Table 3. Us and Japan's International competitiveness of the financial system rankings

Year	2003	2004	2005	2006	2007
US	1	1	1	1	1
Japan	24	21	19	16	24

Source: "Global Competitiveness Report 2008" http://www.imd.ch/index.cfm?nav1=true&bhcp=1

Table 2, Table 3 shows that the U.S. financial system, regardless of the comprehensive international competitiveness or the economy as a whole ranks is the first in international competition, and the two is symmetrical. From the sub-rankings of the international *competitiveness*, the United States have the most dynamic economic performance, operational efficiency is low; Japan, Germany, the comprehensive competitiveness of the financial system higher than their overall economic strength, but the efficiency of Japan and Germany lagged behind in their financial comprehensive ranking system. Thus we can see, market-based is stronger than bank-based in international competitiveness.

3.3 Risk Prevention

The so-called risk prevention capability is the ability to self-adjusting, resolve, avoid and control risks when country's economic system is *facing* uncertainty. The two modes are some differences in risk and risk prevention. We analyze the risk prevention capability of the two modes mainly from dependence on debt and financing. Specific as shown in Table 4:

Table 4. The two modes characteristics of risk prevention capability

	Market-based	Bank-based
Dependence on debt	Lower	High dependence on Internal debt
Financing	Debt - equity financing mix based Expand the company's risk ,but can effectively prevent financial risks	Based relational Financing Can effectively prevent corporate risk, but expanded the potential financial crisis

Table 4 shows that in prevention the national economy, market-based financing by maintaining a certain distance take an advantages; but in *terms* of business or industry, it is bank-based take advantage through the relation financing. As dependence on debt, market-based countries are more easily than bank-based countries. Therefore, from a country's risk prevention perspective, market-based stronger than the bank-based; but for the management of an enterprise, bank-based is better than market-based in the risk prevention capability.

4   The Enlightenment of the Two Modes

The integration of finance with industry is inevitable and objective laws under the market economy condition. Different countries should have their own modes of finance with industry's integration, according to the differences of the economic, social and historical traditions, even in the same country, due to different stages of development, the finance with industry's integration also have obvious differences. Carefully study the two modes' differences, the pros and cons will have important implications to our *country* to explore the way.

The integration of finance with industry is one important way to the rapid development of one country's economy. The realization of integration of finance with industry is direct impact on a country's economic performance. It can promote our country's economic growth mode from extensive to intensive, which is a basic path to adjust economic structure, improve *the* international competitiveness.

The integration of finance with industry has great meaning to the economic restructuring and reform. To achieve the integration of finance with industry, a new enterprise system and financial system should be established, which can create a new *mode* of economic operation.

From the above analysis we can see, under the conditions of market economy, we should closely follow the evolving economic situation, make efforts to adapt the general rule of the finance with industry's integration. However, in our country the market economy system has not been fully established, so there are greater risks to conduct exploration of the finance with industry's integration. To realize the integration of Chinese characteristics, the risk prevention mechanism must established, at the same time: financial companies should be required to further standardize according to the market economy, banks and various financial institutions to deepen the reform; state-owned enterprises should have clear property rights, solving the deficient problems, improving   the effectiveness and restricting the risk of investment decisions. Otherwise, the integration of finance with industry just to increase the financial risks and affect the autonomy of decision-making, it cannot play the purpose of alliance between giants. The integration model of finance with industry choose that should be associated with the condition of our country, promoting the financial capital and industrial capital effectively integration and growth to accelerate the development of market economy in our country.

Acknowledgment. The author would like to thanks for the great support and assistance obtained from teachers and some friends in the process of writing this paper. In addition, the authors extend sincere gratitude to the authors whose words are cited or quoted. Finally, the *authors* are profoundly grateful to the continuous support, understanding and selfless dedication of the family.

References

1. Li, Y., Wang, Z.: Industrial and financial capital integration, benefits and the international tendency. Journal of Dalian Maritime University 5(4) (December 2006)
2. Xu, T., Shan, X.: Integrating Industrial Capital with Financial Capital: Reasons, Methods and Effects. Journal of Xiamen University (5), 107–112 (2005)
3. Meng, J., Liu, Z.: Combination of Industry and Finance in the Perspective of Rotation of Capital. Journal of Xiamen University (2), 37–42 (2010)
4. Zhen, W., Gou, W.: The study of the integration of finance with industry mechanisms in china. Economic Research (3), 47–51 (2000)
5. Fisman, R., Love, I.: Trade credit, financial intermediary development, and industry growth. Journal of Finance 58, 353–374 (2003)
6. Peterson, M.A., Rajan, R.G.: Trade credit: theories and evidence. Review of Financial Studies 10, 661–691 (1997)
7. Huang, M.: New Theory of the integration of finance with industry, pp. 50–71. China Economic Press (2000)

The Transfer of Rural Labor's Analysis

Hu Qiangren

School of Economic & Management, Shanxi Normal University,
Shanxi Normal University, SXNU
Linfen, China
huqiangren@163.com

Abstract. The present situation of rural labor employment is very grim. Now, there are two hundred millions village surplus labor forces in our country. The paper illustrates factors of labor transfer and provides a framework and method of analyzing. In conclusion, it brings forward corresponding strategies.

Keywords: Surplus farm labor, Transfer of rural labor, Urbanization.

1   Introduction

For each country, it is commonly and unavoidable that workforce will transfer from traditional sectors (such as agricultural industry) to modern sectors (such as modern industry and services).This is also an only way to realize "modern economy growth". Rural labor migration is not only a crucial issue we must face on the way to industrialization but also a basic solution to solve issues concerning agriculture, countryside and farmers in China. China's labor force is related to realize the national people well-off, and the difficulty is to solve the problem in rural areas. Only real solution the employment of the rural labor force and achieve a reasonable transfer of rural surplus labor force, so to avoid waste of labor resources and to avoid creating other social problems. Therefore, "Pleasant, agriculture and countryside" are the key to the question whether the social target of well-off can be realized on time. But in these three problems, it is very important to realize the transfer of surplus farm labor.

2   Factors Influencing Labor Migration

Factors of rural labor transfer are variety and complex, which caused the process or outcome of transfer. The rural labor transfer macro-level factors include: the urban-rural income gap, the level of non-agricultural industrial development, labor market structure, urbanization, unemployment and others; the micro-level rural labor transfer elements include: farm family-owned agricultural natural resources, educational level of labor, family property status. Transfer of rural labor force of various factors and the relationship shown in table 1.

J. Luo (Ed.): Soft Computing in Information Communication Technology, AISC 158, pp. 437–442.
springerlink.com   © Springer-Verlag Berlin Heidelberg 2012

Table 1. The scale of rural labor transfer factors

The role of these factors on the direction of the discussion as follows:

2.1 Agricultural Natural Rresource Conditions

Agriculture is highly dependent on natural resources sectors, and natural resources are an important revenue stream of agriculture, especially in China's agricultural infrastructure and technical progress are relatively backward conditions, which is even more. If under the conditions of observed the agricultural natural resources are better, farmers engaged in agricultural labor can get higher income, and natural risks of agriculture is relatively small, so the labor force give up agriculture to non-agricultural the opportunity costs are higher. Therefore, we can think that if the number of rural households have more, better quality of agricultural natural resources, be able to derive more income, they would be willing to be more labor time allocation in agriculture.

2.2 Fixed Assets of Farmer Household to Production

Fixed assets of farmers to production by type can be divided into fixed assets for production of agricultural and non-agricultural use of productive fixed assets. The scale of fixed assets for production agriculture is the main symbol of agricultural equipment, in theory, better agricultural equipment conducive to increase agricultural output, increasing the opportunity cost of labor force and reducing the probability of labor transfer out of the agricultural sector. However, only the fixed assets of non-agricultural production together with family labor resources be used an appropriate of non-agricultural economic activities to play its full role, that is, farmers have non-agricultural fixed assets contribute to the promotion of non-farm labor allocation.

2.3 The Human Capital of Rural Labor

Theoretically, in a highly developed labor market, human capital is the key variable to determine the labors income and employment. Education is seen as a kind of human capital, because the capacity of education includes both production capacity and configuration capacity. Production capacity is the workforce with higher level of education who can produce more products, under the same condition combination with other factors of production, if expressed by the marginal production capacity, which is reflected in the same of other elements, an additional unit of education can increase level of product output. Configuration capability means that education has the ability to find opportunity, seize the opportunity and set the most efficient allocation of resources so can increase output capacity. Production capacity together with configuration capacity plays the role, but configuration capacity take place before than production capacity. So, only when resources are reasonable allocated, higher production capacity will bring more revenue. The workforce with higher level of education is better able to discover and seize the job opportunities and can better avoid the uncertainty risks of employment. On the other hand, in the case of asymmetric information for job seekers and employers, education-lever can be used as screen, the employer choosing from candidates who recognize with high capacity, so highly educated workforce has priority get employment opportunities, and most of the rural labor force with poor human capital remain in the agricultural sector, if they choose flow, may have to bear greater risks. Therefore, in theory, the scale of rural labor transfer is positively related to human capital of the rural labor force.

2.4 Industry or Regions Income Gap

If the flows of transfer costs do not be considered, the income disparities between industry or regions will induce the employment with the homogeneity mobility from lower income to higher income industry or geographic. Therefore, it is assume that the bigger the income gap between agriculture and non-agricultural, rural and urban and the larger the transfer scale of rural labor to non-agricultural industries and to urban mobility.

2.5 The Distance between Labor Outflow and Inflow

If the transfer of rural labor accompany with labor migration from one area to another, the distance between areas of labor migration has become an important factor. If the move is permanent, then the path of migration costs can not as a major factor affect the labor movement. Because the permanent labor migration means that individuals and their families after migration, as long as stability of employment and residence in the new place, it will not move out and move in between the two places frequently travelers. However, if migration is temporary, mobile labor force may often need to back and forth between the two places. Thus, the road cost of the transfer associated with the migration distance may become an important factor affect the transfer flow cost. For the mobile workforce, migration distances have a certain connection with the psychological stress and whether ease to access employment opportunities. Generally, the farer the distance migration, the greater migration population exposure to psychological pressure, and the greater uncertainty and risk of they move into the employment. This means that long-range transport heavier than the close migration of the migration cost. So, just the local or neighboring areas have access to non-agricultural employment opportunities, the rural labor force will be selected as non-agricultural employment in the nearby region.

2.6   Non-agricultural Industrial Development

Non-agricultural industrial scale is not only an important indicator to measure the degree of industrialization, but also reflect economic development to accommodate the transfer of the labor force from the agricultural sector. The larger scare of non-agricultural industries, the faster development, and the more labor force people demanded. Therefore, the transfer of rural labor impacted by the non-agricultural industrial development, and the scale of rural labor transfer has a positive correlation with non-agricultural industrial development.

2.7   The Level of Urbanization

As an empirical conclusion, urbanization is positively related to job opportunities. This is usually due to the higher level of urbanization, the greater the demand for services, while the service sector compared to other industries, typically have higher labor absorption. The transfer of rural labor force is impact by the levels and development of urbanization. Therefore, the level of urbanization is an important factor to affect the rural labor transfer and employment patterns.

2.8   Labor Market Structure

In a fully competitive labor market, labor flows between departments or regionals are guided by the supply-demand relationship of labor market. However, China in a period of economic transition not yet formed a unified labor market, so, the economy inevitably exist the two systems department, a government-controlled sector, a market-oriented sector. There are still employment protections to the urban labor force and the more stringent restrictions to absorption the rural labor force on government-controlled sector. The market-oriented departments in accordance with market rules to abort and use the labor, labor employment in the non-discriminatory practices, rural migrant workers are mainly enter these fields which are mainly market dominated sectors. Therefore, the level of market-oriented sector development is important factors to affect the scale of rural labor transfer and employment patterns.

3   Rural Migration Mode

According to China's household registration system, the basic population is divided into two parts: rural and urban population. Assumption that the two parts caused the annual natural growth rate is same and equal to α (i.e., in the k+1 year, the population number is the number of α times of the k year population), assuming the best of the rural population is the country's total population's γ times ($\gamma < 1$), with the national population distribution between rural and urban migration of population. Migration rate β is affect by the national food security to meet the basic needs of agricultural production, rural-urban income gap, the national macro household registration system and so on, which is equivalent to the ration of the amount of from rural to urban migration to rural population of more than the best ratio of basic rural population.

Under the above assumptions, rural migration can create a simple dynamic model as follows:

$r(k+1) = \alpha r(k) - \beta\{r(k) - \gamma[r(k) + u(k)]\}$

$u(k+1) = \alpha u(k) + \beta\{r(k) - \gamma[r(k) + u(k)]\}$

Model variables that: r (k), u (k) k-the k, u year's rural and urban population;

α - the natural population growth rate;

β - the mobility of the rural population;

γ- the ideal of the rural population to total population ratio.

In fact, the various variables are a function of time k, to simplify the model α, β, γ is set to a constant. The model can be viewed as a linear dynamic system model, to write the state vector form:

$$x(k+1) = Ax(k)$$

$$A = \begin{bmatrix} \alpha-\beta(1-\gamma) & \beta\gamma \\ \beta(1-\gamma) & \alpha-\beta\gamma \end{bmatrix}; \quad x(k) = \begin{bmatrix} r(k) \\ u(k) \end{bmatrix}$$

According to scholars D • G • Run Berger's analysis, the table 2 shows the migration model:

Table 2. Rural Migration

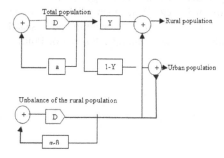

According to brief description of rural migration, the rural population's should meet the basic needs of food security factors coordination, timely and appropriate transfer to the city. Appropriate transfer of surplus rural population and mobility, will ease the tension contradiction between people and land in rural areas, increase the employment rate of farmers, increase farmers income and improving their scientific and cultural quality, it will effectively promote the coordinated development of rural economy and society.

4   Conclusion

As the existence of surplus labor force is concealed, in a short time, it can't bring society turbulence. But we can't look down on it. The surplus labor force not only is a human resource wasting, but also will turn into seriously social problem. So we must consider how to transfer these labors. First, we should commence from the system. We must reform the system, allow the farmer going into the city, and give them the equal employment opportunities. Second, we invest education; improve the farmer's educational level. Third, we must optimize the industrial structure; extend the agricultural independence digestive power. Last, we must develop the big city; exploit the big city's advantages to the full. We also consider the international exportation of the labor force. In the strategy of "many outlets flows, various forms transfer", we expand the employment in many ways, do our best to increase the employment opportunities and promote availably the village surplus labor force transfer, will not bring great social upheaval, but we cannot be taken lightly. Existence of surplus labor in rural areas is not only a waste of human resources, and over time, but also evolved into a serious social problem. Therefore, we must consider how to transfer.

References

1. Cai, D., Wang, M.: Household Registration System and Labor Market Protection. Economic Research (12), 41–49 (2001)
2. Jiang, Y.: Research on Theories of Rural Migration in China: A New Explanation to Shor tage of Rur al Labor. Economics (4), 41–49 (2006)
3. Zhao, Y.: Leaving the Countryside: Rural to Urban Migration Decisions in China. Journal of Development Studies 35(3), 105–133 (1999)
4. Hu, A.: Motivation and Obstacles of the Transfer of Chinese Rural Labor. Journal of Changchun University of Science and Technology 3(4), 153–156 (2008)
5. Wu, S., Zhao, D.: Study on Strategic Adjustment of Rural Economic Structure. Shijiazhuang Vocational College of Law and Commerce Teaching and Research 4(2), 23–26 (2008)
6. George, S.: Information in the Labor Market. Journal of Political Economy 70(5), 94–104 (1962)

A Reflection on the Mental Health Education of College Students

Qi Zongli

School of Information and Communication Engineering
Tianjin Polytechnic University
Tianjin, China
83131590@qq.com

Abstract. Based on the importance of mental health education, I do an analysis of the mental health of college students at present and the research on the way of building up university students' health psychology.

Keywords: Mental Health, Education.

1 Introduction

With the perfection of market economy system in socialist society and the development of modern science and technology, people live a faster and faster life and social competition becomes fiercer and fiercer. Under the fiercer competition environment, university students, who are in youth, do not adapt the environment day and day and their psychological burdens are heavier and heavier resulting in the increase of psychological problems. Therefore, universities are supposed to pay attention to their students' mental health and encourage the development of the education of university students' mental health.

2 The Meaning of Mental Health

According to the opinion of modern health, one's health is not only physical health but also includes his/her psychological health. World Health Organization defines health as: a perfect situation where neither physical disadvantages nor diseases exist and there is an ability of adapting society physically, psychologically and socially. Thus, only when one is under good physical and mental conditions is he or she real healthy.

What is mental health? And what is the standard? Humans' understanding of this problem is deeper and more developed as time goes by and society develops. Different researchers have different opinions. At present, a general opinion holds that mental health is a relationship between human and human as well as human and social environment which can inspire one's biggest potentiality. Based on the college students' specific reality in our country and combining the suggestions of the experts at home and abroad, we should consider the following standards to judge the mental health levels of university students:

- normal intelligence.
- healthy feeling.

J. Luo (Ed.): Soft Computing in Information Communication Technology, AISC 158, pp. 443–448.
springerlink.com   © Springer-Verlag Berlin Heidelberg 2012

- perfect will.
- integral personality.
- right self assessment.
- harmony interpersonal relationship.
- normal adaptability in society.
- mental behavior coinciding a university student's age.

3 The Significance of Doing Mental Health Education to University Students

First, strengthening the education of university students' mental health provides a strong base for roundly accelerating quality education and cultivating high quality people. University has a responsibility of cultivating high quality people and mental health education has a tight relationship with the work of professional cultivation. High quality people should have good moral qualities, scientific and cultural qualities, physical qualities as well as mental qualities. History tells us that a nation can't be one member of the independent nations in this world without stimulating spirits and strong wills. If one does not have stimulating spirits and a strong will, he can not be a high quality person. From many successful people, we can discover that what they have in common is that they not only have a good knowledge of their professions, but also have a good mental quality. However, those failures in working or life usually have weak wills and can't go through difficulties, failures and the challenges and tribulations of success. Therefore, the topic of psychology and mental health education attracts more and more social attention, which is known to young students. In order to cultivate high quality people, it is an urgent need to strengthen the mental health education of college students, cultivate their good personal mental qualities, improve their abilities of adapting society, endurance and feeling adjustment, and to promote a harmony development of their mental quality, moral quality, scientific and cultural quality and physical quality.

Second, to strengthen university students' mental health education is a student-oriented education and an urgent demand of helping students to grow up physically and mentally. Students are the objects of the cultivation of school education. To orient humans, satisfy many kinds of needs of students' development, promote students to improve themselves in all aspects and become qualified builders and reliable successors of the socialist cause with Chinese characteristics is the starting point and ultimate goal of universities' all jobs and is the ultimate demand and concrete reflection of the important thought of implementing "Three Represents". At the present age, most of college students are the only child in their families, who shoulders high expects from their families and society. Usually, they have a high self-orientation and are eager to succeed, but they lack enough social experience and mental   development, which cause them to have mood fluctuations easily. With the development of economy and society, the social environment which they face is more and more complex, which brings them more pressure in study, occupations, economy and emotions, therefore many kinds of mental problems will turn up, which need guidance and adjustments urgently. Therefore, universities should strengthen and improve the educational work of their students' mental health.

Third, to strengthen university students' mental health education is a useful way to improve the ideological education work in universities. College students are being the key period of the development of life, when their worldviews, views of life and views of value are shaping. To a university student, every difficulty, contradiction, worry and conflict in the process of their growing up may develop into a psychological problem. And these problems often go with the shaping of their worldviews, views of life and views of value shape. Psychological problems are the reflections of worldviews, views of life and views of value in psychology. Thus, the settlement of a psychological problem must be based on right worldview, view of life and view of value. In return, the existence of psychological problems definitely will affect the establishments of one's worldview, view of life and view of value. Therefore, to strengthen and improve the ideological education work of university students, we must expand all aspects: ideal and faith, ideology and morality, behavior cultivation, mental health, and so on, which make the relation between ideological education and mental health education is complementary and mutually improved. We should catch university students' thoughts pulses and psychological characteristics tightly and combine the ideological education with the mental health education strictly, following the rules of these two educations. During the work of daily ideological education, we should make a difference between students' ideological problems and mental problems, be good at guiding or counseling with clear aims and provide the students who have psychological perplexities and psychological barriers with necessary helps in time. During the work of mental health education, we should follow the guidance of dialectical materialism and historical materialism, on one side we must know the importance of mental health education completely; on the other side, we must avoid overstating the function of it and make sure this work develops in the right direction.

4 The Present Situation of College Students and an Analysis of the Reasons

In recent years, many experts and scholars made use of various ways to research into university students' mental health. The result shows: the situation of university students' mental health in our country is not optimistic and there exist lots of bad reactions and adaption barriers. For example: down mood, weak emotion, strong sense of inferiority, insufficient confidence, weak psychological endurance ability and social adaption ability, the increase of interpersonal communication barriers, and so on. Nowadays, mental health problem has become the main reason of university students' reductions of class and leaving school.

University students' various problems in mental quality are linked to the process of the development of their mentalities, as well as the social environment around them. Usually a university student is at the age of 17 to 23, which is the middle period of youth when one's psychology is the most changeable in his life. Because one's psychology is not mature and his emotion is often unstable, when facing a series of physical, psychological and social adaption problems, mental conflicts may happen sometimes, such as the conflicts between dream and reality, sense and sensibility, independence and dependence, self-esteem and self-abasement, thirst for knowledge and bad discrimination ability, competition and seeking stabilization, and so on. If these conflicts and contradictions can not be guided efficiently and solved reasonably, they will become mental barriers after a long time, especially at present when students nearly devote themselves to studying so as to succeed in the competition of National College Entrance Examination. Then, as a result of parents' over-protection, the examination-oriented education at high school, a lack of life experiences, these students are mentally weak, weak minded and short of frustration tolerance. So when meeting some small frustrations in study, life, making friends, love, occupations, and so on, some of them may have psychological diseases, even may leave school for a long time or kill themselves. Considering the factor of environment, as competition becomes stronger and life speeds up, one may feel the tension and pressure of time; his choice of life goals increases so much that his mental conflicts become severer because he cannot get both, which makes him more anxious about everything. Without adapting the changes of environment, confusions, puzzles, anxieties, and tensions are increasing and the revolution of society gives college students stronger and more complex mental shock than ever before. Various factors, physical factors, mental factors and social factors, are combined together, which easily lead to college students' unbalance in the development of psychology. A person of low mental quality certainly cannot adapt the environment of high speed, high technology and fierce competition, so this person will suffer from various mental diseases. All these show that the education of university students' mental health needs to be carried out immediately.

5   The Measures of Strengthening Mental Health Education of College Students in University

At present, college students' mental health has become an important factor of affecting the quality of cultivated people, which restricts the realization of cultivating goals. Thus, it is necessary and urgent to strengthen mental health education of college students and improve college students' psychological quality.

5.1   Penetrate Mental Quality Education into the Teaching of All Subjects

Penetrating mental quality education into the teaching of all subjects is not only the way of carrying out mental education in school, but also the demand of the development of each subject teaching. Each subject's teaching includes mental education, because teaching process takes cultures, moral standards, values in history as its content and mainline.

When imparting knowledge, if a teacher considers students' mental requirements, inspires their interests in studying and dicks out the educational meaning inside, he or she can change historical knowledge, experience, and skills into his own spiritual wealth, that is to say, afterwards, this wealth will be changed into his or her students' psychological points, life values and good mental qualities, which may stay with them forever.

5.2   Develop the Various Activities of Mental Quality Education, and Popularize the Knowledge of Mental Health

Universities should provide selective courses, required courses or lectures on special topics to help students to acquire the knowledge and ways of keeping good psychology such as" College Students' Mental Health" "Health Psychology". The specific content should include: studying mental education, emotion education, instructions of interpersonal relations, cultivating healthy personality, cultivating the ability of enduring frustrations and students' self-cultivations. In the class of mental health education, we should not only pass theories on, but also combine theories with practices and develop mental training activities, by practicing social identity, self identification and role playing to improve students' abilities of enduring frustrations and confidences. For example, aiming at the problem of freshmen's adaption, the content of class can be designed as: interpersonal communication, the first effect, the art of communicating, self- exploring, the expression of emotions and control, cooperation, and so on. Students can acquire new feelings in the class training and role playing, and they will bring the feelings back to life to strengthen themselves, so that their mental adaption can be improved.

By giving students lectures, universities can make students know the ways of preventing and curing some mental problems and diseases that young university students may suffer from. Besides, wall newspapers, radios, TV sets, newspapers, brochures should be used to tell students the basic knowledge of mental health. Anyway, we should help students know his or her psychology and acquire perceived coping competence and the ability of asking for help.

5.3 Strengthen the Work of Mental Health Research and Study

In order to handle every student's mental situation, they must divide students into some groups according to different grades, different genders and do researches into students' mental problems, then they should conclude, arrange, make use of active ways to overcome problems. For examples: they should research into freshmen's adaptations in life and study environment; to sophomores and juniors, they should be concerned about their problems during study and communication after adapting environment; as for senior students, they should focus their attention on the instructions of social adaption and future occupations. As for characters, generally speaking, boys are strong and impulsive, while girls are relative mild and introvert. University teachers and directors should study the differences between boys' and girls' psychology. In addition, they should strengthen the information communication and cooperation between brother colleges and between colleges and colleges. They can study and explore the measures and ways to deal with mental problems. Meanwhile, they should improve the way of discovering problems and dealing with problems. Then, they can set up an information network which treats the department of information inquiry as the center, the management people in the department as hubs, backbone students as the structures of the information network.

5.4 Make a Good Campus Cultural Environment

A University campus is the main place where university students study and live. It should offer college students a good and loose mental environment. Some serious but active schedules, wonderful and interesting amateur cultures and harmony healthy interpersonal relationships…all these can be combined into a psychological set to add interests and colors to students' study and life. Otherwise, the suppressed environment with a mechanical life speed is not good for cultivating students' mental quality.

Make full use of the psychological inquiry centers in university. The psychological inquiry centers in university are important approaches on improving students' mental health and quality, as well as an important part of mental quality education. With the development of society, psychological inquiry has been admitted and accepted by more and more people. Today, more and more colleges and universities, even some middle schools, have set up psychological inquiry centers. Therefore, we should make full use of the psychological inquiry centers in those universities, improve and perfect the teams of psychological inquiry workers. Make sure that consultants train and guide visitors to reduce the visitors' contraptions and conflicts, settle their troubles and develop their potentials. At the same time, the consultants help students know themselves better and grasp their own fortunate and adapt the outside environment.

References

1. Duan, X., Zhao, L.: Mental Health Education of College Students. Science Press (2003)
2. Liu, H.: A recognition on the concept and Criteria of Mental Health. Psychological Science (4), 481 (2001)
3. Tu, X.: A Course of Mental Health Education of College Students. Chinese medical sci-tech Press (2006)
4. Yang, M.: Healthy Personality and Shape of College Students. Chinese Youth Press (1999)
5. Zhang, X., Liu, X.: Developmental psychology. Northeast Normal University Press (2000)

Analysis of Social Change of Neutral Fashion

Lili Xu and Yuhua Li

Institute of Art & Fashion, Tianjin polytechnic university, Tianjin, China
{Xulili8688,liyuhua}@163.com

Abstract. "Neutral" clothing plays an important role as a social phenomenon prevalent in contemporary society. In order to explore "neutral" clothing space of development, it embedded "neutral" clothing to social version. Analyzed, researched the style development theory from the view of social exchange so as to point out neutral fashion direction affected by social theory and also convey inner feelings of wearers.

Keywords: social exchange, neutral, clothing.

1 Introduction

Recent years, neutral clothing as a social cultural phenomenon has been presented in contemporary society, but from today's mainstream culture, neutral clothing is still in the scope of the study sub-culture. It hits with the mainstream culture and interaction, continue to define itself, maintains a certain distance to mainstream culture, the distance is exclusive of personal expression and sub-culture power.

"Neutral" in the garment industry as an aesthetic fashion, in just ten years time, its markets is growing step by step, the future will have greater development potential space. "Neutral" clothing is getting totally out of shape, forming an independent consumer groups, grasping the law of the development of fashion, expanding the commercial space, for garment industry it means a lot.

2 Summarize

2.1 The Concept of Social Change

Sociology as one of the core issues of social change, from the time of creation of sociology, social scientists began to discuss on it. Broadly speaking, social change refers to changes in all social phenomena; narrowly, mainly refers to the social structure and function of the major changes.

2.2 The Concept of "Neutral" Clothing

"Neutral" clothing which is "no gender trend clothes," is applicable to both men and women, no significant gender characteristics of the clothing style. Such clothing has its wide adaptability, people of different genders, different ages, different occupations, and different sizes can all find their types.

J. Luo (Ed.): Soft Computing in Information Communication Technology, AISC 158, pp. 449–453.
springerlink.com   © Springer-Verlag Berlin Heidelberg 2012

2.3   The Development Process of "Neutral" Clothing

In traditional society, women are basically attached to men, which definitely is a weak position. At that time, the aesthetic value of the social trend was basically consistent with the aesthetic point of view of men. Women should recognize, consciously or unconsciously by these ideas and action. With the advent of the industrial age, women are increasingly involved in social life. However, traditional women's tedious and long robe accessories do not meet the needs of labor, so women's simplification is inevitable.

In the early 20th century, more and more women began to participate in outdoor sports. Due to the need of movement, men's pants appeared in women's clothing. After that, because of women's requirements to improve the social status, promote gender equality and the development of the feminist movement, consciousness of modern women has undergone tremendous changes, which can be expanded to two clues, one of which is to take the "neutral" of the trail. After World War I, Chanel was the first one, who put men wool knitted fabric for underwear into women used, designed the men's knitted fabric female suit.

After decades, gender clothing learn lessons from each other in various arts, men and women's jeans of same style, T-shirts, casual wear began to widely exist in people's everyday clothing. Neutral fashion, clothing appeared the phenomenon of gender convergence. Today's society is the information age, this "no gender trend" clothes will be an important topic for clothing industry.

3   Analyzing Neutral Clothing Problems from the Social Change View

3.1   Changes in Social and Political Status

With the development of productive forces of society and the trend of cultural diversity, women's status is higher and higher so that most of them are out of the family cage, women are not appendages of men. This has a subversive function of men's status, also shakes clothing view of male-centered. Thus, basic division of clothing gender has changed, so as to change the way of making amphiprotic clothing of modern fashion. Women acquire the access to survival, development and aesthetic power, due to their own strength, while fully in accordance with their own aesthetic, cultural taste freedom to dress themselves, having the sense of autonomy and independence. So neutral clothing styles appeared, it is no longer for please men, but a tool for women highlights their individuality.

3.2   Changes in Social Roles

It is well known that the achievements women made in pursuit independence. In business, politics and other areas, women politicians, women entrepreneurs, scientists are everywhere. This shows women have gotten rid of men and have their own world. Women got freedom on the basis of politic and social roles began to change, so the clothing production, manufacture and dissemination's owner is no longer only for men, clothing and fashion started to become real women exclusively.

When the traditional gender roles become more and more neutral trend, the boundaries between masculinity and femininity are more obvious. Modern male body break away tradition nature of subject, which has become like a woman being turned into objects to be appreciated, need dedicated care items. In fashion roles, started to play a similar role with women, so that is how neutral develops.

4   Social Changes on the Neutral Selection of Fashion

In the sociological sense, it refers to all social changes not only changes in social phenomena, but also specific major changes in social structure; not only refers to the process of social change, but also result of social change. With the change of society, neutral clothing will change accordingly as a social and cultural phenomenon. It can be said that this change is the inevitable result of neutral clothing adapts society development.

4.1   Population Structure of the Evolution of "Neutral" Clothing

Age characteristics: As widespread of neutral, consumers with "neutral" clothing is increasing and started with popular features. "Neutral" clothing for different age groups, up to the old, down to the young, reflects the wide range of adaptability.

Gender characteristics: In ancient times despite the existence of gender differences, many styles of men and women are common. After 20th century, women propose greater social status, promote gender equality and the feminist movement started, so that consciousness of modern women has undergone tremendous changes, which reflect women from the "charming" to "handsome." Women pants, suits, coats, etc., made they are the same as men, able and mature. On the men's side, affected by hippie, broke the Western tradition of men's clothing tough masculine style since the 19th century, subverted traditional "gender clothing identification system", thus promoting the development of neutral clothing.

Social class characteristics: According to sociological theory of social stratification, social structure can be divided into the upper, the middle, the lower three levels. In the past, the upper tend to have huge salaries, dividends and commissions, they have money and time sufficient to ensure, so most of them are more likely to pursuit of individuality. The middle is a social knowledge and its own professional and technical services in the community and get paid groups, work stability, wage increase every year, living standards steadily, one could take some investment money for clothing. The lower classes, whether social wealth, power or education, compared with the upper and the middle, are still lower status. Its subject is the working class and coolies, etc. It is impossible to note clothing without basic needs of living. Thus it can't be so common in the masses, only spread between the upper and the middle. As China's political, economic, cultural and social life of the great development, wearing "neutral" clothing of the population structure is changing, from intellectuals, dignitaries, down to ordinary people, young and old, which reflects "neutral" clothing adapt the society well with the better change of society.

4.2   "Neutral" Value Function of the Evolution of Clothing

People's idea originally on the "neutral" clothing value concept is to make a living and do sports under the background of industry, both of the two have functional value. As time changes, people's idea on the "neutral" clothing values, will also change, and constantly enrich and develop. Women made remarkable achievements in the state of gender equality presented gradually. "Neutral" clothing is responsive to this trend, showing both men and women equal, also means that the era of continuous improvement, it is an extremely wide range of adaptability, of different gender, ages, different sizes, and different occupations that can find suitable styles. In the increasingly modern society, the gradual rise of the modern lifestyle, more and more consumers want to wear different from the others to show their unique style and temperament. Most basic needs of consumers of clothing, which are comfort, economic, aesthetic, are related to self-actualization. Therefore, the "neutral" clothing with its unique personality and temperament is the most acceptable.

4.3   "Neutral" Clothing Evolution of Power Relations

Changes in power relations not only affect people's lifestyles and habits, but also real impact on people's costume and dress behavior. In male dominated society, women's clothes are to please men. The feminist movement of thought on the evolution of the popular clothing plays an important role.

In the first wave of the feminist movement in order to strive for "equality of the gender", a large number of women began to enter society, women began to imitate men, with the obvious characteristics of male trousers and accepted by the general public. This shows women began to achieve gender equality completely; meanwhile, neutral fashion design began to take shape. The second wave of the right to eliminate gender differences, strengthen and balance men clothing elements into women clothing, thus neutral clothing officially appear. In the third wave feminism, the promotion of men and women of the "unity of yin and yang" advocated no "men" and "women" in society, and emphasized the fundamental differences in gender and gender identity, thus "neutral" clothing to develop rapidly. In short, the wave of the feminist movement over and over again, pushed the traditional relationship of right, in this changing right, the "neutral" clothing as a product of the times has been widely popular.

5   Conclusion

Based on the understanding of the theory of social change, this paper researches and studies "neutral" clothing trends of the evolution from the view of society change, in order to better grasp the law of the development of fashion .Mite changes in very age and society will be reflected by the sensitive female image clearly. It not only reflects the era of transformation, social change, but also reflects the thinking of the concept of progress and innovation.

With the advent of information technology, aesthetic and visual aspects of the plot in the respected , clothing is constantly cross-cultural developing; On other hand, the impact of traditional culture, global cultural eclecticism, and cultural inclusiveness will directly affect the gender pattern of clothing, both women's masculine or feminine men, will have a broader space for development of neutral clothing.

References

1. Feng, X.: Social research methods. Higher Education Press, Beijing (2006)
2. Vago, S. (wrote), Wang, X. (trans.): Social change. Peking University Press, Beijing (2007)
3. Zhang, L.: Sociology of modern design. Hunan Science and Technology Press (May 2005)
4. Kaise, S.B., Li, H.: Social psychology of clothing. China Textile Press (March 2000)
5. Ma, L., Zhang, Z.: On the history of neutral clothing space, vol. 19. Xi'an Engineering Science and Technology University (2005)
6. Min, H.: From the "hierarchy of needs" of the history of clothing personalized with neutral phenomenon. Suzhou Silk Institute of Technology (May 1999)
7. Xia, X.: Interpretation of the trend of the prevalence of women in neutral. Changsha Institute (2007)
8. Yang, W.: Anatomy "neutral" movement. University Times (12) (2006)
9. Su, G.: Century Review of neutral clothing. Ornaments (01) (2003)
10. Ma, X.: Clothing neutral Phenomena. Suzhou University (2008)

Brief Study on the Economic Value of Advertising in Great Sports Event

Jiang Yi-fan and Xiao Chang

Institute of Art and Fashion, Tianjin Polytechnic University
candy2834@sina.com, xiaochang_@126.com

Abstract. The paper, taking 2008 Beijing Olympic Games for example, elaborates current situation of sports advertising at home and abroad from macro perspective, and then has a research on sports advertising type and reciprocity between advertising and sports event. Sports advertising type is divided according to characteristics of advertising, which is convenient for sports advertising to study and summarize. Discussion on the economic value of sports advertising entailed in China's sports event can better provide service for companies and their brands.

Keywords: sports event, Advertising, brand, economic value.

1 Introduction

After accession to WTO, China's economy has played a significant role globally; brand of china has received a sharp development, changing with each passing day. Under current circumstances, effects of traditional advertising have gradually declined; enterprises thus begin to show their magic power to find a new advertising medium for their brand publicity. In what kind of communicative form can advertising provide much more original and effective service for brand publicity and its profits yield? Such question undoubtedly has become a new challenge to nowadays' advertisers.

Due to hot broadcast of 2008 Beijing Olympic Games and great passion towards sports from audiences, sports stars like Yao Ming, Liu Xiang etc. became image representatives of some brands. Making use of their popularity and influence, they represented certain products and made advertisements; which showed their celebrity effects; set up brand images; improved brand recognition, and moreover, brought excessive profits for entrepreneurs objectively. Although the upcoming 16th Asian Games will be held in Guangzhou On November 12th, 2010, Guangzhou Asian Games' sponsors club has been organized by Guangzhou Asian Games Organizing Committee on November 12th, 2008. 14 companies singed as the first spate of Guangzhou Asian sponsors.

Sports are considered to be a common topic around the world. No matter which countries and nations you come from, what religions you believe in, and what cultural and economic level you possess, sports are our common language, favorites and ideal. Therefore, sports event marketing evolves as the easiest acceptable communications platform by the audiences. Jiang Jiazhen, doctor of science of physical culture and sports in Beijing Sport University, famous expert in sports event marketing, expressed:

J. Luo (Ed.): Soft Computing in Information Communication Technology, AISC 158, pp. 455–458.
springerlink.com
© Springer-Verlag Berlin Heidelberg 2012

"Compared with hard advertisement, sports event marketing has received better communicative effects because of its public interest, flexibility and hidden implication."

2   Main Types of Sports Event Advertising

Sponsored advertising. Sponsored advertising is divided into three sub-types: tournament title, Garment and equipment provision, and prizes provision. Tournament title can be named before and after the competition so as to increase the degree of brands' exposure and publicity. Garment and equipment provision can help to achieve the goal of gaining more brands' appearance rate before the sports audiences. In the process of broadcasting contemporary sports event, more and more media begin to pay attention to the interactive process with audiences. For that reason, more media shift to provide prizes to the interactive audiences, enhancing their enthusiasm. The interactive process is much more emphasized by the enterprises, not only improving product publicity, but also receiving preferences and attentions from audiences.

Brand advertising of sports event. Entrance ticket designs of competitions, extension products designs etc. all belong to this category. Such advertising mainly aims to publicize cultural connotation of competitions.

Commercial advertising. Commercial advertising refers to a pure kind of commercialized publicity which purchases time and space for advertising, such as commercial breaks in sports event, arena fence advertising etc. Commercial advertising is recognized as the most common type of sports advertising. As figure 1-1shows, "Double Fish" brand advertising around a table tennis stadium.

Fig. 1. "Double Fish" brand advertising around a table tennis stadium

3   Economic Value of Advertising in Sports Event

Taken 2008 Beijing Olympic Games for instance, it has stretched very profound significance and considerably great impact as a world-class event held in china. The difference with World Cup is that Olympic Games have a wide range of different audiences regardless of age and gender. It is no wonder that Olympic Games get the highest attention with remarkable audience loyalty among any other sports game. International Olympic Committee has ever surveyed audience's familiarity of Olympic Five Ring logo with a result that audience's familiarity towards Olympic Five Ring logo surpasses any other logos. That is why so many enterprises intensely competed for qualification as Beijing Olympic Games Partners.

Low cost, high exposure. It takes at least one or two hours to hold a competition, which means thousands of seconds can be used for sponsors to publicize their brands. After long time exposed in public, a brand will impress the audience with certain cognition and favor. Time spent on sports event advertising is many more times than the equivalence of TV advertising. However, the cost of its brand advertising has decreased a lot. Such mode of communication with the feature of low cost and high exposure will bring incalculable profits beyond all doubt. And of course, enterprises will prefer that advertising form in the future.

High brand recognition improvement. Sports event has obvious celebrity effects. When audience acknowledges that some brand has become a part of his adorable sport or sports star, it will galvanize intimacy between audience and such brand. Because sports stars have very high public attentions, enterprises will try their best to create sports stars' advertising in order that brand can "graft" on their public attentions. Like smelling flowers can make people memorize some things of beauty, audience can also think of his adorable sports star and lovable sports, and then recognize the brand he met in the case of not watching competitions.

Brand reputation improvement. The patronage of companies supports and advances sports event significantly. Firstly, sponsors provide funds for the sports competition. Only strong economic strength a high-level sports event can be held and people's daily life can be richened. Reliance only on the tickets' sale of competitions is not that enough. Secondly, culture and sports are correlated. As for people's daily life, sports competition promotes a positive psychology and active philosophy of life. It pursues a concept of unity, diligence and hard working. For example, the Olympic motto "Swifter, Higher, Stronger" expresses a positive message to all people. With the benefit of positive influence from sports competition, sports event advertising always transmit a positive ideal to its audience. Such positive impression can deplete audience's exclusion and reverse psychology toward sports event advertising. On contrary, that advertising can easily and preferably impress audience. Thirdly, compared with traditional advertising, sports event advertising is of little commercialization and utilitarian. From different kinds of sportswear to colorful billboards around venues, sports advertising indeed provides a vision feast, enjoyment, and moreover, public entertainment products to its audience. Therefore, consumers' recognition can be acquired easily and brand reputation be promoted. Sun Liping, Chief Executive Officer of China Olympic Sports Industry Co. Ltd said, "Sports event marketing can help to build steady emotional connection between enterprises setting up brands and consumers, and also increase the emotional value of brands."

Brand culture enrichment. Great sports event can help a brand quickly recognized by audience and win in the crucial market competition. What's more, it can enrich and enhance a brand's culture and connotation. Many invaluable sprits are embodied in sports event. The sports' spirits such as fairness, peace, communication and friendship can merge with sponsored brands to enrich brands' culture and also enhance brands' connotation. Those spirits all human beings accept and admire, will definitely and easily make audience receive and recognize the sponsored brands and products.

Advertising effect improvement. 21 century is regarded as an age of information. However, some advertisements as a kind of information scatter like "trash" everywhere in the era of full information. In current TV programs, people always complain that advertising occupies too much time and start to revolt against it. Sports event advertising does not involve in that issue. Thousands and millions of audiences are attracted by the cliff-hanging competitions. If advertising is plugged in blindly, the intense competition will lose its meaning. The effect of advertising will definitely decrease, and advertisers will never favor sports event as a platform. But no matter how fierce the competition is, athletes need to have a break as well as audience from about a few minutes to several hours, which provides a very good opportunity to the sports event advertising. Audience can relieve from the tension of the game and have a psychological and physical adjustment properly while enjoying the well-prepared advertisements. Accordingly, positive image of a brand will be taken into shape, and audience will even more accept and remember the brand unconsciously.

Product promotion. Sports event advertising can not only improve brands' culture and connotation, but also promote products directly for profits. There are many sports event promotions, such as purchasing products with tickets presented, visiting sports stadium by groups on free etc. In 2008 Beijing Olympic Games, "Yili" dairy was considered as designated milk; while in 2010 the 16th Guangzhou Asian Games, "Kunlun Mountain" mineral water has become its designated water. That undoubtedly will bring the most directive profits to the companies' brands.

China's sports cause has been stirred without question by China's economic up growth, China's sports' rise and also the successful hold of 2008 Beijing Olympic Games. Competitive sport has become the focus of attention and topic of discussion in people's daily life. A great opportunity has been brought to China's enterprises and a healthy platform has been set up and popularized for their brands. How to handle and make use of sports event advertising will be a new theme for advertisers to study.

References

1. Ting, T.: Budweiser and expo marketing—interview vice marketing director of ABInbev Wang Dao. China Advertising (2010)
2. Hu, C.: Research on brand advertising shaping. China Advertising (03) (2004)
3. Xu, S.: Brand advertising exploitation, media brand image shaping. Advertiser (02) (2006)
4. Hu, C.: Brand advertising shaping (2003)
5. Hong, Q., Shen, Y.: Strategy of sports advertising in market economy. Journal of Xi'an Institute of Physical Education (2) (1999)
6. Sun, L.: Study on sports event sponsorship. Group Economy (2) (2007)

Service Quality Comprehensive Evaluation and Research of Community Service Industry

Zhao Jing and Li Xiangbo

College of Management, Tianjin Polytechnic University, Tianjin, China
lengjing3@126.com

Abstract. This paper established hierarchy structure model of community service industry service quality comprehensive evaluation, and with community chain supermarkets as example carried out comprehensive evaluation to service quality through community chain supermarket's service environment, employee quality, the selling goods and after-sale service four aspects and certain indexes reflecting these aspects, to provide reference for improving service quality and adjusting operation strategy.

Keywords: Hierarchy analysis method, Community service industry, Community chain supermarket, Service quality comprehensive evaluation.

1 Introduction

With economy development and life rhythm speeding up, the level of service for the service industries have become increasingly demanding. Especially for community service industry, besides the demands of high quality goods and reasonable price, the consumers have higher demands for service convenience and humanization. Seller can timely adjust operation strategy and make solution according to evaluation results to better meet consumer's demands, enhance enterprise competitiveness and improve economic benefit.

2 Community Service Industry and Community Chain Supermarket

Community service industry refers to some public service character industries with small scale, less employee and roots in the community, such as: snack bar, smoke & wine shops and community chain supermarket, etc. These industries have the characteristics e.g. direct service to community; easy operation and management; can provide convenient and efficient service, etc. With rapid development of market economy, people engaging in operating service industry are relatively more and more, service quality competition is also increasingly fierce. Community chain supermarket, commonly known as "Neighborhood Sopping Center" which originated from South Land Company of America, now is rapidly developing in China. Be different with common department store and supermarket, community chain supermarket's business

J. Luo (Ed.): Soft Computing in Information Communication Technology, AISC 158, pp. 459–464.
springerlink.com   © Springer-Verlag Berlin Heidelberg 2012

area is smaller, near consumer centralized places such as residential areas, traffic artery and so on, the goods are mainly quick food, beverage and small articles of daily use, business time is long, main open-shelf and self-choice and settlement is uniformly carried out by cash register. Community chain supermarket has many advantages such as: commodity is fewer but better, inventory adjustment is flexible, profit is high, easy to expand, purchasing is convenient and close to consumer demands, etc. so it is suitable for our country's economic development level. [1]Meanwhile, as a service nature retail business mode, service quality of community chain supermarket directly impacts on their performance. This paper uses hierarchy analysis method carry out quantitative analysis to community chain supermarket service quality to help managers timely adjust operation to improve service quality and increase market competitive capability.

3   Service Quality Comprehensive Evaluation of Community Service Industry

Service Quality Comprehensive Evaluation Model Establishment. With community supermarket chain as example, the service quality comprehensive evaluation model is divided into 3 hierarchies: Target hierarchy——with "service quality comprehensive evaluation" as supreme target hierarchy and record it as A; Rule hierarchy——choose service environment, employee quality, goods and after-sale service as evaluation criteria and record it as B; Proposal hierarchy——choose evaluation index significantly reflecting target hierarchy as evaluation proposal and record it as C. Service quality comprehensive evaluation system structure and specific meaning of evaluation indexes[2] please see Figure 1.

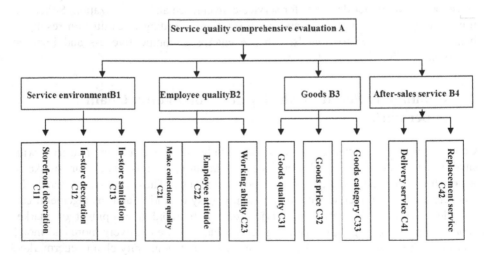

Fig. 1. Hierarchy analysis chart of service quality evaluation

Construct Judgment Matrix. Do binary comparison between each hierarchy element and construct comparison judgment matrix. In order to carry out comprehensive evaluation, randomly sampled Tianjin Hedong District Jingong Supermarket Chain, Huarun Wanjia Community Convenience Store and Jinmaitian Food Supermarket 3 community chain supermarkets as evaluation objectives, respectively sampled 180 consumers from 3 community chain supermarkets and composed jury (including 60 persons of below 30 year-old, 60 persons of 30-50 year-old and 60 persons of above 50 year-old), and invite them give comparison evaluation to each hierarchy element of comprehensive evaluation system by questionnaire survey form. By questionnaire recycling, sorting and statistics construct judgment matrix e.g. Table 1.

Table 1. Rule Hierarchy Judgment Matrix

Service quality comprehensive evaluation	B1	B2	B3	B4	Importance sequencing value
Service environment B1	1	1/3	1/3	1/7	0.068
Employee quality B2	3	1	1/3	1/3	0.145
Goods B3	3	3	1	1/3	0.251
After-sale service B4	7	3	3	1	0.537
λ_{max}=4.140 CI=0.04656 RI=0.90 CR=0.0517					

Hierarchy Single Sequencing and Consistency Inspection. Calculate according to all indexes importance construction judgment matrix [3], the results please see Table 2 to Table 5:

Table 2. Hierarchy Single Sequencing of Service Environment (B1)

Service environment	C11	C12	C13	Importance sequencing value
Storefront decoration C11	1	1	1/3	0.200
In-store decoration C12	1	1	1/3	0.200
In-store sanitation C13	3	3	1	0.600
λ_{max}=3 CI=0 RI=0.58 CR=0				

Table 3. Hierarchy Single Sequencing of Employee Quality（B2）

Employee Quality	C21	C22	C23	Importance sequencing value
Make collections quality C21	1	1/5	1/3	0.105
Employee attitude C22	5	1	3	0.637
Working ability C23	3	1/3	1	0.258
λ_{max}=3.038 CI=0.019 RI=0.58 CR=0.033				

Table 4. Hierarchy Single Sequencing of Goods（B3）

Goods	C31	C32	C33	Importance sequencing value
Goods quality C31	1	1	5	0.455
Goods price C32	1	1	5	0.455
Goods quantity C13	1/5	1/5	1	0.091
λ_{max}=3 CI=0 RI=0.58 CR=0				

Table 5. Hierarchy Single Sequencing of After-sale Service（B4）

After-sale service	C41	C42	Importance sequencing value
Delivery service C41	1	5	0.833
Replacement service C42	1/5	1	0.167

Total Sequencing of Hierarchy. Calculation hierarchically from top to bottom sequentially along progressive hierarchy structure can calculate lowest hierarchy element's relative importance or relative merits sequencing value compared to highest hierarchy (total goal), namely is total sequencing. Hierarchy total sequencing should be carried out consistency inspection which is from higher hierarchy to lower hierarchy, the results please see Table 6.

Table 6. Total Sequencing Table of Small Service Enterprise Service Quality Comprehensive Evaluation

Hierarchy B / Hierarchy C	B1 0.068	B2 0.145	B3 0.251	B4 0.537	Total Sequencing of Comprehensive Evaluation
C11	0.200				0.0136
C12	0.200				0.0136
C13	0.600				0.0408
C21		0.105			0.0152
C22		0.637			0.0924
C23		0.258			0.0374
C31			0.455		0.1142
C32			0.455		0.1142
C33			0.091		0.0228
C41				0.833	0.4473
C42				0.167	0.0945

CI=0.068×0+0.145×0.019+0.251×0+0=0.002755
RI=0.068×0.58+0.145×0.58+0.251×0.58+0=0.26912
CR=0.002755÷0.26912=0.0102<0.10
So we can think that judgment matrix has satisfied consistency.

4   Analysis and Discussion

Through the evaluation result we can find that: after-sale service of related element sequencing in community service industry service quality comprehensive evaluation has located in top important position, as service nature industry, now after-sale service has been key point by consumers attention.

Through above evaluation indexes we can find that: hierarchy analysis method can objectively and truly reflect comprehensive evaluation results, and by mathematical modeling method can make qualitative problems quantify, this method's thinking is clear and calculation is simple, and has certain universality and realistic significance. But still has some insufficiency such as: people subjective factors instability easily produces major impact to evaluation results.

References

1. Sun, H.: Simply Talk About Convenience Store Marketing Strategy. Business Time (10), 19 (2006)
2. Du, D., Pang, Q.: Modern Comprehensive Evaluation Method and Case Selection, pp. 41–42. Tsinghua University Press, Beijing (2005)
3. Wang, H., Zhao, L.: Judgment Matrix Improvement of Hierarchy Analysis Method. Statistical Education (05), 20–22 (2004)
4. Yang, H., Ma, D.: Judgment Matrix Consistency Research of Hierarchy Analysis Method. Modern Electronic Technique (30), 19 (2007)
5. Liu, Y.: New Development of Service Quality Evaluation Theory Research, vol. 02, p. 80. Tianjin Finance Institute School Newspaper, Tianjin (2006)

Research on Policy Fluctuations on Coping with Australia Climatic Change

Xiaobo Hou and Zhenqing Sun

School of Economics and Management, Tianjin University of Science & Technology,
Tianjin, China
paulabobo@126.com

Abstract. The political situation in Australia had changed in July 2010, which Gillard accedes to the 27th premier. The paper takes Australia climatic change policy as a point since 2007. It reviews the changes of climatic change policy for emphasis between Kevin Rudd and the Gillard, and then it analyzes the attributions of the climatic change policy and appraises its significances and the enlightenments.

Keywords: Climatic change, Policy, javascript:void(0)Enlightenment, Australia.

1 Introduction

Since the joint signature on the Kyoto Protocol, the Australia government has taken a positive attitude toward climate change for keeping seeking for better policy. On the one hand, it is the gaming result of various domestic interest groups; on the other hand, it profoundly shows that climate change has become the focal problem in policy of all countries.

2 The Process and Performance of Policy Fluctuations on Australia Climatic Change

From Kevin Rudd Government in the end of 2007 to Gillard Government in July 2010, Australia has been keeping adjust the policy on coping with climate change.

Kevin Rudd Government Period (2007.12-2010.06), government policies on Australia climate change. In Dec. 2007, Kevin Rudd signed Kyoto Protocol immediately after he was sworn in, which obviously assumed Australia Government determination on dealing with climate change. He said, "It is an very important step our country and international society should take together for slowing down climate change."[1] In Feb. 2008, Australia reached a consensus with New Zealand on coping with climate change together[2].

In June 2008, Kevin Rudd gave an opinion that Australia need not use nuclear power to solve the problem on reducing greenhouse gas emissions. In order to realize the target that based on the emissions in 2000, the greenhouse gas emissions will be cut 60% by 2050, the government would pay more attention on renewable energy, including wind energy, solar energy and so on, and make a short-term goal.

J. Luo (Ed.): Soft Computing in Information Communication Technology, AISC 158, pp. 465–473.
springerlink.com
© Springer-Verlag Berlin Heidelberg 2012

In Dec. 2008, Kevin Rudd proclaimed some measures should be taken to reduce greenhouse gas emissions and the anticipated goals are shown as follows: Greenhouse gas emissions will be cut 5% to 15% by 2020 compared with that in 2000; Industrial polluting enterprises can discharge carbon-containing material after getting the permission of the government. However, Kevin Rudd was accused of those measures he proclaimed. The critics insisted that the measures were not adequate. The reduction of greenhouse gas emissions the above goals showed was far away from that of some environmentalists suggested to prevent from horrifying climate change. Compared to the goal European Union raised that the greenhouse gas emissions would be cut 20% (or 30%) by 2020 compared with that of 1990, the goal the Australia proclaimed was inadequate. [3] In the same month, the government proclaimed a legislation draft on renewable energy that the share of renewable energy in the energy structure would account for 20% by 2020 and the corresponding measures including, encouraging every family to install solar system, promoting the renewable energy investment such as, wind, solar, geothermal energy and so on[4].

In May 2009, Kevin Rudd brought forward a plan for building the largest solar power plant by which the whole country covered solar energy net would be formed in Australia and Australia would be the most famous country with renewable and clean energy in the world. At an estimated cost of over A$1.4 billion ($1.05 billion), the government decided to allocate A$4.65 billion ($3.48 billion) for clean energy projects, including the solar energy plant[5].

Since Kevin Rudd Government came to power, he has actively promoted the development of his carbon cap-and-trade plan, which required that enterprises could discharge exhaust gas after getting the government permission that can be exchanged in the market. On August 2009, all the Opposite senators opposed the bills, including carbon cap-and-trade plan, when the bill was put forward for the first time; The revised bill was rejected by the Senate again in the same year December. From the beginning, the bill had received extensive attentions and controversy: public opinion of the Media and the populace showed that this plan would improve the cost of consumers and enterprises and then impair Australia economy development and could not curb climate change effectively. Among the Opposite Party Union serious disagreements arose, which made the Opposite Leader Malcolm Turnbull who advocated pushing forward that bill through bipartisanship, was ousted from the leading position during the inside struggle of Liberal Party, and Tony Abbott who had a negative opinion took Malcolm Turnbull's place[6].

In January 2010, Kevin Rudd appealed to the public in New Year's message that the general public should take some measures on coping with climate change, which is in accord with the national interest. He insisted, "If we do not take actions (to curb climate change) at home and abroad, it will be a shameless betrayal to our future generations who should have the right to enjoy the natural resources including Great Barrier Reef." [7]

In April 2010, Kevin Rudd declared that "Carbon Pollution Reduction Scheme" should be postponed to 2012, which deviated from his promise in the last election in 2007. Kevin Rudd believed that the main cause of the emissions reduction delay is attributed to the disfavor of the Opposite and the slow progress of coping with climate change all over the world. However, the Opposite---- the leader of the Liberal Albert offered his criticism that Kevin Rudd tried to get around the issue on climate change and he should work out some concrete policies on climate change. Shadow cabinet finance ministry spokesman said, Kevin Rudd's decision was a kind of financial trick, which was used to cover the financial deficit through removing the carbon-trading amount from the budget[8].

In May 2010, Australia finance ministry declared a plan on imposing a kind of "Resource Super Profits Tax". According to the plan, from July 1st 2012, the government will impose a very high tax RSPT, even up to 40%, from the whole non-renewable energy field. The plan will make all the effective tax rate of Mining Company like Broken Hill Proprietary Billiton Ltd. goes up from 43% to 57%, which means Australia resources industry will be the industry with the highest tax in the world[9]. Right before the election, "Resource Super Profits Tax" made the Australia Mining Co. market value crashed, and Kevin Rudd's approval ratings had been in an accelerated decline. Therefore, the delay of "Carbon Pollution Reduction Scheme" played a role in the out of power of Kevin Rudd.

Gillard Government Period (2010.06--present), government policies on Australia climate change. Australia Green Stock Exchange officially opened in June 2010 in Sydney. The Exchange was established in terms of the joint venture of Australia Finance and Energy Exchange Group and Australia National Stock Exchange, which provided the special capital market service for clean technology and green energy companies. Australia Indigo technology Co. Ltd. applied to the Stock Exchange immediately to be the first listed company.

In July 2010, Gillard Government proclaimed that RSPT would be cut down from 40% in Kevin Rudd Government to 30% and enter into force in July in 2010.That was a great concession to mining industry made by Gillard and it was also a practical measure taken in election year. Not called RSPT, the revised mine tax was called "Minerals Resource Rent Tax (MRRT)", and the objects of taxation were iron ore and coal ore. According to the new standard, the tax was reduced about A$1.5 billion compared with that of the original plan of Kevin Rudd. [10] In the same year, Gillard tried to play down the issue on imposing carbon emission extraordinary tax and claimed that the carbon cap-and-trade plan would not resume before 2012. [11] Besides, the government made a decision that A$1 billion would be invested into market to build the renewable energy market throughout the country and A$100 million would be put into the renewable energy technology research in the next 10 years; The new power-station would be strictly prohibited coal-fired electricity generation from now on. The new plan also included the establishment of a citizen parliament and a Committee of experts, which can be used to seek for nation-wide consensus on climate change, "Australia need take actions on climate change and also need come up to nation-wide consensus." [12]

In Oct. 2010, Gillard declared that the government would work out a concrete plan on how to impose tax on the large Minerals Industry Companies in the recent. The government has always supported the mine-use tax and other desired adding taxes plan, she said, "we will not agree on the way that local governments in all the states take use of federal government to improve the local tax revenue." [13]

Fluctuations in climate change policies between the two Governments. Form "Resource Super Profits Tax (RSPT)" to "Minerals Resource Rent Tax (MRRT)".

Kevin Rudd Government proclaimed a plan on imposing RSPT on May 2,2010. According to the plan, Australia Government will levy a high tax up to 40% on all the fields of non-renewable energy and the plan will take effect from July 1st 2012, which made Australia Minerals Industry Companies market value crash and the approval rating was in an accelerated decline. Since "Resource Super Profits Tax (RSPT)" was put forward, Australia Minerals Industry Companies market value had shrunk for about A$180 billion in two days. Minerals Industry Companies came out against government for the unsatisfying result that Australia would be a country imposing the highest tax. Complaints are heard all over, not only in the Minerals Industry Companies, but also from a recent survey covering the views of Australian main institutional investors who blamed the plan as a "clumsy conception".

On June 24th 2010, under the pressure of Australian Labor Party, Kevin Rudd had to make way for Vice-Premier Julia Gillard who became the first woman Prime Minister. On July 2, Gillard Government announced the revised "Minerals Resource Rent Tax (MRRT)", making concessions to Minerals Industry Companies. Compared with former plan proposed by Kevin Rudd, the new plan focused on the tax on iron ore and coal ore, and the decline of tax rate from 40% to 30%. However, the former 40% rental tax rate of offshore oil and gas resources would expand to onshore oil and gas resources, including coal-bed methane (CBM), which received an enthusiastic ovation from most Minerals Industry Companies.

From actively promote "Cap And Trade" Emissions Plan to Australia Delays "Carbon Emission Reduction Program" to play down the issue on imposing "Carbon Emissions Extraordinary Tax"

Australia Government was scheduled to reduce carbon dioxide emission to 5% before 2020, which compelled almost 1,000 large-scale enterprises who should exhaust fumes to buy a special permit of waste gas emission. In Sep. 2009, all the members of the Opposite in Senate objected "cap and trade" emissions plan, which was proposed by Australia Government; The Senate vetoed the revised proposal again in Dec. 2009; Because of the obstruction in parliament in April 2010, Australia Government will not push forward the "cap and trade" emissions plan until at the expiration of Kyoto Protocol in 2012. The decision on the delay of carbon emission reduction plan played a role on the out of power of Kevin Rudd. As an important part supported by the center-left Labor Party Government, the progressive voters were disappointed at the result that Kevin Rudd changed his proposal from considering climate change as "the most important moral problem in our times " to delay the emission reduction plan.

The newly appointed Prime Minister Gillard prepared to play down the plan of imposing carbon emission temporary tax and proclaimed that the carbon cap-and-trade plan would not be resumed till 2012. She promised that she would make much more comments on the issue in the near future, and she declared that government would make assessment on the relevant carbon trading social consensus and the international carbon pricing process, and then government would decide whether resume the plan or not, "In my view, carbon pricing should be completed through market mechanism, that is Carbon Pollution Reduction Scheme, and the timeframe of 2012 lies there." "We want to reduce carbon emission in our country, so we should hold a discussion on the plan and what we should do before 2012."[14]

Table 1. Goodbye Resource Super Profits Tax, hello Mineral Resource Rent Tax

	Mineral Resource Rent Tax	Revised Petroleum Resource Rent Tax
Headline rate	30%	40%
Date of application	1 July 2012	1 July 2012 (for projects not currently subject to PRRT)
Coverage	All iron ore and coal mining projects	All oil, gas and coal seam methane projects, onshore and offshore Australia
Capital Expenditure	New capital expenditure immediately deductible	New capital expenditure immediately deductible
Transferability of expenditure between projects	Yes	Limited transfer of losses
Extraction allowance	25% of otherwise taxable profit	No additional allowance
Uplift Rate	Long term bond rate plus 7%	Eight uplift rates for capital expenditure
Allowance for existing project expenditure	Yes - choice of book value or 1 May 2010 market value. Where market value elected, no uplift available and amount written off over a period up to 25 years	Yes - choice of book value or 1 May 2010 market value
State Royalties	Fully creditable - no refund if royalties exceed MRRT liability and not transferrable between projects	Fully creditable - no refund if royalties exceed PRRT liability
Exemptions	Miners with resource profits below $50 million per annum will not be subject to MRRT	No exemptions currently contemplated
	Mineral Resource Rent Tax	

Source: 《Goodbye Resource Super Profits Tax, hello Mineral Resource Rent Tax》
[2007.07.08]
http://www.mallesons.com/publications/2010/Jul/10403773w.htm

3   Attribution Analysis on the Fluctuations of Australia Climate Change Policies

Through straightening out the climate change policies fluctuations in the recent three years and drawing the reasons form the fluctuations, the author makes an analysis in the following key points.

The attitude of domestic voters. Kevin Rudd won victory over Howard and got the support of voters in 2007, which was mainly attributed to Kevin Rudd's active reaction on climate change that he signed Kyoto Protocol as soon as he was sworn in. He actively pushed forward various policies and measures on climate change during the three years in power; However, right before the election, when "Resource Super Profits Tax (RSPT)" was put forward, Australia Mining Industry market value crashed and the supporting rate was in an accelerated decline, and the delay of Carbon Pollution Reduction Scheme played a role on the out of power of Kevin Rudd, and the progressive voters were in disappointment onu the result that Kevin Rudd changed his proposal from considering climate change as "the most important moral problem in our times " to delay the emission reduction plan.

As for Gillard, she announced one important point in her policies when she was sworn in, that is restarting the carbon emission trading plan, pricing for the emission and promoting the application of green and renewable energy for climate change. Therefore, she gained a high supporting rate from the majority of the people, which was mainly credited to her various new policies, including the tax change Form "Resource Super Profits Tax (RSPT)" to "Minerals Resource Rent Tax (MRRT)", the tax rate change from 40% to 30%, which received an enthusiastic ovation from most Minerals Industry Companies.

Based on the need of Australia own interests. Australia carried out from beginning to end the principle of setting national interests before all the other interests to work out climate change policy. Because of the complexity and uncertainty of the climate change issue, being confused about the developing trend, Howard Government believed that the scientific study of climate change was not necessary and the economic loss would be huge for the performance, so he refused the permission; While Kevin Rudd Government, on the basis of national interests, adapted to the trend and signed the agreement. In the 2010 New Year's message, he declared that the climate change policies were in accord with the national interests; And a series of adjusting policies made by Gillard Government in the same carried out from beginning to end the principle of setting national interests before all the other interests.

On the one hand, many kinds of international mechanism for adaptation and slowing down of climate change are increasingly developed; The uncertainty of many fields is slowing down; The cost and profit for the performance can be accurately count out, and long term benefit will be gained by taking part in the international mechanism. On the other hand, the various of international developing mechanism for adaptation and slowing down of climate change are in progress now, which realized win-win of economy development and environment protection, illustrated an attracting bright future. Besides, with the fierce competition of technology coping with climate change, it is an excellent opportunity to develop and strengthen economic competitiveness, getting the advantages on climate change policies for the huge opportunities for investment and interests[15]. As shown in a latest report from a research institution in Australia, taking active measures on climate change can offer about 56 opportunities to promote economy development in Australia. The emission reduction index will be achieved in a form with relatively low cost. Otherwise, the more you delay the measures, the bigger loss and cost you will pay for. [16]

The results of game among domestic interest groups. Each adjustment of Australian climate change policies will get a fastest reaction from all the domestic Parties and interest groups. Kevin Rudd declared some measures on cutting greenhouse emissions in Dec. 2008, which got an immediate criticism: the measures were not adequate, the spokesmen of Green Party, Christine Milne insisted that the policy of Government was a "total mistake" and the reduction goal of minimal 5% greenhouse emissions was a "global embarrassment".

When the news of "Resource Super Profits Tax (RSPT)" distributed, both politics and business world were shocked. Australian Mining Industry put the policy under a boycott; so Labor Party pushed Gillard to take the place of Kevin Rudd as PM under the pressure of the election.

"Carbon Emissions Charge Policy" proposed by Kevin Rudd Government had gone through three times hardships and the new resource tax also objected by the Opposite whose resources affairs spokesman Mcfarlane said that hundreds of companies would be affected and the policy would scare the international investor.

4   Enlightenment for Our Country

Seen from the policies change in Australia, the climate change have been a important problem not only in the environment field but also a serious politic issue in the international and domestic situation, even involving national development. Each country can in no way ignore the issue.

National interests come before everything else. Seen from the fluctuations in policies of coping with Australia climatic change, all the change are made according to national interests. The fact that Kevin Rudd delayed the carbon cap-and-trade plan and the new resources tax proposed by Gillard, are totally concerned from the view of nation development in the future. Any measures we take should in accord with domestic economy and social development.

To coordinate the interest groups and set up a system concept. Climate change is a complicated problem, involving many sovereign states, special fields and industrial departments, so system concept is required when we dealing with the problem. From the above statement and analysis, we can conclude that once all the interest groups can get their desired benefit through coordination, the adjustment of climate change policies can achieve success; If the conflict of Interest existing among the interest groups, the policies will fail. Although our Government controls the majority resources in the country with strong policy enforcement, we still need comprehensive concern many aspects such as energy, environment, economy, politics and so on and coordinate the interests of various governments, industries and enterprises. Otherwise, the phenomenon that the apparent support but actually taking negative attitude, will cause resources wasting and miss a good chance for economic transition.

Increasing research funding and achieving technique innovation. Kevin Rudd Government offered A\$4.65 billion (\$3.48 billion) to a plan on building the global largest solar power plant for providing clean energy, which will cost more than A\$1.4 billion (\$1.05 billion); Gillard Government will invest almost one billion Australia dollars to establish the renewable energy market around the country in the next ten years and about A\$100 million for the renewable energy technology research. In our national eleventh Five-Year technological plan (2006-2010), the available fund for energy conservation and emission reduction is about 7 billion Yuan. Seen from the experience in Australia, China is required to raise funds from multiple sources and attract more social investment for the technology research of climate change. Although we have strengthened our investment on climate change in the recent years, we should pay more attention to the fact that the activity of enterprises R&D has not cultivated. During the process of coping with climate change, we should improve our comprehensive competitive strength through straightening system mechanism, stimulating the enthusiasm of enterprise R&D, and bringing the mainstay of enterprise technological innovation into full play.

In a word, climate change is a common issue faced by people around the world. As a developing country, China should take the responsibility of dealing with the problem. However, China is a developing great power is the factor what we must first take into consideration. There are still quite a large number of impoverished people and economic development is not balanced around the country. Only focusing on the economy development can we have enough energy and ability to deal with climate change and win the war of coping with climate change at last.

References

1. Australia Signs Kyoto Protocol (December 04, 2007), http://news.sina.com.cn/o/2007-12-04/090013017434s.shtml (October 28, 2010)
2. Australia Forms a Partnership with New Zealand on Coping with Climate Change (February 29, 2008), http://env.people.com.cn/GB/6941890.html (October 28, 2010)
3. Australia Proclaims New Measures on Greenhouse Gas Reduction (January 18, 2008), http://lvse.sohu.com/20081218/n261282884.shtml (December 28, 2010)
4. Australia Announces the Legislation Draft of Renewable Energy (December 19, 2008), http://www.hbzhan.com/News/Detail/7770.html (October 28, 2010)
5. Australia Plans to Build the Largest Solar Power Plant in the World (May 17, 2009), http://news.163.com/09/0517/23/59I7C9300001121M.html (October 28, 2010)
6. Australia Government Gets a Resistance Again on Carbon Emissions Trading Plan (December 02, 2009), http://gb.cri.cn/27824/2009/12/02/2225s2693477.htm (October 28, 2010)
7. Rudd, K.: Appeals to the Public for Emissions Reduction for Australia gets the Server Damage (January 03, 2010), http://news.qq.com/a/20100103/000041.htm (December 28, 2010)
8. Australian PM Claims to Shelve Emission Reduction Plan for the objection of the Opposite (April 28, 2010), http://news.zman.cn/dibvguoji/adlyzlcy_ijbbff.htm (December 28, 2010)

9. Recent Development and Foreign Media Comments on Australia Resources Tax Reform (August 24, 2010), http://www.penmeiwang.com/show.php?contentid=18500 (October 28, 2010)
10. Li, J.: Gillard's Second Action—— Cut the RSPT proposed by Kevin Rudd (July 02, 2010), http://world.people.com.cn/GB/57507/12043038.html (October 28, 2010)
11. Australia PM Not Consider Imposing Carbon emissions Temporary Tax Right Now (July 8, 2010), http://www.hnys.gov.cn/Read.asp?IC_ID=89458 (October 28, 2010)
12. Australia Proclaims New Policy on Climate Change (July 23, 2010), http://finance.ifeng.com/news/special/cxcmzk/20100723/2439742.shtml (October 28, 2010)
13. Australia Will Announce Detail Tax Rules on Large Scale Mining Industry Company (October 20, 2010), http://www.dayoo.com/roll/201010/20/10000307_103748274.htm (October 28, 2010)
14. Australia PM Not Consider Imposing Carbon emissions Temporary Tax Right Now (July 8, 2010), http://cdm.ccchina.gov.cn/web/NewsInfo.asp?NewsId=4597 (October 28, 2010)
15. Li, W., He, J.: The Reading and Evaluation on Australia Climate Change Policy. Contemporary Asia-Pacific Studies (01), 108–123 (2008)
16. Climate Works Australia.Low Carbon Growth Plan for Australia (March 2010), http://www.climateworksaustralia.org

Problems in the Ideological and Political Education in Contemporary Colleges and Their Countermeasures

Luquan Zhang

School of Economics, Tianjin Polytechnic University, Tianjin, China
159160816@qq.com

Abstract. Ideological and political education is the battle position of moral education of colleges and universities and an important way to enhance the thought quality of students. The investigation shows that there is a certain gap between actual results of the course and our expectations. So the reform must be carried on from the aspects of direction, effectiveness, and subjectivity, etc.

Keywords: Ideological and political education, problems, countermeasures.

1 Introduction

Stick with the combination of teaching and educating people; stick with the combination of education and self-education; stick with the combination of political theory education and social practices; stick with the combination of solving ideological problems and solving practical problems; stick with the combination of education and management; stick with the combination of inheriting and improvement & innovation. The "six combinations" principle is the integrity practical conception system of enhancing and improving the ideological and political education of college students'. The contemporary students should hold and implement the "six combination principles", they should widen their eyes, advance with the times and seek truth.

2 The Contemporary Situation of Ideological and Political Education in Colleges

Although the teaching reform has achieved some results in recent years, in the results of the questionnaire survey on "the impression or influence of teaching on my practical thought", 46.7% freshmen and 31.6% senior students agree definitely, indicating that some problems still exist.

First, there is a big gap between teaching content and ideological practice of students'. The survey shows, 87.6% freshmen and 75.4% sophomores, junior students and senior students approve that it is essential to go on ideological and political education in colleges, besides, 73.2% freshmen and 61.4% other students believe that it relates to their ideological practice tightly. However, almost all the students suggest making the course tie to the real life. The contradictory indicates that the students realize the importance of ideological and political education, but not to agree with the teachers' practical instruction. Only 46.4% freshmen and 31.6% other students think the instructions have an effect on their ideological practice. That strongly proves that students are still eager to the instructions to morals and ethics after entering the college, which has a difference with given instructions. Many teachers pay more attention to theory, ignoring the importance of the combination of theory and students' real life. Thus, the students can't relate the content to real life and the teaching effect is discounted.

Second, colleges and universities pay inadequate attention to the ideological and political education and the ideological and political work team staffs are not enough. At present, not few colleges and universities pay inadequate attention to the ideological and political education. Due to the ignorance of ideological political workers, the counselors and communist youth league cadres have not high statures, low pays, difficulty in professional evaluation, housing and other problems. Besides, not few increase the pays of ordinary teachers, which makes teachers' pay are more than political work cadres'. As its results, some teachers who work for ideological and political education turn to major education, which leads to insecure political work cadres and instable team. Due to the serious effect, we must highly value this problem[1].

Third, practice is lack. The basic principle of education is the principle of unity of knowing-doing, with reference to ideological and political education, which refers to the combination of theory education and practice, united knowing with doing to make ideological and moral conceptions internalized as ideological morality. Knowing is the premise of doing, knowing is a process of conscious awareness. Practice is the essential way, which includes two connections: one is advocated by moral education that values should become students' belief, pursuit and moral desire, to form behavior motivation of morality; the other is moral's "practical spirit", which requires that the moral conception finally acts on one's activities, through the experiences of moral emotions, forming and consolidating moral ideas during practices to become conscious moral obligation, thus finishing "unity of knowing-doing". But since many years, no matter the theoretical instructions or major learning, teachers do a lot of work in the aspect of "knowing", however, the truly internalized sector of moral ideas is not much. The final reason is to lack of the moral practice. All in all, the phenomenon that knowing breaks with doing generally exists in the ideological and political education work, featuring as "non-knowing-non-doing", "knowing-non-doing" and "knowing but wrong doing". The plight highly dilutes the effectiveness of ideological and political education in colleges[2].

Fourth, the methods of education lack of flexibility. Ideological and political education is used to rigid indoctrination all the times, forming education model with the style of executive order, lack of effective counseling education, inspired education, instruction and practice education, ignoring the main need of students and spiritual communication, to look on improvement of ideological and political qualities and cultivation of moral quality as acceptation and understanding of science and technology. The conceptions of people oriented, respect for the students' characters and statures, understanding students, respecting students, loving students and focusing on students' real need are not throughout the process of education. The situation that students accept education initiatively according to their need and social development is not formed so that students receive education in passive positions, without consciousness and initiative. The methods of education are mechanized. In practical teaching, teachers are used to teaching by classroom teaching, traditional report, lectures, etc, which could not attract students. They do weak in linking theory with practice and are short of targeted methods such as lectures, debate, discussion, observation and experience. In terms of classroom teaching, only "a book, a pen", which seldom uses modernized teaching methods such as multimedia. Such is the main reason for students to cut classes, sleep much and do other things. Other ways like meeting and activities, are only echoing what the books say, reading paper or materials, the activities are all the time alike with no new meaning or originality, lacking of attraction. Due to tense funds, worrying about security problems, some favorable activities are not well functioned. So the effectiveness of ideological and political education is difficult to be distinguished.

3   Basic Solutions and Suggestions

First, renew ideas, consider moral to be fundamental and establish an open view of the ideological and political work system. Enhance and improve the ideological education work. Improve the efficiency of students' ideological education work. The basis is to renew ideas. In the new period, these new ideas must be set in the work. Teachers must keep pace with times with new education ideas and thought and have a deep understanding in the ideological and learning states, preventing from big theories which make students bored and unrealistic. Teachers can't pay too much attention to teach theories but neglect its relation to real life. They had better take examples for the real things happened in real life, combine with self practical condition, explore new trend and problems of ideological education, to make ideological and political workers to renew knowledge, exchange experiences, widen their eyes, keep up with changing situations and propel the ideological and political education effectively.

Second, the colleges and universities have to place ideological and political education in a prominent position and construct the scientific team mode. Strengthen leadership and look on construction of the instructors and class teachers as the important landmark of qualities and standards of schools and important criterion on assessment leaders. According to the more anticipation for teachers' qualities, some work is efficient to do. One is to enhance the teachers' selection, employment and capability of management to configure better teachers for ideological moral courses, and strengthen the management by observation and evaluation. The other is to create more chance for teachers to go on learning. The last is to stimulate teachers to improve their own qualities initiatively by methods of selecting or evaluating teachers. Improve the effect of education and train professionally constantly. If instructors and class teachers want to adapt to the work reality of ideological and political education, they must improve their comprehensive qualities and accept training and education in political and professional qualities.

Third, social practice is the essential section of ideological and political education, which plays an irreplaceable role in promoting students to know about society and national conditions, enhance their abilities, delicate to society, exercise perseverance, cultivate qualities and enhance their social responsibility. In the process of practice, students can know about the development and change of country since reform and open-up. In the process of practice, I have a deep sense in the masses' guileless, kindness and hard working, besides I know more about the fine traditions of Chinese nation, to enhance my national pride and national confidence and improve my moral standards and self qualities. These comrades wish students to set up ideas and emotions, which has a good effect by social practice. All in all, practice helps students know and understand society, establish the correct concept of career, solve the doubts and difficulties in ideology and enhance their sense of historic mission and social responsibility, so that various practical activities truly become the process of students' self-education and find a new way to enhance the efficiency of ideological and political education.

Fourth, insist complex instilling principle must combine with various education modes, methods and ways, besides the purposed, planned and organized teaching of scientific theory of Marxism and socialism. Instilling education must be combined with inspired grooming, complementing and promoting each other, "Cramming education", "Gulping down a whole date", "Saying what the books say" must be avoided in order to perfectly integrating philosophy and art. The level of indoctrination must be paid attention to, according to their ages, receptivity and self- needs, teach according to their abilities. Stop "one size fits all", "looking down from high". Enrich the teaching content, conform to the times, follow the law of development, enrich creativity and combine theory with practice, refusing to get in a groove and quote out of context. Choose a good vehicle for instilling, change the steady education mode such as class meeting, reports, etc. Bring practical education, network education, cultural education and investigation education into education mode and establish a democratic mode.

Students are precious human source, national hope, national future and builders and successors of the socialist cause of distinct Chinese characteristics, whose qualities directly affect the realization of the goal of building a well-off society in an all-round way. So various colleges and teachers must be fully aware of the importance of the ideological education, combination with practical situation of colleges and explore the new trend and new problems of ideological education to make students to renew knowledge, exchange experiences, expand minds. According to changing situation, promote ideological and political education effectively.

References

1. Yang, X.: To strengthen college students' ideological and political education in an effective way. Research on Education of Shanghai University of Engineering Science (2), 49–51 (2005)
2. He, B.: About college students' ideological and political education validity of thinking. Heilongjiang Science and Technology Information (2), 215 (2009)

Study on Cause Analysis and Educational Solutions of Learning Disabilities

Luquan Zhang

School of Economics, Tianjin Polytechnic University, Tianjin, China
159160816@qq.com

Abstract. "Education of learning disabilities" is vital work of colleges, with expanding range of higher education in recent years, the educational problem of "learning disabilities" causes more and more attention. This paper puts forward the corresponding education policy, on the basis of analysis of causes of learning disabilities , mainly including no respect of persons, guidance to different types of areas, psychological consultation, cultural construction and creating an atmosphere of mutual help and common progress, reviewed to provide a convenient reference for theory.

Keywords: colleges and universities, learning disabilities, causes, education countermeasures.

1   Introduction

"Education of learning disabilities" of colleges and universities has become a hot educational topic attracting attention from all walks of life in society. The common aspiration of The National Student Loan implemented by State Council, admission green channel and tuition waiver implemented by State Education Ministry, award, loan, assistance, compensation, avoiding and other measures conducted by all the universities, social patronizing and school charitable fund is "never let a student drop out of school due to poverty". "poor students" initially refers to students whose families are poor. With the updating and development of education and training conception and practice, the educational workers of colleges and universities are enriching the content of poor students continually. Students with lower adaptiveness, thinking backward students and students with poor self-care ability are classified as different "poor students" groups.

In recent years, especially after the college expansion plan, with the development of higher education from "elite education" to "mass instruction", more chances to receive higher education have been given, which caused diversification of the quality of students. So a new group generates in the mass---learning disabilities. Due to its universality and particularity, to solve this problem, we must clearly understand causes and suit one's method to the situation.

2   The Analysis of Causes of Learning Disabilities

Learning disabilities generally refer to students with normal intelligence and physical qualities. Due to learning methods, learning motivation, learning psychology, willingness

J. Luo (Ed.): Soft Computing in Information Communication Technology, AISC 158, pp. 479–483.
springerlink.com   © Springer-Verlag Berlin Heidelberg 2012

and other reasons, who have low integrated level of learning, academic performance significantly behind the other students and not meet the basic requirements of college students. The causes are complicated (shown as Figure 1), the main causes are as follows:

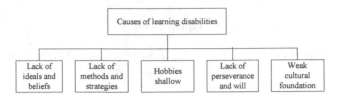

Fig. 1. The analysis of causes of learning disabilities

Near ideals and power shortage. Most learning disabilities have not enough motivation because of not having lofty ideals, who don't know why they have to study in colleges. The situation is causes by family education and primary education. During the process in high school and before, their ultimate goal is to go to college. However, teachers and parents never teach them what has to do after entering colleges and they never think the problem by themselves. After experiencing "ten years of sweat and swot", "the hordes of troops and horses lead only wooden bridge", they generate the ideas of "resting your feet", which is the common saying "idealistic stage of failure". Some students relax themselves completely. Some students are hard to endure days, not knowing how to spend their college time or how to plan their careers.

Wrong methods, be at loose ends. The teaching mode of college leads to the big differences between learning methods of college students and those of primary and secondary students. The college learning emphasizes more on autonomy, flexibility and exploration. Having not hold this rule, learning disabilities how nothing to plan their study, can't learn by themselves and fritter away the reliance on teachers. They are working hard, but having no effect. They learn in library every day, but with low learning efficiency. They can't find the suitable learning methods and strategies.

Lukewarm interest and here today, gone tomorrow. Some learning disabilities have no interest in their own majors. The reason is that some students' majors are decided by their parents' will; some are decided by their teachers according to the entrance scores of various colleges; some students are blind to pursue the hot majors. Such makes them passive after entering colleges, which makes them difficult to learn well, even thinking about day to day.

Weak willed, going with the stream. The mode of colleges and daily management give more time for students to freely arrange. Some learning disabilities are weak in controlling themselves, with no idea about how to manage their spare time. Some are indulging in online games, some are learning and practicing with people who have special skills because of the thought to be hot, some are making up cheerful times they have missed in primary and middle schools, who spend most of their time on playing, which has resulted in the situation that failing in many courses.

Weak basis, having a heart but no strength. Some learning disabilities come from remote and backward districts, whose educational standards decide their knowledge level and learning capacities' big distance to that of students in other districts, especially in English and math. To make up the shortage, they have to spend much time and energy on learning. Although with desires to make progress, the unfavorable results have a bad effect on other courses.

3   The Educational Countermeasures to Learning Disabilities

According to causes of learning disabilities, as a college educator, we should start from practical situation of every learning disability, giving the right medicine. Meanwhile, build the good learning atmosphere in general, grasp the construction of the learning style, help and conduct learning disabilities to solve their problems, to lead them to the road accessing to life. The main educational countermeasures are as follows (Figure 2):

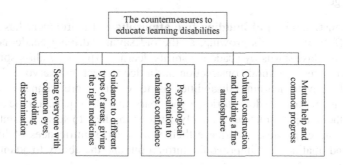

Fig. 2. The educational countermeasures to learning disabilities

Seeing everyone with common eyes, avoiding discrimination. As college educators, firstly, we should treat learning disabilities with right attitudes, set a conception of overall cultivation, promotion to overall development. While facing learning disabilities, we should not have bias or discrimination, but to treat them with developing eyes, with the firm belief that everyone can succeed and the difficulties are temporary and can be avoided. In common life, they have to care for, help and love students warmly. The counselors and classroom teachers must meet the good, enhance the communication and exchange with learning disabilities, create conditions and conduct them to exert their advantages to promote and develop their maximum capacities and potentials.

Guidance to different types of areas, giving the right medicines. Because every learning disability is different, we must insist the method of making a concrete analysis of a concrete problem. To different learning disability, teachers should be fully aware of their performances, hobbies, familial environment, grown background and interpersonal relationship. Holding the various factors of their difficulties in learning, conduct and help students to self-analysis, change their thoughts of attributing their weak performances to the uncontrollable external factor, make the suitable plans and methods, exert their advantages, so as to change the motivating system of learning and improve their learning performances.

Psychological consultation to enhance their confidences. Confucius summarized the psychological conditions of success in learning as "wisdom is wise at moral. They who know the truth are not equal to those who love it, and they who love it are not equal to those who delight in it. 'Xin' is believing in and loving ancient, that is to say, confidence, will and belief. 'heng', that is students should study perseveringly, referring to perseverance. 'xu' , that is having but seem to nothing, referring to an open mind. ". It can be seen that learning disabilities have psychological obstacles and difficult in treating their facing difficulties, with the situation of anxiety, inferiority, collision, solitude even disparity. The work must stress on everyone to help them analyze causes, find their own shining points, set confidence and cultivate them to overcome difficulties, never fearing hardships and having fine qualities of sacrificing. Become their friends and source of absorbing motivation by enhancing communication and exchange.

Cultural construction and building fine atmosphere. Environment has a huge and potential effect on people's growing, as the old saying "driving academic spirits by school spirits, class spirits by academic spirits, learning spirits by class spirits". It can be seen that campus cultural construction has a deep effect on everyone. Enhance the educational help to learning disabilities is to create more chances for learning disabilities to go out of dormitories, cyber bars and their psychological enclosure, encourage them to join practice positively. Encourage them to serve students with their knowledge and capacities, experience their values and significances of life, exercise their will and qualities and improve the throughput and the ability to adapt to the new environment.

Mutual help and common progress. More successive problems happen in students' life and intercourse because of learning difficulties, they will become more inferior, solitude even commit suicide. As responsible educators, we are not allowed to be forgotten by this group. So colleges should pay attention to enhance constructions of class and dormitory, exert overall on student groups, develop mutual psychological help and mutual learning help, form and establish the policy of helping and assisting learning disabilities. Through teachers' instructions, solidarity and friendship among students and mutual helps, let learning disabilities feel warm, valued, loved and the true motion among people, cultivate their positive attitude to life.

In general, learning disabilities is an important group in colleges, an increasing group. No matter in the aspect of cultivating talents or management, promoting learning disabilities' transformation scientifically and effectively can reflect the improvement of educational profits. Moreover, enhancing and valuing education of learning disabilities should have been one of the important contents of students' work of colleges. Therefore, we should catch a right sight at work ideas. While propagandizing prototypes and learning from talents, we should adopt scientific work methods, promote the transformation from learning disabilities and cultivate them to become the needy talents.

References

1. Yan, Z.: Penetrating into studying weariness of college and university students. Journal of Xichang Teacher's College (02) (2002)
2. Chen, X., Lei, W.: A summary on the recent studies of student with learning disabilities in China. Journal of Henan Vocation-Technical Teachers College (Vocational Education Edition) (01) (2006)
3. Shao, L.: Analysis and countermeasures on psychological behavior features of college and university students with learning disabilities. China Science and Technology (05) (2006)
4. Zhu, Y., Zhu F., Ren, Y.: Students' learning psychological crisis in changing surroundings and intervention ways. Journal of Chuzhou Vocational and Technical College (05) (2006)
5. Wang, M., Jiang, L.: Discuss on the psychological crisis of freshman and the intervention countermeasures. China science and Technology Information (02) (2007)

References

Management of Multilevel Students after University Merger

Luquan Zhang

School of Economics, Tianjin Polytechnic University, Tianjin, China
159160816@qq.com

Abstract. University merger and enrollment expansion in higher education has given rise to the prevalence of multilevel students studying within the same university. Thus the student management is faced with a new problem and measures must be taken accordingly. The paper analyzes and reflects on this situation and proposes the corresponding measures.

Keywords: university merger, multilevel, student management.

1 Introduction

The merger and restructuring of Chinese colleges and universities has been carried out since 1992 to meet the development need of Chinese higher education in the 21st century. In 1997, Vice Premier Li Lanqing put forward on the practical basis a policy of "joint development, adjustment, cooperation and merger" to urge the reform of management system and adjustment of layout structure in higher education in a bid to create a new landscape with comprehensive universities, multi-subject universities and single subject colleges being well proportioned. And this has been the biggest reform in Chinese college layout ever since the "college adjustment" in 1952.

Yet while enjoying its benefits, we should also realize the new challenges that the merger has brought to the student management. The greatest challenge in terms of student management is the increase of student levels, which requires the corresponding management philosophy and management approach. The author analyzes and reflects on the specific situation of his own college in hopes of proposing feasible measures so as to manage the multilevel students much more scientifically, reasonably, and smoothly and provide a pleasant environment for the growth of students.

2 Current Situation of University Merger

Since the early 1990s, the management system that colleges and universities are run by the departments of the State Council under the planning system has been reformed in accordance with the policy of "joint development, adjustment, cooperation and merger". Most of the colleges and universities have been adjusted, merged, and the administration of them has been handed over to the local authorities.

J. Luo (Ed.): Soft Computing in Information Communication Technology, AISC 158, pp. 485–490.
springerlink.com   © Springer-Verlag Berlin Heidelberg 2012

Four hundred and six colleges and universities have been merged into 171 institutions since 1996, and comprehensive universities with large scale and specialties of arts, science, engineering, agriculture, and medical science have thus taken shape, for example the new Zhejiang University. Meanwhile, 360 colleges and universities of the central government have gone through the reform of "the central government working with the local authorities while the latter dominating the administration". At present, 71 colleges and universities are administered by the Ministry of Education, and above 40 more are managed by other departments of the State Council, so the total number of the institutions of the central government has decreased to about 110.

3   New Problems of Student Management That Brought by University Merger

The greatest problem of student management after university merger is that multilevel students are studying within the same university. First, their educational backgrounds are multilevel. That is to say, after the universities that mainly confer bachelor's degree and master's degree have taken in the adult education institutes and higher vocational colleges, students with various educational backgrounds are mixed together. Second, due to the transformation from the elite education to the mass education caused by the merger and enrollment expansion, the cultural and ideological awareness of the students is multilevel. Third, the tuition fees are increased by a big margin after the merger and restructuring, so that the multilevel household incomes are highlighted. Fourth, the enrollment expansion and the increasing number of students that are brought up in one-child families with distinct environments and living experiences will bring out the problem of multilevel mental adaptabilities. Fifth, different age levels are produced as a result of different requirements of majors and educational backgrounds. Take the author's college as an example: there are 2180 students in total, among which the graduate students number 45, the undergraduate students 883, the PE students 294 (28 of them entered this college in 2007 with diplomas of higher vocational colleges), the students trying to upgrade the diplomas of junior colleges to bachelor's degrees 49, and the higher vocational students 909 (355 of them entered this college in 2007 to study in the five-year consistent system and 3+2 program). In general, the students in this college are multilevel in the diplomas they get, the cultural and ideological awareness, and even only from the perspective of student source. They entered this college with diplomas of universities, junior colleges, senior high schools, technical secondary schools, junior high schools, and primary schools and they are of different ages ranging from 14 to 26. So we have here almost all the varieties of students.

The fact that multilevel students are studying within the same university made us realize in our work that if we only adhere to a single level yardstick, the student management will be biased and unfair because some special groups will be neglected, so that we will not reach the overall goal of the well-rounded development of all the students.

4   Reflection on Management of Multilevel Students and Corresponding Measures

First impressions keep a strong hold—regulate multilevel students with a unified system. Different students have different learning and living experiences, and different schools have different cultural environments and management modes. So after the merger, students will bring various habits and customs to the university. Good customs deserve to be inherited, but bad ones must be banished. In terms of study, the university will set down respective requirements for graduate students, undergraduate students, and higher vocational students on their status management; but the requirements for behaviors are basically the same, namely the provisions in the Student Handbook. A unified code of conduct needs to be obtained by the first impressions to keep a strong hold. On one hand, students should learn the provisions of the Student Handbook at the first time they attend the orientation; on the other hand, the problems that are founded during the admission process and at the beginning of their university life should be handled in time and known to everybody. Thereby, the students will understand what are permitted and encouraged and what are forbidden and need banishing, and the bad customs and habits will not develop and spread in the newly merged university. Otherwise, the student management will be plunged into chaos and become out of control.

Treat all on the same footing—create an equal, fair and harmonious environment. In the ideological and political education and daily regulation of multilevel students, the most important is to treat all on the same footing, because the special groups usually would rather not be treated specially, just like the handicapped dislike others' peculiar and sympathetic eyes.

First, show them the same respect.

It is a common sense that every human being is eager to be respected during the interaction with other people. Besides, individuals of different student levels tend to have different psychological states. In the author's college, undergraduate students are more or less prone to admire the graduate students and even feel inferior before them. The same thing goes for the PE students, the higher vocational students and the students trying to upgrade the diplomas of junior colleges to bachelor's degrees in the face of undergraduate students and graduate students. Yet all of them hope to be respected by other students, and they particularly don't want to see any contempt in teacher's eyes as a result of their weakness.

Therefore, we've learned from our work experience that respect is more effective than reasoning, criticizing, and punishing. A note written with "higher vocational students and dogs not admitted" was once founded at the door of the public teaching building of a university. So an increase of higher vocational students violating the disciplines can be easily imagined. Thus, we need to respect our students ideologically, and then show it to them, especially to their personalities with concrete actions. For instance in our college, we should not say how naughty the PE students are to other students, or call the students studying in the five-year consistent system and 3+2 program "little five" - we can name them "Customs Clearance 07" or "International Business 07" instead according to their majors and the year they entered this school.

And we should also comply with the principle that "serious without contempt, critical without blasphemy" while punishing the students.

Second, provide them with equal chances.

As educators, we have to teach the students according to their aptitudes; while as managers, we shall never divide them into different ranks. We need them to have equal chances in participating in every activity that is conducive to the growth of them. Anyway, this is written in the Student Handbook as a right enjoyed by every student. In our college for example, every academic lecture or opportunity of postgraduate recommendation is open to students of all levels, including the five-year students and the PE students; excellent PE students and higher vocational students also enjoy equal chances of gaining high scholarships; the presidents of the student union in three consecutive terms were PE students; the higher vocational students missed the chance of taking part in the university's Autumn Sports Meeting of 2007 because the date they arrived in the school is later than others, yet we still arranged them to perform in the flag array and aerobics team.

Third, rewards and punishments should be objective and fair.

Every student is equal in the face of rights and obligations. Therefore, we should make sure that the students are treated equally and fairly when we are doing the rewards and punishments.

Encouragement—help the poor students to live self-reliantly. With the advance of reform and opening up, the income gap among people is deeper and deeper. In addition, the tuition fees are increased by a big margin after the merger and enrollment expansion in 1990s. Hence a new group, or a new level of students has come into being—the poor students. The party, state, governments at all levels and educational institutions have devoted much attention to them and the student loans and subsidies have been gradually raised especially in recent years, which have eased to a large degree their living pressure and strongly guaranteed their fulfillment in the studies.

We firmly support the national policies, and we have also seriously done lots of work accordingly. Yet the student tutors should thoroughly think about the work on the poor students, especially the education work after the subsidies are dispersed. Inertia is the character that humans are born with. Most of us will not let go the chance of making gains without doing any work. Being poor is not the fault of the students, and they didn't beg for the subsidies, but it will be horrible if the poor students begin to simply rely on the help of others. Thus the inspirational education afterwards is indispensable. We must encourage them to live self-reliantly and let them know that the spiritual poverty is worse than the economic poverty.

Focus on the mental health—advance with major efforts the improvement of young students' personalities. The cultivation of young students includes two aspects—the study and the personality. The most important norm of personality is the mental health. The merger and enrollment expansion goes hand in hand with the increase of students that are brought up in one-child families. The change of studying and living environments have produced different levels of mental adaptabilities like strongly adaptable, adaptable, basically adaptable, inadaptable, and adaptive disease.

The study on mental health in China was carried out later than other countries, and the mental health didn't draw any attention until some relevant problems occurred in recent years. Yet the current work on mental health is confined to the evaluation, consultation, and individuals, to be more specific, the treatment of unhealthy students. However, from the perspective of the cultivating the students' personalities, every student should receive psychological guidance and training. The unhealthy students need consultation and correction, and the healthy students need to improve their personalities via the group tutorship and training in each academic year.

Rely on the increase of the party members - unite the students with the core value of socialism. We often talk about the work on the "three difficulties" students, namely the students with difficulties in living, studying, and ideological awareness. Compared with the elite education, it is true that due to the merger and enrollment expansion, the students' ideological awareness is different from person to person. Nevertheless the students still have good wishes for themselves and most of them are willing to join the CPC, which is a good way for us to improve their ideological awareness. The author's college has been stick to an idea that the increase of party members should take priority, which has been proved effective by the practical work. We've been persisted for a long time in the "party member in charge of the class" program in order to advance the progress of the class via the influence of the student party members. In the second half of 2006, we created and put into practice the cultivation and judgment system of the student party branch to cultivate and judge the student party members, activists, and applicants on the basis of the "record of cultivation and judgment". Now the students in this system account for 70% of the total number of the college. From the practice of more than one year, we have realized that this system not only improved the quality of party members, but also became an effective joint between the increase of the party members and the improvement of the students' ideological and political awareness. To add some remarkable entries on the record, students increased their requirements for themselves. Besides studying hard and abiding by the laws, students were also proactive in the activities organized by the college and the university. For instance in 2007, the admission of higher vocational students happened to be on the National Day, but nearly 300 students still volunteered to work on holiday, and some of them even returned their train tickets.

Develop harmoniously by encouragement and narrowing down the gap. The major difference between managing people and managing things is flexibility. In the work of student management, the most effective mechanism may lose its effectiveness. Therefore, after the idea has been formed, and the plan has been worked out, the most important thing is the management and control of the whole process as well as the summary. Thus our work in every part has to be complete and consistent. In the process of daily management and control, we should attach importance to the approach of notification and dissemination. The students that violated the disciplines should be notified in the monitors' meeting at an early date, and the excellent students' deeds and awards should be disseminated in time through various channels. Consequently, students' concept of honor and disgrace will be intensified by learning from both others' faults and achievements. In the summary of the previous work and activities, we should lay stress on comments and encouragements. Commendation to the less advanced students is scarce and thus important to encourage them to improve themselves. Only by doing so, can the students develop themselves and help others in a harmonious atmosphere.

References

1. Ying, Y.: Multi-level comprehensive vocational colleges student management pattern to explore. Health Vocational Education 25(15), 20–21 (2007)
2. Zang, X., Li, J.: Multi-level comprehensive strengthen student management work. Journal of Jiamusi Education Institute (2) (2002)

Study on Leading University Student Party Members to Pioneers for Party Flag Luster

Wen Zhang

Institute of Art and Fashion, Tianjin polytechnic university, Tianjin, China
awen_129@163.com

Abstract. University student party members to be pioneers for party flag luster, is not only the internal demand of promotion to university party construction and campus cultural development, but also the object requirement for times of economic knowledge and realization of the rejuvenation of China. This paper objectively analyzes its necessity, the problems facing us and the basic methods and ways to conduct university student party members to be pioneers for party flag luster.

Keywords: university student party members, pioneers, party flag luster.

1 Introduction

Most university student party members are the most dynamic and energetic groups of party groups. It is necessary for university party organization and political workers to have strategic views, positively instruct students party members to be clearly aware of their historic responsibility, firmly establish the mind of party, emancipate their mind, look for the truth and be practical, work with utmost concentration, be energetic and promising, fully show the party's advantage in promoting development of science and social harmony, and to be pioneers for party flag luster.

2 The Necessity of Conducting University Student Party Members to Be Pioneers for Party Flag Luster

University student party members are not only representatives of future social advanced productivity but also important propagandas for social advanced culture. Take concrete steps to enhance the education has the great and deeply significance of realizing university's educational target, ensuring prosperity of Socialism with Chinese characteristics and successors.

Comrade Hu Jintao proposed that a far-sight people always pay attention to youth and a visionary political party always push forward youth as an important force for social progress. Since 1980s, our party has paid great attention to develop party members among university students and made a quite distinguished performance. Let alone the recent developed party members have been the indispensable talents for party

J. Luo (Ed.): Soft Computing in Information Communication Technology, AISC 158, pp. 491–494.
springerlink.com © Springer-Verlag Berlin Heidelberg 2012

and country, more important, under the leading by this advanced group, hundreds of students have been grown up to be the important power for various fields.

Cultivate a lot of university party members with high political consciousness, good professional qualities, strong fighting capacity, who can become pioneers and models in the process of growth and success of students, which relates to the party's development and significant socialist construction. It is the currently important political task for the party construction of students in colleges to do well in educating students, cultivate right political direction, root party's ideology in the young students and cultivate successors for party, because they are the backbone and successors of socialist cause. It is imperative to lead student party members to be pioneers for party flag luster. In terms of student party construction in our school, we have made beneficial try and exploration in constructing system of cultivating student party, completing educational content and playing party members' vanguard and exemplary roles. Especially in aspects of improvement of party's comprehensive qualities and function, taking Three Represents as the representative of research, we have explored the new measures to strengthen and improve university student party's education and have achieved some progress.

3   The Main Problem for University Student Party Members to Be Pioneers for Party Flag Luster

Besides paying attention to the progress we have achieved, we must also be aware of some problems can't be ignored are existing during the course. We have to pay highly attention to it and take concrete steps to solve it. Some party constitution in campus lack of attention they should have and they are short of deeply exploration and effective countermeasures to new situation and problems. The situation of party construction and ideology political work team can't meet the demand of work. Some student party members have not been aware of duty and responsibility as a party, who are weak in conscious of advancement and play role in pioneers and models. The specific aspects are as follows:

Part of new student party members' self-discipline are not high. The mainstream of student party members is upward, which has high recognition in the society. But parts of new student party members have not definite motivations, with their laziness, they don't demand much of themselves. They can't act as pioneers and models.

Relax re-education after student party members to be party. The phenomenon that much attention is paid to development before joining party but less attention to re-education exists in the student party construction. Before development, the improving candidates have to face time-to-time inspection and multi-level education, with the need to join constitution study group and the party school's training, etc. But after becoming party members, they think their tasks have been accomplished, ignoring the re-education work, which leads to drive through before joining the party but loose half of energy after. The vanguard and exemplary roles are no longer their pursuing standards, some deep problems in ideology having been exposed, they paralyze in ideology, act on the loose, have poor performance, even have committed violations of the law and breaches of discipline, which had a bad impact on figures of party.

Student party members' education lack of epochal character and direction. With the deeper step to reform and open-up, it is evitable for student party members to suffer various ideological trends. The party education's key is to help student party members establish and insist right political directions. At present, with the increasingly fierce competition of employment, not few student party members are busy with NETEM, learning software, learning foreign languages and dealing with kinds of exams in order to enhance the employment competitiveness, but not interested in study of political theory. So, how to face new situation, on the basis of inheriting and carrying forward fine traditions, striving to innovate and improve the student party members education from contents, forms, methods and ways, etc, with the purpose to enhance the epochal character   direction, activity and timeliness of student party construction is the important task for party construction work to face.

Student party education lacks of effective stimulation and limit system. Stimulation and limit promote student party members self-improvement and to enhance party qualities from two contradict aspects. To give students proper encouragement and right comments can make them face the performances rightly and to be more definite in ideals, moreover, they will cultivate their good qualities of modest, courtesy and guarding against pride and haste. Two aspects are included in it, one is supervision, the other is to hold activities of criticizing and self-criticizing. In our practical work, we always pay much attention to the developing work of party, but neglect effective supervision and the party's self-supervision and self-education, so that weaken the limit system to party members.

4   The Ways and Methods to Propel University Student Party Members to Be Pioneers for Party Flag Luster

At present, all party and people are conscientiously implement the guiding principles of the Seventeenth National Party Congress and the third and fourth plenary sessions of the Seventeenth Central Committee, striving to construct overall better-off society and create new situation of Chinese social characteristics. New situation gives more requirements to current student party members. The main methods to instruct and promote student party members to be pioneers for party flag luster in the educational work are as follows:

Lead student party members to pave on the road of Chinese characteristic socialism firmly according to Marxism. Party branches should develop multi-level and multi-channel training of theoretic knowledge of party to lead student party members to pave on the road of Chinese characteristic socialism firmly. Combine broad ideal of communism and current historic tasks of party and transform spoiled enthusiasm to the strongly motivation to become a good self, playing as a role of pioneer and models.

Enhance the practical education on the principle of serving people wholeheartedly. By means of organizing various campus and social volunteer activities, promote student party members to practice the basic object of serving people wholeheartedly, relate being responsible for superiors to being responsible for people with loving the people, for the people and benefit people, do things according to laws, fairly and justly and realize the life pursues in the college and university in the process of uniting with students, serving students and good to students. In the activities such as donation for poor students and earthquake-stricken areas, and unpaid blood donation, educate student party members to go ahead, lead students to show their loving hearts and enhance the social responsibilities.

Improve capacities and standards of serving the people by learning with no end. It is essential to educate student party members to be models of industrious learners, to strive to learning innovative theory of party, major knowledge and essential science, culture. In the daily life, insist rigid demand, rigid management to student party members. Instruct and help them improve their abilities in the process of summarizing experience and holding up regularity, purify their motions in the process of gaining wisdom and receiving educations, hammer party qualities in the process of trying to equal one better than themselves and improving themselves and improve the abilities and standard of serving the party and people by continuous study.

Motivate student party members to create new phase with insisting reform and innovation, and to pay much attention to do solid work. It is important to make student party members insist emancipating the mind, seeking truth from facts, advancing with times, always to be upwardly dynamic, wise in trying and courage of pioneer and keeping forward ahead. Encourage them to be facing difficulties and contradictory, to be courage to take up duties and responsibilities, to create new phase in the course of innovation and reform and doing solid work. Let student party members to be positive practitioners and disseminators of constructing campus culture of democracy and rule by law, fair and justice, honesty and love.

Enhance the moral influences of student party members. Party organizations have to instruct student party members to insist enhancing their moral and ethic in the daily life, to strive to implement social core value system, to cultivate highly moral sentiment, to encourage fine moral qualities, to figure lofty moral images, insist eight honor and eight shame, to cultivate themselves and defeating the people by moral, and to always keep the moral influences of party.

As rooted power of student groups, university student party members have good radical effect on most young students. Their striving to be pioneers for party flag luster will promote the scientific development of campus culture and harmonious development; moreover, it will impact the deep implementing the spirits of party Seventeenth, the overall improvement of educational qualities and cultivation of comprehensive development in moral, intelligence, sports, aesthetics and labor education for constructors and successors for social construction.

References

1. Xi, J.: Party members to be pioneers for flag luster. Party Construction Articles (13), 3 page (2003)
2. Chen, P.: The communist party member must strengthen the sense of honor and mission. Party Construction Articles (17), 33 page (2009)
3. Yan, Z.: Historical choice, realistic choice. Party Construction Articles, 26 page (July 13)
4. Zhuye: Clear about the relations between scientific outlook on development and harmonious socialist society. Cards World (01), 226 page (2010)

Study on Mercerizing Technology of Sliver

Yiping Ji[1,2], Rui Wang[1,2], and Xiuming Jiang[1]

[1] School of Textiles, Tianjin Polytechnic University
[2] Key Laboratory of Advanced Textile Composites, Ministry of Education, Tianjin, P.R. China
thymeping@163.com

Abstract. The processing device and technics of sliver mercerization was introduced. The device includes sliver feeding device, alkali lye dipping device, mercerizing device, rinsing and drying device. The main technological parameters of sliver mercerizing were selected in this paper; depending on these parameters, design of a mercerizing machine and product development were achieved. And also this mercerizing technology showed excellent practical value.

Keywords: sliver, mercerization, self-made device.

1 Introduction

The aim of mercerizing is to obtain fibers with improved and silk-like luster, higher tensile strength after wash and wear finish, better dimensional stability and increased dye uniformity and dye yield.

The mercerized finish is always carried out on cloth, yarn or garment, but with the diversification of product materials, there are more and more bi-component and multi-component products. In order to produce mercerized multi-component blended yarn, the sliver should be mercerized. The advantage of mercerized sliver is that it can directly blend with stock or blend with other fiber slivers during drawing process. Thus, blended yarns with mercerized effect are produced. This not only increases yarn variety but also improves product added value.

Currently, there is no report on research of sliver mercerized finish, and there is also no finished technics and equipment. The difficulty of sliver mercerizing is to solve tension control of its fibers. Fibers in cotton sliver are always arranged in parallel and put in order, and adhere force between fibers is poor. So the development of sliver mercerizing is hindered.

This paper dealt with experimental research of sliver mercerizing on self-made sliver mercerizing machine, in order to find out the optimized technics and measures to reasonably control sliver tension, also provide instructions for equipment parameters.

2 Basic Principles of Cotton Mercerizing [1,2]

Cotton mercerizing, normally means a process to deal with cotton fibers in strong sodium hydrate solution and under certain tension, in order to gain dimensional stability, lasting luster and improved dyeing absorbability. The fundamental reason of obtaining favorable mercerizing effect is that cotton fiber undergoes irreversible and

J. Luo (Ed.): Soft Computing in Information Communication Technology, AISC 158, pp. 495–499.
springerlink.com   © Springer-Verlag Berlin Heidelberg 2012

acute swelling in strong sodium hydrate solution under proper tension. When cotton fiber strongly swells, fiber section turns from flat kidney or auricular shape to round. Its cells start shrinking, and almost become one point for complete mercerized fibers. Meanwhile, longitudinal natural torsion disappears and wrinkles on fiber surface go away. The fiber turns into a slick cylinder, and reflects light regularly, so luster appear. After cotton fiber undergoes strong alkali, the reactivities would be (1) and (2):

$$C_6H_7O_2(OH)_3+NaOH \rightleftharpoons C_6H_7O_2(OH)_3 \cdot NaOH \qquad (1)$$

$$C_6H_7O_2(OH)_2OH+NaOH \rightleftharpoons C_6H_7O_2(OH)2ONa+H_2O \qquad (2)$$

Fig.1 is the transformation of cross sections in cotton fiber during process.

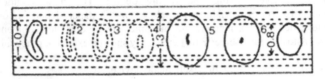

1—non-mercerized cotton;   2, 3—after dipping into lye;4—after neutralization;5,6—after scouring off lye, 7—after drying.

Fig. 1. Transformation of cotton fiber during process

As shown in Fig.1, fiber section swells with dipping hours in lye passing, and the shape of section changes a lot after neutralization, especially after scouring off lye, the section swells and turns into round, cells almost close down. There is some shrinkage in section after further scouring, but there is obvious shrinkage after drying and cells disappear completely.

3   Experiment on Sliver Mercerizing

Raw material: Combed drawing sliver. Normally, the primary condition of mercerizing is under certain tension, and dealing with cotton fibers in strong sodium hydrate solution. The mercerized finish is always carried out on cloth or yarn, because of the easy control of fiber tension. But with diversification of raw materials, it is impossible to mercerizing on blended-spinning products. According to this, we will discuss the mercerizing technology of cotton sliver. In order to obtain certain control of fibers during cotton mercerizing and carry out the process under certain tension, we chose combed drawing sliver which has good fiber arrangement.

Equipment: Self-made sliver mercerizing device. As shown in Fig.2, the self-made sliver mercerizing device mainly includes sliver feeding, lye dipping device, mercerizing installation and scouring setting. Oven drying or natural drying is adopted due to limited condition. Among which, revolution of each driving roller is actively. In order to ensure certain tension during sliver mercerizing process, multi-slivers are merged together, false twisted and fed into maceration tank, the maceration tank is equipped with grooved driving roller. And then, feed slivers underwent lye into mercerized groove which can apply false twist to them. Fibers in sliver gain further control due to application of false twist, and through certain tension and drawing, the sliver shows excellent mercerized effect.

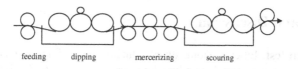

feeding   dipping   mercerizing   scouring

Fig. 2. Sliver mercerizing device

Settlement of main technological parameters. Lye concentration, temperature, time of operation and tension applied on sliver are main factors to influence mercerizing effect.

Considering mercerizing technology of cotton cloth and yarns, and through the primary experiment on sliver mercerizing, we choose lye concentration as 240g/L～280g/L. Since the action between sodium hydrate and cellulosic fibers is an exothermal change, and in order to make fibers dissolve and swell sufficiently, the lye temperature during mercerizing should be low. Also taking account of energy consumption and enough soaking of sliver, the reaction is always carried out under room-temperature. It takes very short time to carry out the reaction between lye and cellulose during mercerizing, time recommended in traditional mercerizing technology is 35s-50s. But if we want the lye evenly penetrate into sliver and undergo the reaction, time of operation needs to be optimized. This process is closely related to sliver structure, lye concentration and temperature, etc.. Even obtaining same mercerizing effect, the operation time of lye also varies according to different mercerizing technology and equipment. In order to guarantee enough time of operation, the productivity inevitably decreases or length of device increases, thus, cost of mercerizing also goes up. There is contradiction. So if even soakage of sliver is guaranteed, time of operation should be reduced as short as possible to improve productivity and save cost. According to current condition, we changed running speed of the device through inverter to control the operation time of sliver dipping and mercerizing. Slivers of different operation time are tested.

4   Tests on Performances

In order to understand performance changes in mercerized cotton, and also with the intention to optimize the processing technology, we carried out performance tests on mercerized cotton fibers under lye concentration 260g/ L, room-temperature, and different operation time in dipping and mercerizing. Tests include bundle strength, equilibrium moisture regain, specific resistance and electronic microscope photos on transect and lengthways surface of mercerized fibers. Among which, bundle strength is carried out on Y162 bundle strength device. As absorption of mercerized fibers to water, iodine, barium hydroxide and dye stuffs are all increased, we may objectively determine the degree of fiber reactions in lye if we understand equilibrium moisture regain and specific resistance of mercerized fibers. Thus, equilibrium moisture regain and specific resistance of mercerized cotton fibers are tested here. Equipments used are Y802A 8-basket conditioning oven and YG321 fiber specific resistance device. And the electronic microscope photos are taken under KYKY-2800 SEM.

5   Results and Discussion

Bundle strength test. Fibers become evenly during mercerizing, thus bundle strength is increased. As sown in Fig.3, bundle fibers are processed under lye concentration 260g/ L, room temperature, operation time as 1min, 1.5min, 2min, 2.5min, 3min, 3.5min, 4min and 4.5min. As illustrated in Fig.3, the bundle strength of slivers are increased after mercerizing, this is because internal structure of cotton fibers changes due to strong lye. Moreover, under certain concentration, the bundle strength of mercerized sliver increases with the operation time of lye passing, and the change is not obvious after 3.5min. Observing from the experimental data, processing time of sliver is longer than fabrics during mercerizing. This conclusion is significant in primarily choosing the operation time of lye and development of device.

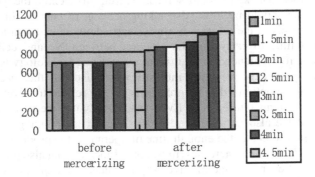

Fig. 3. Bundle strength before and after mercerizing

Equilibrium moisture regain, specific resistance. The data of equilibrium moisture regain and specific resistance on mercerized cotton fibers are shown in Table 1, and the fibers are treated under lye concentration 260g/ L, room-temperature, and 1min, 3min respectively.

Table 1. Equilibrium moisture regain and specific resistance of cotton fibers, before and after mercerizing

Samples	equilibrium moisture regain (%)	specific resistance ($10^6 \Omega$)
cotton wool	6.69	$2.16*10^3$
Mercerizing 1min	6.72	$1.79*10^2$
Mercerizing 3min	6.78	$1.81*10^2$

As listed in Table 1, adsorption performance of mercerized fibers is increased, and the specific resistance is decreased. They will make fiber processing easier.

Electronic microscope photos. The SEM photos of mercerized cotton fibers are shown in Fig.4, and the fibers are treated under lye concentration 260g/ L, room-temperature, and 1min, 3min respectively. Observing from the photos, when mercerizing 1min, fiber cells and longitudinal fiber distortion have changed, but not sufficiently. While mercerizing 3min, fiber cells and longitudinal fiber distortion almost disappeared.

Fig. 4. Mercerized cotton fibers under operation time 1min (left) and 3min (right)

6   Conclusion

The primary experiments of sliver mercerizing on self-developed sliver mercerizing device prove that preferable mercerizing effect can obtain through sliver mercerizing;

The sliver mercerizing device needs to be further improved to gain mercerized sliver with better effect, so that excellent materials to develop high added value product are supplied.

Acknowledgment. The authors thank the Tianjin Municipal Science and Technology Committee for a Science and Technology Support Key Project Plan (09ZCKFSH02000).

References

1. Zhang, X.: Introduction to Dyeing and Finishing, pp. 22–24. China Textile and Apparel Press (June 2000)
2. Yang, J.: New Textile Materials and Applications, pp. 63–65. Northeastern University Press, Harbin (2004)

The Furniture Design Strategies Based on Elderly Body Size

Qiuhui Wang and Aihui Yang

School of Mechanical and Electronic Engineering, Tianjin Polytechnic University,
Tianjin, China
{wangqiuhui,yahfighting}@126.com

Abstract. How to meet the elderly people's needs with their body size is an important furniture designs direction for population aging. So, according to physical characteristics and their body size of the elderly, this paper analyzed the national standard of furniture size, and the market furniture size. And it was focused on the high definition for the furniture. It was found that high-scale of furniture is a major problem for the elderly furniture designs. According to the body measurements of the elderly people and theoretical analysis about the scale, this paper proposed furniture design strategies about the bedroom, living room, kitchen, bath room and study room for the elderly people.

Keywords: The elderly people, Body size, Furniture design.

1   Introduction

The world aging population is already a global problem, and the aging are integrated into national development strategies of social and economic, put in the work of the main agenda of governments[1]. The elderly as a specific user groups of environmental space, and the barrier-free design in living space should be meet the physical and psychological needs of older persons, the elderly. And it can be more convenient to travel home and daily life. It is the social development needs of aging [2]. Home space furniture is essential appliances for old people everyday work, study, rest and other activities, and it is an important component of indoor facilities factors. The size is an important part of the furniture design, the design must meet the physical requirements of users, especially vulnerable groups and to meet the requirements of little user.

Furniture design based on a reference measurement size is inseparable from the human body, so the human size of furniture to meet the users can meet people's requirements. In the social environment of the aging population, how to meet the elderly living space furniture requires physical size of the furniture design is an important research direction in future. So, this paper built model based on the body size changes, and analyzed national standard furniture and facilities of the aging. And put out the elderly living space furniture design strategies meeting the aging.

J. Luo (Ed.): Soft Computing in Information Communication Technology, AISC 158, pp. 501–509.
springerlink.com   © Springer-Verlag Berlin Heidelberg 2012

2   Methods

2.1   Body Scale Analysis

2.1.1   Changes in Body Functions of the Elderly

The body size data are important for the work space design, machine equipment and the control devices. It is directly related design rationality to the spatial layout and tool. It should be reduced fatigue to improve work efficiency, so that the operator can be safe, comfortable. According to surveys, when the people were older, the physiological function is recession [3]. Due to physiological function, with the elderly age increases, the body's height continue to lower and shorter5% -6% than the youth. Due to contraction of the spine, the aging started hump, and the body became smaller and smaller. While the weight and lateral size become bigger, sensitivity decreases, arm strength and physical decline, visual loss, hearing loss, the body organs is lower.

Body size of the elderly. Body size is an important basis for the older elderly living space furniture design. But the GB10000-88 only provides a body size of Chinese adults [4]. In 1989, the GB10000-88 shows body size based on data standards about the 18-55 years male and the 18-60 years female adult. They are height, weight, arm length, forearm length, thigh length, and calf length. And it is given out the standing and sitting body size data.

This old body size data from the static size of the sample test results [5-6] and related literature on changes in body size of the elderly. Because the time, economic and human problems are relative lack , the project only collected 100 samples from the elderly people, all the dress sizes after the data is static size. In the investigation, there are also individual body size is a big gap between the theoretical analysis. The individual size increases nearly 10-20cm, than the theoretical analysis. And the maximum abdominal circumference body up to 127cm. In fact, the greatest change in the elderly because the contraction of the spine, so the most obvious changes in the elderly is body size, as shown in figure 1, figure 2.

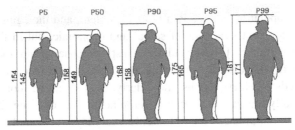

Fig. 1. The male body size changes

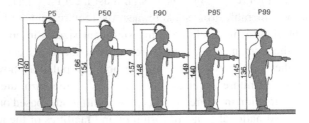

Fig. 2. The female body size changes

According to GB10000-88 body size, the elderly contraction height, body characteristics, the paper analyzed the standing and sitting middle-aged body size, as show in table 1, table 2. This paper selected body size for environmental space data, listed percent 1, percent 5, percent 50, percent 90 and percent 99 of the older male and female standing, sitting and static size of the body changes [7].

Table 1. The older male body size

	Old male (mm)				
	P1	P5	P50	P90	P99
standing	GB10000-88 6% contract size				
height	1450	1488	1577	1649	1705
Eye height	1349	1386	1474	1544	1602
Shoulder height	1169	1204	1284	1349	1404
Cubits height	869	897	967	1014	1060
Hand height	617	639	697	740	778
	sitting				
Sitting height	786	807	854	890	920
Eye height	685	704	750	786	816
Shoulder height	507	524	562	593	619
GB10000-88 10% increase size					
bust	83.8	87.0	95.4	103.8	111.9

Table 2. The older female body size

	Old female (mm)				
	P1	P5	P50	P90	P99
standing	GB10000-88 6% contract size				
height	1362	1396	1476	1542	1595
Eye height	1257	1289	1367	1431	1501
Shoulder height	1096	1123	1195	1253	1302
Cubits height	821	845	902	948	987
Hand height	592	611	662	701	731
	sitting				
Sitting height	742	760	804	838	865
Eye height	637	653	695	727	755
Shoulder height	474	487	523	550	573
GB10000-88 10% increase size					
bust	789	820	908	1011	1106

The used human body size in furniture design. According to person for engineering principle, furniture scale determines whether man-computer system safety and efficiency [8]. Human body size is main basis for old furniture dimension design. Therefore, all kinds of equipment height, such as furniture, must base on the user's body scale to design. Operating side height is the most important scale relations to furniture design. And it related directly to the factors system [9]. The body height, eye height, calf height, cubits height, sitting height, sitting depth are the most commonly used furniture design, and also are the important human body size factors. Cubits height determines the operating side, desktop height; the cur's height determines the seat surface height.

According to old people standing hand activity space analysis when old people no loading, the old male size percent 1, percent 5, percent 50, percent 90 to percent 99 are1850 mm, 1840 mm, 1980 mm, 2080 mm and 2130mm. And older women hand lifted in turn for 1730mm, 1740mm, 1860mm, 1940mm and 2000mm, as shown in table 3.

Table 3. Old people hand lifted height (mm)

percent	P1	P5	P50	P90	P99
Old male	1840	1850	1980	2080	2130
Old female	1730	1740	1860	1940	2000

According to old people sitting activity space analysis, the old male size percent 1, percent 5, percent 50, percent 90 to percent 99 are350 mm, 360 mm, 390 mm, 420 mm and 440mm. And older women hand lifted in turn for 310mm, 320mm, 360mm, 380mm and 390mm, as shown in table 4.

Table 4. Old people sitting height (mm)

percent	P1	P5	P50	P90	P99
Old male	350	360	390	420	440
Old female	310	320	360	380	390

3   Results

National standards and the market furniture scale analyzing. About furniture scale problems, national standards and market existing products not unified. GB3326-3328-82 showed the standard furniture scale, the ministry 04J923-1 showed the old residential building standard atlas "[10] gives toilet and kitchen furniture scale, as shown in table 5. GB3326-3328-82 formulate: the work desk height 780mm, the bed height 480mm, chair height 440mm: 04J923-1 Atlas provides the standard measure of home space facilities: the toilet seat height 450mm, chair height 450mm, bathroom wash basin height 850mm, width 450-550mm, basin height of wheelchair disabled 800mm. The market furniture Operation height as follows, computer desk height is 790mm, dining table height is 790mm, sofa height is 450mm, bed height is 450mm, tea table height is 470mm, and bedside cabinet is 450mm, as show in table 6.

Table 5. National standards furniture (mm)

| | GB3326-3328-82 | | | | GB04J923-1 | | |
	length	*width*	*height*	*depth*	*length*	*length*	*width*	*height*
desk	1000-1500	550-850	780	—	—	—	—	—
chair	400	400	440	420	—	—	—	—
Wardrobe				200	600	—	—	—	—
Double bed	2000	1500	480	—	—	—	—	—
single bed	2000	900	480	—	—	—	—	—
Cupboard	—	—	—	—	—	—	—	—
Toilet	—	—	—	—	.600	500	450	—
Work table	—	—	—	—	—	—	800	—
Wash basin	—	—	—	—	—	—	850	—
wheelchair Wash basin	—	—	—	—	—	—	800	—

Table 6. Market furniture (mm)

	length	*width*	*height*
Computer desk	1400	700	790
Bookcase	2200	470	2200
Dining table	1300	800	790
Wardrobe	2430	640	2220
Double bed	2190	1510	450
Big tea table	1400	700	470
sofa	2210	880	450
Dining table	550	500	450
Bedside cabinet	710	470	450

From the little older men and older women's body size analysis showed that the national standards and the market scale of existing furniture can not fully meet the needs of the users. Especially since furniture seat height and high-handed operation height, increasing the difficulty of older operations. As an example, the toilet 450mm height exceeds the limit the ability of older users. From the perspective of spatial analysis using the toilet, 50 percentile for height toilet old male is 390mm, suitable for older women toilet height is 360mm. According to ergonomics principle, low design can effectively reduce the pressure inside the disc users, in line with the requirements of barrier free design. So the old furniture should be designed to accommodate the elderly limit the ability.

Furniture size design strategies. According to the elderly scale modeling evaluation and analysis about the old people body size changes in physical condition, and the home of the national standard atlas space for facilities, this paper proposed furniture scale strategies.

Bedroom furniture

As show in table 7, bedroom double bed, single bed, bedside cabinets designed for height 390-400mm, large wardrobe depth of 650mm, so this can easily accommodate large fat coat, and double and single bed width to retain the existing size.

Table 7. Bedroom furniture (mm)

	length	width	height	Seat height
Double bed	2000	1500	750-800	390-400
single bed	2000	9000	750-800	390-400
Wardrobe	2000	605	1700-2000	-
Bedside cabinet	400	400	-	390-400
TV cabinet	1200	700	120-400	-

As shown in table 8, according to percent1, percent 5, percent 50, percent 90 and percent 99 old people using large closet space analysis, the hand touching height of wardrobe were 1550,1650,1750,1850, 1860mm. The size wardrobe elderly women were 1500mm, 1550mm, 1700mm, 1750mm and 1780mm.

Table 8. The Wardrobe height(mm)

percent	P1	P5	P50	P90	P99
male	1550	1650	1750	1850	1860
female	1500	1550	1700	1750	1780

Living room furniture. Living room furniture is clean single, it includes sofa, coffee table, cabinets, corner cabinets, and TV cabinets. The sofas number is three sofas, two sofas or corner sofa, coffee table, TV cabinet and corner cabinet depth should be according to the spatial decision.

As show in table 9, in living room, the sofa, seat height were 390-400mm, total height should be controlled between 750-800mm. So it is easy to rely on for the elderly, but also it should avoid the environment space feeling crowded. The corner cabinet height is 710mm, and the table height is 400mm. The height of TV cabinet should be designed for the elderly based on the eye height and TV size. The new large-screen TV is usually large, so the TV cabinet height should be designed 120-400mm.

Table 9. Living room furniture (mm)

	length	width	height	Seat height
Sofa	2000	820	750-800	390-400
Tea table	1000	600-700	400	-
Conner cabinet	-	500-600	710	-
TV cabinet	1200	500-600	120-400	-

Kitchen furniture. The Kitchen furniture typically includes cabinets, wall cabinets, vegetables pool, console, and kitchen stoves, general requirements should be set neat and orderly, as show in table10. Because the elderly body size reducing, the kitchen cabinet and console should be reduced 50-100mm than the young people using. And it should be reduced 100-150mm than the national standard, and always control in 650-710mm. Refrigerators, washing vegetables pool, console and stove should be placed in turn within the region in recent. So it is reduced the times of elderly people move to reduce their fatigue. To meet the 50 percentile older women's needs, a separate design should be considered their own house size.

Table 10. Kitchen furniture (mm)

	length	width	height
Cabinets	1000-	600	650-710
Wall cabinets	1000-	450	600-700
Vegetables pool	700	400-500	650-710
Console	1000-	600	650-710

Dining room furniture. Dining room furniture main is the dining table, chairs, buffet. It is should be consideration of its scale of the older user's body height, especially elderly women of smaller body size. To satisfy older people of percent1, percent 5, percent 50, the chair height should be designed 750mm, seat height 400mm, table height 710mm. The buffet height as long as the hand can reach both scales, as shown in table 11.

Table 11. Dining room (mm)

	length	width	height	Seat height
Dining table	100-120	70-80	71	-
Dining chair	40-42	40	75	40
Dining buffet	100-120	50-60	120-150	-

Bathroom furniture. Toilet facilities and furniture mainly consists of vanities and countertops, toilets, vanities and countertops, bath. According to body size characteristics of the elderly, proposed facilities, bathroom furniture scale strategies, such as shown in table 12, the basin installation height is 710mm, toilet installation height is 360-400mm, bath height is 360-400mm.

Table 12. Bathroom furniture (mm)

	length	*width*	*height*
Basin	100	45-60	71
Toilets	70	40	36
bathtub	150-180	70-80	36-40

Study room furniture. Study furniture includes desks, office chairs and bookcase. According to the old physical size we design the furniture, as shown in table 13, the desk height is 710mm, office chair seat height is 400mm, for more convenient, office chairs can be designed to be adjustable, between 360-450mm in height.

Table 13. Study room furniture (mm)

	length	*width*	*height*	*Seat height*
Desk	1000-1200	700-850	710	-
office chairs	400-420	400	750	400
Bookcase	1500-2000	350-450	1700-2000	-

4   Conclusion

This paper analyzed to compare the national standard of furniture and the market existing furniture based on the elderly physiological characteristics and changes in body size. And it is main focus on the furniture height. It is found that it is a problem for the furniture height. Because that the furniture size is too high to the older people. According to body size theories of older people, this paper established vector model to evaluate and compare the rationality, and proposed the design strategy of the living space furniture size for the older people. This paper proposed the furniture size design strategies about the bedroom, living room, kitchen, dining room, bath room and study room. It is presented low height design size than youth furniture to meet the requirements of older people.

References

1. Luo, C.: From aging to aging: a population-based Perspective exploratory study: A Probative Study Based on Demographic Perspective, pp. 35–38. China Social Sciences Press, Beijing (2001)
2. Mora, E.P.: Life cycle, sustainability and the transcendent quality of building materials. Building and Environment 42, 1329–1334 (2007)
3. Dickinson, T. L.: User Acceptability of Physiological and Other Measures of Hazardous States of Awareness. Sponsor. National Aeronautics and Space Administration, Washington, DC. Report: ODURF-101441, p. 64. Old Dominion Univ., Norfolk (2001)
4. Ding, Y.: Ergonomic, pp. 14–15. Beijing Institute of Technology Press, Beijing (2005)

5. Nairen, Z., Hui, W., Hui, Z.: Beijing Accessibility investigation report. Beijing Institute of Technology, Beijing Municipal Science and Technology Commission, Beijing (2007)
6. Tilley, A.R., Dreyfuss, H.: Office (U.S.). Ergonomic diagram - human factors design, p. 33. China Building Industry Press, Beijing (1998)
7. Wang, Q.: Beijing elderly people living space accessible for system analysis and design of countermeasures. Beijing Institute of Technology, Beijing (January 2009)
8. Araki, I.: Atlas of foreign construction design details - accessible buildings. China Building Industry Press, Beijing (2001)
9. Phillips, L.: Road users can grow old gracefully-With some help. Public Roads 69(6) (May/June 2006)
10. GB04J923-1-2004. Elderly residential building. Ministry of Construction, China Building Standard Design & Research Institute published, Beijing, pp. 4–15 (2004)

New Characteristics of Clothing in the Culture of Social Gender

Guanglin Chen and Jing Sun

Institute of Art and Fashion, Tianjin Polytechnic University, Tianjin, China
chenguanglin_2010@163.com, 124489550@qq.com

Abstract. This paper mainly focuses on clothing's constructional process under the influence of the culture of social gender and demonstrates the different characteristics of clothing culture result from different cultures of social genders while the whole society is searching for low-carbon life. It will further analyze the new tendency led by neutralized clothing culture, the clothing's characteristics of male and female has produced a kind of cultural ambience and complementary and reinforced appeal, which has new clothing language effect with the decreasing boundary of social gender to clothing style.

Keywords: social gender, neutralized, clothing language.

1 The Clothing's Constructional Process under the Influence of Social Gender

As human's psychological signal and mental expression, clothing is psychological product of social gender group in the culture of social gender. In social gender, "gender" is used to distinguish biological primal instincts of "sex". Social expectation, requirement and evaluation towards gender and their relationship has an effect on gender's constructional process under the influence of inanimate elements of gender's development, and during this process, civilized social gender consciousness has affected clothing profoundly. Different social gender's construction has led to the difference of social gender consciousness and this difference refers to the biological difference between male and female result from social system's strength and consciousness in social construction. The unbalanced development of social gender has generated a series of social behavior standards, social status and the imbalance between male and female. So the difference of social gender has led to clothing difference in different cultures.

2 Different Characteristics of Clothing Culture Resulted from Culture of Different Social Gender

In the society where male is main source of productivity, the aesthetic standard is slender tender figure and lightsome posture to cater to male's aesthetics. Female's beauty is their competence, and many women even turn to sacrificing health to search for this kind of beauty to abstract male's eyes and female's admiration. In China, foot binding is some beauty; the little "lovely" feet are beauty's sign. In contrast, in west, corset has something common with foot binding; the female prefer to suffer the pain of internal organs from corseting to pursue slender waist. Both foot binding and corset demand a wealthy family otherwise the young lady will have little chance to bind small feet and slender waist. Because neither the small feet nor the slender waist is good for physical labor, and it's hard for an ordinary family to offer their daughters such a chance. At that time and even today, little feet or slender waist is regarded as the symbol of wealth. This is the result of invisible gender inequality and different social genders. In the history of clothing, it's rare to see such kind of abnormal aesthetics towards male, and this desire has always been female's mental deformity to meet male's aesthetics.

From the prospective of Europe, ancient west is based on The Mediterranean civilization. West culture of social gender centers on Nilotic civilization in North Africa, Mesopotamia civilization in West Asia, Aegean civilization, and ancient Greek and Roman civilization in South Europe and all these civilizations are around Mediterranean. The countries of West Europe have totally different culture from China's civilization after adventuring medieval civilization and Christian civilization. There are many races in West Europe, especially in the area of Mesopotamia civilization in West Asia, their cultures collide, penetrate compromise with each other to form a special national culture and active social gender characteristics. While in New Guinea-the so-called Living Fossil, the gender difference in some ancient races is utterly different. For example, Mundugamor is a race of male chauvinism, where female even sisters or daughters can be exchanged with each other like goods. Kola is on the contrary, and female is the leader of the race. Female works and plays male's role in modern civilization and male make up themselves to please their wives. However, western traditional culture belongs to the combination of the ancient Hebrew (the main source of Christian civilization) and ancient Greek cultures. But the ancient Hebrew civilization holds negative attitude towards body, and the ancient Greek positive. Such conflict makes the sexual awareness more sensitive in western culture and western countries began to literature social gender in a new way and pursue physical and psychological liberation after the ascetic medieval era. Shaping female's figure curve and exposure of skin are new characteristics after the medieval era, and sex becomes one of clothing's sources. Clothing is more public than reserved and represents self-consciousness and self-aesthetics. Mini trend began after English designer Mary Quant's work-the skirt hem 15 to 20mm upper than the knee. Modern clothing requires more on personality that on exaggeration and decoration to break the restrictions of social gender and reflect the natural beauty of low-carbon life.

In modern China, the development form of gender culture is like this: information technology centers on male, and female is in collective unconsciousness, which is reflected in the early period of liberation, almost all the Chinese people's clothes are grey, black and green, and the style stresses more on utility. In 1979, Pierre Cardin held an internal fashion show in Beijing National Culture Palace, which made a stir in China and was looked on as a new starting of China's fashion. In China, women got liberation and began to pursue self and express self with the build of socialism. From this, we can see that, Pierre Cardin has brought China not only western clothing culture but also enlightenment to China's fashion.

3   Analysis on Causes of Male and Female's Common Liberation

Traditional view of social gender first faded in fashion world, which is decided by fashion's characteristics and also the inevitable outcome of improved science and technology, culture and overall quality. At the same time, gender difference under the influence of social gender is relative, and it will change under some condition.

First, we will analyze it from the prospective of sociological human rights. In the information era, social productivity and cultural diversification makes intellectual work the dominating working factor. Women can do the intellectual work what men can, and women try to improve their quality to prove that they are no worse than men. So it's inevitable to collide with traditional psychology of social gender. Women are eager to be independent out of kitchen and get public recognition. But they have unconsciously admitted male's authority which is reflected in their pursuit of hale style and male's dressing. Although on the whole, the traditional view of social gender won't have any creative change in a short period of time, female's aesthetics has begun to break male's view. The feminism's development makes female gradually get out of male's control and male is deprived of traditional role. With the reformation of social gender, male begin to pay more attention to external self-aesthetics.

Secondly, the author will explain it from the evolution of social status and role. In modern politics, it's not strange any more that female take part in it and own power. Female begin to enjoy the same individual autonomy and independent economic, politic and cultural opportunities and resources. Women's success and contribution to the society offer them equal rights, social status and responsibilities as men, and women have gained the autonomous rights of spiritual aesthetics and given up making up themselves by male's aesthetic standard, which is reflected that women turn to unique taste in clothing. Female's liberation in politics and economy pushes forward the role change of social gender. It's has not been male's unique rights that the production of clothes, the transmission of media and the domination of fashion and fashion begins to evaluate fashion in female's prospective. When the neutralized tendency of social gender becomes more and more prominent, the traditional disposition of male and female is becoming lighter and lighter and melt with each other. Male the traditional social subject starts to play equal role as female does, and the role change of social gender is directed to neutralization.

Thirdly, the paper will interpret it from the psychology of aesthetics. It's psychological main cause of modern neutral clothing to break through the self and break the traditional boundary. Development is one of the themes of the times, and worldly boundary and restrictions have to be given up and freedom and creation have to be the designers' pursuit in order to develop. Fashion also mirrors this idea that both male and female's clothes try to break the limitation of traditional gender and pry into opponent's style, so neutral beauty becomes more and more popular. The psychologist Carl Jung thinks that there exists a heterosexual archetype subconsciously and take it as a reference of association. Human beings are androgenic in sub-consciousness and intensity of gender tendency decides the diversity of characters, but it's good for them to accept with each other in spirit.

4   Breaking the Cultural Orientation of Gender and Making Reunion

Fabric, color and composition are the three main aspects of fashion design. According to the market research and the popular low-carbon boom, we can break cultural orientation of gender from the following three aspects.

First is the choice of fabric. Color and composition depend on fabric to stand out, and all kinds of fabrics can show the characteristics of gender full of modern feelings, such as frivolous fabric and stretch fabric setting off figure curve, fabric of mental light and matt, nonwoven fabric and mesh fabric, knitting fabric, leather, buckle, cleverish mental decorations and so on.

Second is the colorific collocation. Color is your first visual impression, low lightness, low purity color scheme, cold mental tone (cold grey, blue and green with mental luster on surface) and tone with dream-like colors are all typical colors of neutral style.

The last is the style. The transfer of clothing elements between male and female and disposition of the fabrics by polishing, embroidering and drawing make the ordinary fabric fashionable and reduce the development cost of senior fabric to realize low-carbon life.

Fig. 1. Works of Limi Yamamoto

Figure One shows a group of female clothes by Limi Yamamoto, and loose blouse with horizontal streaks, long narrow tie, reserved hue and tough lines highlight modern women's personality. Figure Two is a series of male clothes by Rei Kawakubo. This design employs female's regular color, pattern and decoration collocation to shape

Fig. 2. Works of Rei Kawakubo

different male and display male's charm and grace. Clothing has turned out to be the direct language of your emotion how to make you standing out in association and show your affinity. For example, a tall man needs to increase his affinity when he intends to take part in a business conference. At this time, there is no good to be in formal clothes because they transfer a message that the man is solemn but also serious and difficult communicator. So how does clothing transfer one's feelings? First, soft cold and cotton mixture can increase easiness, randomness and intimacy, and tight style can show tidiness and seriousness and transfer fashion. In the choice of trousers, close-fitting trousers give people the impression that the wearer is lively and capable. Also fashionable accessories can improve life taste. For female, stiff and smooth wool fabric can help change the traditional image of women, and reveal elegance and dignity. However, there is a long history that both male and female wear similar clothes and low-carbon idea push forward neutralization of clothing indirectly. At the same time, neutral dressing is a new way to break through the limitation of social gender.

5   Conclusion

You With the development of globalized economy, world culture collides and blends with each other, so are Chinese and western clothing culture. China's clothing industry is gradually integrating with world fashion and is complying with world fashion. People in east and west have different shape, view of social gender and cultural background, but they have the same aesthetic pursuit in clothing. Moreover, social gender has less and less restrictions on clothing style, and both male and female's clothes will blend better and reflect positive sentimental aspect. This development complies with low-carbon culture, fashion culture and personality culture, transmitting fresh charm of era.

References

1. Zhou, S.: Analysis on Clothing's Emotional Design. Domestic and Oversea Textile (2), 20–21 (1999)
2. Lurie: Interpretation of Clothing, Li, C.(trans.): China Textile & Apparel Press, Beijing (2000)
3. Li, D.: The History of Western Clothing. Higher Education Press, Beijing (1995)
4. Tong, X.: China's Women in 30 Years /Sociological Research on Gender. Collection of Women's Studies (3) (2008)
5. Hua, M.: Psychology of Clothing. China Textile & Apparel Press, Beijing (2004)

A Research on University Teachers' Information Use Behavior Analysis Model

Wu Liming[1], Zhang Hui[2], and Yang Xiudan[1]

[1] College of management, Hebei University, BaoDing, China
[2] Personnel, Hebei Finance College, BaoDing, China
{wlming2000,huizi536,poshyang}@126.com

Abstract. Based on the theory of TAM and TTF model, the paper proposes an analysis model of university teachers' information use behavior, points out the factors of university teachers' information use behavior include external factors, internal factors and core factors, and the relationship among the above-mentioned factors. The paper uses the survey methods and chooses 220 teachers of 5 universities as the sample, through the analysis of sample data, discusses the relationship among the main factors and the information use behavior, and finally verifies the model.

Keywords: university, teachers, information use behavior.

1 Introduction

With the information resources showing diversity characteristics, university teachers' information behavior is changing in the networked environment [1].When or after getting information resources, there is an implicit judgment on whether to use it or not for university teachers' further actual using of information. The paper adapts the idea from the theory of Technology Acceptance Model (TAM) to study the information use behavior, where users judge on the usefulness and usability of an information technology and the further affects of accepting and using of information technology [2]. Since the use of information resources is needed by their work, users have to check what is needed in the work to decide whether to use information resources, when using information resources, there must match up the need of work with the task. The paper points out that either TAM or Task-Technology Fit (TTF) model is not enough to researching university teachers' information behavioral separately, therefore, the paper learns from the core theory of TTF, and makes an integration of these two models, in other words, adding technical suitability to the TAM as the external factors to study its effect on the usefulness, usability and actual use of user's behavior. Based on the former study, the paper proposes the framework of analysis model of university teachers' information behavior and furthers to research the components of the model.

2 The Framework of Analysis Model of University Teachers' Information Behavior

In the teaching practice, university teachers' acceptance and use of information resources is a technology acceptance model. In order to finish their works better, university teachers

J. Luo (Ed.): Soft Computing in Information Communication Technology, AISC 158, pp. 517–526.
springerlink.com   © Springer-Verlag Berlin Heidelberg 2012

use information resources, which is affected by tasks, external objective conditions, target audience and teachers themselves. Teachers' information behavior analysis model is an integration of TAM and TTF [3]. Teachers' use of information resources is affected by different factors, such as teachers themselves, characteristics of information resources, characteristics of tasks and other external objective conditions, which are the key factors that affect and limit teachers' use of certain information resource. These factors could be summarized as internal factors, external factors and core factors. Based on the researches both at home and abroad, combined with the preceding analysis, the paper creates a new and more comprehensive model by integrating TAM and TTF [4]. The model is mainly made up with core factors, internal factors and external factors; its framework is showed in figure 1.

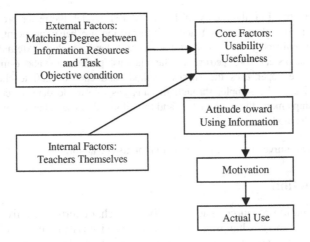

Fig. 1. Teachers' information use behavior analysis model

3   Hypothesis of Analysis Model of Teachers' Using Information Behavior

Teachers' information use behavior analysis model constitutes of core factors, internal factors and external factors, which affect the users' attitude. Based on the extensive research of TAM, usability and usefulness have a significant impact on attitude, which further affects the users' motivation [5]. Based on the model framework, we could see there are three relationships, which are relationship between core factors and attitude, internal factors and core factors and external factors and core factors.

A. Relationship between Core Factors and Attitude. In order to analyze the university teachers' use of information resources, the teachers' information use analysis model takes usability and usefulness as reference. The behavior of teachers' using information implies the process of teachers' making a judge on information resources, in other words, users have to make screening and judge on the usability and usefulness of information. In judging practice, users make judges based on characteristics of information and come to a conclusion whether it is usability or usefulness, which further affects the users' attitude on information resources. The process is the core part of the above-mentioned model. Usability means teachers feeling on using certain information resources, whether it is easy or difficult to use, namely usability means how much energy and stamina they have to give in order to use certain information resource. Usability means teachers thinking certain information resource is helpful to their task or themselves and how much information resources have been helpful [6]. Certain information resource is useful usually shown as enriching the teaching contents, teaching methods and teaching means or improving the teachers' quality level.

Based on the previous analysis, the paper proposes the relationship hypothesis between core factors and attitude:

H1: Core factors have a positive impact on attitude.

B. Relationship between Internal Factors and Core Factors. In the model, internal factors mainly mean teacher's personal factors, which mainly include personal intellectual learning ability, cognitive ability, and cultural background, teaching experience, gender, age, education background, interests and hobbies. Generally speaking, teachers' judge on usability and usefulness of information resources is affected by these factors.

Based on the previous analysis, the paper proposes the relationship hypothesis:

H2: Internal factors have a positive impact on core factors.

C. Relationship between External Factors and Core Factors. Teachers' use of information resources is connected with teaching work, which means whether information resources can be applied to teaching task decides their judge on usability of use and usefulness on information resources, namely the match degree between information resources and task affects core factors, In addition, teacher's judgment on the usability of use and usefulness of information is related to external condition (teaching conditions, target audience) which belongs to external factors.

Information Resources and Task Suitability

The matching degree between information resources and task depends on whether information resources are suitable for task; it contains three concepts which are information resources characteristics, task features, suitability between information and task.

a)Information resources characteristics: Information resources are abstracted as the information technology to be considered in our research. There are many kinds of information resources, which are not specific and clear if they are not downloaded. Since research on information resource features needs a common feature summary, the paper does a case study mainly on academic information in order to do the study on teachers' using information behavior in their teaching work. Generally speaking, information resources have certain forms, such as title, publisher, content, time, source and so on. Based on the study, we can summarize the features of information resources as follow: authority, accuracy, novelty, carrier forms and restricted access.

b)Task Features: Task features are delimited as teachers' task demand characteristics, which contains requests and conditions needed in their teaching work. The paper shows that teachers' requests are simple and clear and their expectation on teaching task is rarely changing too much by interviewing university office of academic affairs and teachers. Therefore, we decide to do a research under the conditions of fixed task demands.

c)Suitability between Information Resources and Task: Goodhue proposed classic task—technology suitability measurement through testing task –technology suitability model. Goodhue put it into twelve dimensions: confusion, right level of detail, meaning, locatability, accessibility, assistance, usability of hardware and software, system reliability, accuracy, currency, compatibility, currency, presentation, and the total number is 24 items [8]. However, as task features and technical features in classic models, the measurement method is just a general one. The paper simplifies the original 12 dimensions into 3 through ask features, information resources, technical features and specific application environment. The 3 dimensions are relevance of information, the degree of help, portability.

Objective Environment.

Objective environment refers to target audience and teaching environment which can be divided into hardware teaching environment and software teaching environment. Besides, teachers have to judge the difference of students' knowledge acceptance among college students, undergraduates and graduates in teaching practice. Therefore, teachers' information resources selection is affected by students' cognitive level and difference in knowledge structure.

Based on preceding analysis, the paper proposes the hypothesized relationship between external factors and core factors:

H3: External Factors Have a Extend Impact on Core Factors

4   Questionnaire Design

Questionnaire Design. In this part, we analyze the components and the relationships among them, and then determine problem description. Problem design includes problem design of external factors, internal factors and core factors.

Problem Design of External Factors

In the designing process, we proposes problems as follows in table 1.

Table 1. Problem Design of External Objective Environment

Project	Corresponding Questions
External factors	Q1. your instruction level
	Q2. Teaching conditions provided by school

The suitability questionnaire design of information resources and task includes design of task, information resource choice and degree of suitability between them, the paper divides information resources into three categories which are paper resources, internet information resources and database resources. Relevant dimensions of suitability degree measuring the information resources and task are information relevance, available help level and portability. Problems proposed in design are as table 2.

Table 2. Questionnaire design of suitability of external information resources and task

Project	Corresponding Questions
External factors	Q3. Which kind of information resources do you usually use?
	Q4. What is the reason for choosing paper resources (book, material and so on)?
	Q5. What is your reason for choosing internet resources?
	Q6. What is your reason for choosing academic papers in database resources (CNKI, Articles and so on)?
	Q7. What will you do if you have a problem in getting information resources (download remote query, paying for the information and so on)?
	Q8. What need to be considered for this issue of suitability between information resources and task in order to using information into work?

Question Design of Internal Factors

Internal factors are mostly teachers' individual factors, mainly including the level of teachers' knowledge and culture, academic background, the ability of cognizing things, the ability of learning things , the users' interests and hobbies, the users ' age, the users' gender, and the users' experience and so on[10]. The corresponding questions are raised in questionnaire and shown in Table 3.

Table 3. Questionnaire Design of Internal Factors

Project	Corresponding Questions
Internal factor	Q1. your gender
	Q2. your age
	Q3. your highest education
	Q4. your academic background
	Q5. your teaching age
	Q6. You feel easy to use the information resource, analysis from yourself

Question Design of Core Factors

The paper insists that core factors which affect teachers' information use behavior are usability and usefulness. The connotation of the two factors as well as the influence relationship between the two factors and use manner are shown in Table 4.

Table 4. Questionnaire Design of Core Factors

Project	Corresponding Questions
Core Factors	Q1. Which aspects do you consider that the information resource can help yourself or your work
	Q2. Which aspects shown as follows do you feel that information resource is useful (stand on yourself)
	Q3. Which aspects shown as follows do you feel that information resource is useful (stand on information resource)
	Q4. The reason you plan to use information resource(usability or usefulness)

Sample Selection and Questionnaire Issuing. Sample Distribution

The object of the questionnaire is based on the university teachers in Baoding, involving Hebei University, Hebei Finance University, North China Electric Power University, Agricultural University of Hebei and Baoding University. The questionnaire involves the comprehensive university, the professional features prominent university, but also relates to the key university in China, the key university in province (affiliated with Hebei Province and the Ministry), the common university, the university just promoted to the undergraduate level. The subjects' fields involve Arts, Science, Engineering and so on. The questionnaire involves more complete subject areas, such as: Hebei University, the only key comprehensive university in Hebei Province, covers Literature, History, Philosophy, Economics and Management, Mathematics and Computer, Machinery and Electronics, and other disciplines; Agricultural University of Hebei is a Agricultural science-related University; North China Electric Power University as a key university is mainly based on electric power engineering; Hebei Finance University is strong in Finance and Accounting; Baoding University is a normal university. The author believes that this questionnaire relates to the major universities in Baoding, and it almost completely covers the university types and subject areas. Therefore the result of questionnaire is a certain representation.

Regarding the universities in Baoding, 130 pieces of questionnaires are delivered to Hebei University, 30 to Hebei Finance University, 20 to North China Electric Power University, 20 to Agricultural University of Hebei and 20 to Baoding University.

The specific situation of questionnaires in Hebei University has three aspects: 30 pieces to Management, involving Science of Business Administration (Accounting, Business Management, Human Resources Management, and Tourist Management), Science of Public Management (Labor and Social Security, Public Administration),

Science of Library, Information and Archival(Library Science and Archives Science), Information Management and Information Systems; 20 pieces to Economics, involving Public Finance and Economics; 20 pieces to Mathematics and Computer, involving Fundamental Mathematics, Applied Mathematics, Computer Science and other disciplines; 20 pieces to Art; 20 pieces to Journalism; 10 pieces to Public Foreign Language; 10 pieces to Architectural Engineering; 10 pieces to other disciplines. The other universities have different conditions. North China Electric Power University mainly involves Power and Electric; Agricultural University of Hebei mainly involves Agriculture, Baoding University mainly involves Normal Education; Hebei Finance University mainly involves Finance.

5   Delivery and Return of the Questionnaire

The questionnaires are delivered in two ways: face to face and E-mail.180 pieces of questionnaires are delivered by face to face, and 40 pieces are delivered by E-mail. Fig. 4 shows the delivery form of questionnaires. 220 pieces of questionnaires are delivered and 211 pieces of questionnaires are returned. 7 pieces of them are invalid questionnaires, including incomplete answers and invalid answers. 204 pieces of questionnaires are valid, and the effective rate is 96.7%. The author believes that the survey data are valid.

6   Data Analysis Methods and Verification of Model Assumptions

Data Analysis Methods. In the paper, various factors, designed in the questionnaire, mainly use the form of a given type, the data analysis mainly uses Crosstabs Analysis in SPSS software. The data analysis is mainly on the Frequencies Analysis and Percentage Analysis. Some data use Chi-square Analysis to compare the close relationship between two factors. For example, in order to analyze the relationship between various factors, the paper separately compares the specific components of external factors, internal factors with the part of the core factors (usefulness and usability), consequently obtain the corresponding relationship. In the analysis of the relationship between core factors and use attitude, the paper uses Crosstabs Analysis firstly, and then uses Chi-square Analysis to verify analysis, e.g. the paper tabulates core factors and use attitude crossways, and the Chi-square Analysis shows Pearson Chi-square value 2.428, Progressive Sig. (Bilateral) 0.119, for a given type of data that illustrates close relationship between column and row, i.e. there is close relationship between core factors and use attitude.

The verification analysis about the model assumption. The verification of assumption relationships between the external factors and core factors
    It mainly analyzes from two aspects through the data analysis of the relationships between the external factors and core factors in the model. a) Objective external conditions. The analysis about Teaching Conditions is as follows. The frequency of choosing "the classroom" and "usefulness, usability" at the same time is 22, accounting for 10.8%. And the frequency of choosing "ordinary classrooms, multimedia classrooms" and "usefulness, usability" at the same time is 136 which accounting for

66.7%. The data shows that the proportion of "usefulness, usability" increases with the improvement of teaching conditions. The analysis about Teaching Levels is as follows. The frequency of choosing "specialist undergraduate" and "usefulness, usability" at the same time is 24 which accounting for 11.8%. And the frequency of choosing "undergraduate, graduate" and "usefulness, usability" at the same time is 180 which accounting for 88.2%. The data shows that the proportion of "usefulness, usability" increases with the improvement of teaching levels. The external factors have some positive influence on the core factors according to the analysis of the above data. b) The analysis is about suitability between information resources and task. The analysis about suitability between Information Resources and Task is as follows. The frequency of choosing "available help level" is 161, accounting for 78.92%. The frequency of choosing "information relevance" is 157, accounting for 76.96%. And the frequency of choosing "portability" is 85, accounting for 41.67%. College teachers think that "available help level" is the most important, "information relevance" followed, and less than half of the teachers chose the "portability". Overall, the frequency of thinking dimensions of the core factors influencing on the core factors is 204, accounting for 100%, but in which the frequency of choosing "usefulness" is 184, accounting for 90.2%, the frequency of choosing "usability" is 148, accounting for 72.5%, and information resources and the fitness of the task has a greater impact. The assumption H3 "external factors have certain impact on the core factors" sets up through the data analysis and judgments.

The verification of assumption relationships between the internal factors and core factors

As the internal factors is affected by many subjective factors, the data analysis about the variables such as gender, age, educational background, and academic background is analyzed as follows: 92.63% of males choose "usefulness" and 88.07% of women select "usefulness", the percentage of men is higher than women; 69.47% of males and 75.23% of women choose "usability", the percentage of men is lower than women. The data shows that men tend to choose "usefulness" and women tend to "usability". The proportion of choosing "usefulness" is 90.2%, and choosing "usability" is 72.55%, overall the choice tends to "usefulness" in all age stages. During the three age stages, such as 25-29 years old, 30-34 years old and 35-40 years old, about 91.89% of 25-29 years old, 89.47% of 30-34 years old and 86.49% of 35-40 years old teachers tend to choose "usefulness". It presents the changes of inverse proportion in choosing "usefulness" with ages increasing. 71.58% of 25-29 years old, 72.97% of 30-34 years old and 75.68% of 35-40 years old teachers tend to choose "usability". It presents the changes of direct ratio in choosing "usefulness" with ages increasing. During education levels, the ratio of undergraduate students, graduate students, and doctoral students choosing "usefulness" is separately 81.48%, 89.71%, and 97.14%. This indicates that the higher education tends to the usefulness. The ratio of undergraduate students, graduate students, and doctoral students choosing "usability" is separately 81.48%, 75%, and 51.43%, the ratio reduces gradually. This indicates that the higher education, the weaker of the tendency to "usability". For the academic background, the ratio of liberal arts, science, and engineering is separately 88.71%, 94%, and 89.66%. And it shows that science and engineering mostly focus on the choice of usefulness than literature and history. The ratio of liberal arts, science, and engineering is separately 75%, 68%, and 68.97%. It shows that literature and history mostly focus on the choice of usability than science and engineering. These internal factors including gender, age, education, academic background, and so on have different degrees impact on the core factors through the data analysis. In addition, 50% of users believe that the individual level of cognitive ability, learning ability and personal preferences (interests) which belong to the variable difficult to measure. The assumption H2 "internal factors have certain impact on the core factors" sets up through the data analysis and judgments.

7   The Verification of Assumption Relationships between the Core Factors and Use Attitude

This model describes the attitudes of users using information resources. There are two kinds of attitudes that must use and may or may not use if the users hold the information resources is valuable, otherwise do not use if worthless. The frequency of choosing "must use the valuable information resources" is 160, accounting for 79.2%, which is the highest. The frequency of choosing   "may or may not use the information resources if it has some value" is 42, accounting for 20.8%. The data shows that college teachers tend to choose "must use the valuable information resources". The frequency of choosing "must use the valuable information resources" and "usefulness" at the same time is 136 accounting for 85%. The frequency of choosing "must use the valuable information resources" and "usability" at the same time is 24, accounting for 15%. The proportion of choosing "usefulness" is far higher than choosing "usability", this shows that choosing "must use the valuable information resources" more because of "usefulness" rather than "usability". And the chi-square test about the attitude of the core factors and cross-table using shows that Pearson chi-square value is 2.428, progressive Sig. (Bilateral) 0.119, for a given type of data there are closer relationships on the ranks, namely there are closer related relations between the core factors and use attitude. The assumption H1 "the core factors have certain impact on the use attitude" sets up through the data analysis and judgments.

8   Conclusion

The paper advances the information use behavior model of college teachers based on their cognition about information resources, and drawing on the research theory which information science and technology is accepted. The assumption sets up through the data analysis and judgments of the model, namely the external factors have impact on the core factors, the internal factors have impact on the core factors, and the core factors have impact on the use attitude.

The objective external conditions of external factors mainly include the teaching conditions and teaching level through the data analysis of various factors. For the three factors of analyzing suitability between information resources and task, the proportion of "available help level" is 78.92%, "information relevance" is 76.96%, but the lowest is "portability", only 41.67%. The first two factors have a high degree of recognition. The main viewpoints of the internal factors include gender, age, education, academic background and so on, and the teaching experience has a weak impact. The proportion of cognitive ability is 68.4%, learning ability is 60.70%, and personal preferences (interests) are 60.2%, which belong to the characteristics difficult to measure. Therefore cognitive ability, learning ability and personal preferences (interests) have a high degree of influence. The proportion of "usefulness" is 85% and the proportion of "usability" is 15%, which indicates that "usefulness" is the most important observation point of core element and "usability" is the weaker observation point.

Acknowledgment

Project source: Soft-science subject of Hebei province

Project name: Research of Hebei province government information technology application evaluation model

Project number : 074572239.

References

1. Cao, J.: The effectiveness of college teachers information behavior and realization acts. The Science of College Education, University Education Science, pp. 79–80 (May 2007)
2. Cao, P., Zhao, Y., Xu, Y.: An empirical study about the use behavior of government websites based on TAM model. New Technology of Library and Information Service, 77–80 (February 2008)
3. Ye, X.: Analysis of Network Information Resources based on the integration model TAM and TTF. Library and Information Service, 39–41 (March 2008)
4. Lei, Y., Ming, L., Li, X.: An model study about the efficiency of Network information of resource to use—Based on TAM / TTF model. Document, Information & Knowledge, 78–82 (April 2008)
5. Lu, Y., Xu, H.: An empirical study review of technology acceptance model. R & d Management, 93–98 (March 2006)
6. Davis, F.D.: Perceived usefulness perceived usability and user acceptance of information technology. MIS Qurterly (September 1989)
7. Yan, W., Deng, X.: The basic problems of Internet users information behavior. Library and Information Service, 37–39 (June 2009)
8. Goodhue, D.L.: Development and Measurement Validity of a Task–Technology Fit Instrument for User Evaluations of Information Systems. Decision Sciences, 105–138 (January 1998)
9. Ren, X.: A research about information technology acceptance model of teacher based on TAM and TTF model. Distance Education in China, 64–65 (September 2009)
10. Li, G.: The influence factors of information Behavior of College user. Academic Library and Information Service, 51–54 (March 2009)

Author Index